Oracle9*i* Database Administrator II: Backup/Recovery and Network Administration

Lannes L. Morris-Murphy, Ph.D.

THOMSON

COURSE TECHNOLOGY

Australia • Canada • Mexico • Singapore • Spain • United Kingdom • United States

THOMSON
™
COURSE TECHNOLOGY

Oracle9i Database Administrator II:
Backup/Recovery and Network Administration
by Lannes L. Morris-Murphy, Ph.D.

Senior Vice President, Publisher:
Kristen Duerr

Executive Editor:
Jennifer Locke

Senior Product Manager:
Barrie Tysko

Development Editor:
Jill Batistick

Production Editors:
Kristen Guevara
Melissa Panagos

Associate Product Manager:
Janet Aras

Editorial Assistant:
Amanda Piantedosi

Product Marketing Manager:
Jason Sakos

Cover Designer:
Betsy Young

Cover Art:
Rakefet Kenaan

Senior Manufacturing Coordinator:
Laura Burns

Compositor:
GEX Publishing Services

Contents

TABLE OF

Contents

CHAPTER NINE
Incomplete Recovery with Recovery Manager 259

CHAPTER TEN
Recovery Manager Maintenance 299

CHAPTER ELEVEN
Recovery Catalog   321

Preface

The last few decades have seen a proliferation of organizations that rely heavily on information technology. These organizations store their data in databases, and many choose Oracle database management systems to access that data. The current Oracle database version, Oracle9*i*, is an object-oriented database management system that allows users to create, manipulate, and retrieve data. In addition, Oracle9*i* has increased database security and includes new features that ease the administrative duties associated with databases.

The focus of this textbook is on the procedures necessary to recover an Oracle9*i* database in the event of failure and to ensure accessibility through a network environment. In a "real world" setting, experienced DBAs tend to use the command-line approach to perform these procedures, while novice DBAs prefer a GUI environment. Therefore, for maximum flexibility, students will be exposed to both command-line and GUI interfaces to perform these procedures.

In addition, the book has been organized to correspond to the published certification objectives for Exam 1Z0-032: Oracle9*i* DBA Fundamentals II. You can find these objectives at *www.oracle.com/education/certification/index.html?dba9i_track.html*.

The Intended Audience

This textbook is designed for students in technical two-year or four-year programs who need to learn how to back up, recover, and set up network configuration for an Oracle9*i* database. Students should have a basic understanding of database architecture and a fundamental knowledge of networking concepts.

Oracle Certification Program

This textbook covers the objectives of Exam 1Z0-032: Oracle9*i* DBA Fundamentals II. A grid presenting the Oracle exam objectives and the chapters of this book that address them can be found on the inside front cover of this book. Successful completion of Exam 1Z0-032 can be applied toward certification as an Oracle9*i* Oracle Certified Professional (OCP) Database Administrator (DBA). Information about registering for this exam, along with other reference material, can be found at *www.oracle.com/education/certification*.

The Approach

The concepts presented in this textbook are based on a hypothetical credit union, Janice Credit Union, and revolve around the tasks learned or performed by Carlos, one of the database administrators. Concepts and commands are first introduced in the chapter and then applied in the context of production or a test database used at the credit union. Students are encouraged to perform each task demonstrated in the textbook.

Although the setting and storyline in this book are designed to simulate a network environment, each example can be performed using a stand-alone installation of Oracle9*i*. This approach provides flexibility for instructors in the lab and allows students to practice tasks at home using the default configuration for an Oracle9*i* database.

To reinforce the material presented, each chapter includes a chapter summary and, when appropriate, a syntax guide of the commands covered in the chapter. In addition, at the end of each chapter, groups of activities are presented that test students' knowledge and challenge them to apply that knowledge to solving various problems.

Overview of this Book

The examples, assignments, and cases in this book will help students to achieve the following objectives:

- Change the archiving mode for an Oracle9*i* database
- Back up an Oracle9*i* database through the operating system
- Perform complete and incomplete recoveries through the operating system
- Use the RMAN utility to backup and recover an Oracle9*i* database
- Create and maintain a recovery catalog
- Transport and load data using the SQL*Loader and Import/Export utilities
- Configure clients and servers for network access
- Configure the Oracle Shared Server to support a larger number of concurrent connections with additional physical resource requirements

The textbook is divided into two parts: Backup/Recovery and Network Administration. The Backup/Recovery portion of the textbook is further subdivided into the following sections: User-Managed Procedures and Recovery Manager. The chapters within each section build on concepts covered in the previous chapters and should be completed in sequence.

Chapter 1 provides an overview of Oracle9*i* database architecture and backup/recovery objectives. **Chapter 2** is focused on changing a database's archiving status and how to enable automatic archiving. **Chapter 3** demonstrates the procedures for backing up database files for open and closed databases through the operating system. **Chapter 4** includes performing complete recovery for ARCHIVELOG and NOARCHIVELOG mode databases through the operating system. **Chapter 5** presents the steps for performing an incomplete recovery of an ARCHIVELOG mode database through the operating system. **Chapter 6** provides an overview of the Recovery Manager (RMAN) utility. **Chapter 7** shows students how to create a backup of various database files through RMAN. In **Chapter 8**, the process of performing a complete recovery through the RMAN utility for ARCHIVELOG and NOARCHIVELOG mode databases is presented. **Chapter 9** covers the types of incomplete recoveries that can be performed through RMAN and the consequences of each approach. **Chapter 10** includes maintenance of the RMAN utility to ensure backups of the database are available and valid. **Chapter 11** discusses the benefits of using a recovery catalog as the target repository for procedures performed through RMAN. **Chapter 12** examines how to import and export data using various utilities available in Oracle9i. **Chapter 13** provides an overview of the Oracle9i networking architecture and server-side configuration. **Chapter 14** focuses on the client-side requirements to access a database through an Oracle9i network environment. **Chapter 15** discusses use of the Oracle Shared Server to enable concurrent user connections without increasing an organization's physical resources.

Features

To enhance students' learning experience, each chapter in this book includes the following elements:

- **Chapter Objectives:** Each chapter begins with a list of the concepts to be mastered by the chapter's conclusion. This list provides a quick overview of chapter contents as well as a useful study aid.

- **Running Case:** The databases utilized by Janice Credit Union serve as the basis for demonstrating the procedures presented in each chapter.

- **Methodology:** As new concepts are discussed, the major steps are demonstrated. When appropriate, both command-line and GUI approaches are included in the chapter examples.

- **Notes:** These explanations, designated by the *Note* icon, provide further information about a concept or where further documentation can be located.

- **Chapter Summaries:** Each chapter concludes with a summary of chapter concepts. These summaries are a helpful recap of chapter contents.

- **Syntax Summaries:** A Syntax Guide table is supplied after each Chapter Summary, where appropriate. It recaps the commands and parameters presented in the chapter.

- **Review Questions:** End-of-chapter assessment begins with a set of 20 review questions that reinforce the main ideas introduced in each chapter. These questions ensure that students have mastered the concepts and understand the information presented. The questions consist of true/false, multiple choice, fill-in-the-blank, and discussion type formats.

- **Hands-on Assignments:** Along with conceptual explanations and examples, each chapter provides hands-on assignments related to the chapter's contents. The purpose of these assignments is to provide students with practical experience and can be completed in either a network or stand-alone environment.

- **Case Projects:** Two cases are provided at the end of the chapter. In the backup and recovery chapters, one case specifically relates to the development of a procedure manual. This case is designed to have students summarize the major procedures performed in the chapter with the dual purpose of reinforcing the chapter material and serve as a reference manual throughout the term and later in the work environment. The remaining cases in the book are designed to help students apply what they have learned to real-world situations. The cases give students the opportunity to independently synthesize and evaluate information, examine potential solutions, and make recommendations, much as students will do in an actual business situation.

In addition to the book-based features, *Oracle9i Database Administrator II: Backup/Recovery and Network Administration* offers the following software components:

- The Course Technology Kit for Oracle9*i* Software, available when purchased as a separate bundle with this book, provides the Oracle9*i* Enterprise Edition, Standard Edition, and Personal Edition database software, Release 2 (9.2.0.1.0) on three CDs, so that users can install on their own computers at home all the software needed to complete the in-chapter tutorials, Hands-on Assignments, and Case Projects. The software included in the kit can be used with Microsoft Windows NT, 2000, or XP Professional operating systems. However, during installation, only certain editions of Oracle 9*i* Release 2 will be available. The edition available will depend on the operating system installed on your computer. The installation instructions for Oracle9*i* and the login procedures are available at *www.course.com/cdkit*. Look for this book's title and front cover, and click the link to access the information specific to this book.

- Boson Software and Course Technology have partnered to extend students' Oracle learning experience. We have created a powerful multimedia practice test engine to help students prepare for Oracle certification exams. The CD containing an extended demo of this tool is located in the back of the book. This self-assessment tool demo includes 50 interactive questions, much like the questions on Oracle Certification Exam 1Z0-032, DBA Fundamentals II. After the student has completed the demo practice test, he or she can find out what questions were answered correctly and incorrectly by just the click of a button. For those questions answered incorrectly, the Test Engine includes references to the page(s) in this book where the student can find and review the topic(s).

  If students would like to obtain additional questions, they may upgrade from the demo to a more robust version of the tool for a fee. This engine offers approximately 150 additional exam practice questions. To upgrade, visit the Boson Web site at *www.boson.com*.

Teaching Tools

The following supplemental materials are available when this book is used in a classroom setting. All teaching tools available with this book are provided to the instructor on a single CD-ROM.

- **Electronic Instructor's Manual:** The Instructor's Manual that accompanies this textbook includes the following elements:

  — Additional instructional material to assist in class preparation, including suggestions for lecture topics.

  — When appropriate, information about potential problems that can occur in networked environments is identified.

- **ExamView®:** This objective-based test generator lets the instructor create paper, LAN, or Web-based tests from testbanks designed specifically for this Course Technology text. Instructors can use the QuickTest Wizard to create tests in fewer than five minutes by taking advantage of Course Technology's question banks—or create customized exams.

- **PowerPoint Presentations:** Microsoft PowerPoint slides are included for each chapter. Instructors might use the slides in three ways: As teaching aids during classroom presentations, as printed handouts for classroom distribution, or as network-accessible resources for chapter review. Instructors can add their own slides for additional topics introduced to the class.

- **Data Files:** The data files referenced within the chapters and in the end-of-chapter materials are provided through the Course Technology Web site at *www.course.com* and are also available on the Teaching Tools CD-ROM.

- **Solution Files:** Solutions to the end-of-chapter review questions, Hands-On Assignments, and Case Projects are provided on the Teaching Tools CD-ROM. Solutions may also be found on the Course Technology Web site at *www.course.com*. The solutions are password protected.

ACKNOWLEDGMENTS

The following reviewers provided helpful suggestions and insight into the development of this textbook: Allen Dooley, Pasadena City College; Alla Grinberg, Montgomery College; Gary Knotts, University of Arizona; Sergey Ivanov, George Washington University; and David Welch, Nashville State Technical Community College.

Read This Before You Begin

DATA FILES

To work through certain examples demonstrated in this book, you will need to load the data files created for this book. Your instructor will provide you with those data files, or you can obtain them electronically from the Course Technology Web site by accessing *www.course.com* and then searching for this book's title. Example files are found in the Data folder under their respective chapter folders (for example Chapter01) on your data disk and have files names that correspond with the instructions in the chapter.

When you download the data files, they should be stored in a directory separate from any other files on your hard drive or diskette. You will need to remember the path or folder containing the files, because the file name of each script file must be prefixed with its location when it is executed.

In chapters where the configuration of the database will be changed, you will be instructed to create a backup of the database files. In various chapters, different drives were used to store the backup copies of the database files. If your computer has only one hard drive available, you can partition the drive to different logical drives, or simply create a folder on your hard drive and store the files in that folder. In either case, you will need access to the backup files to restore the database in the event an unrecoverable problem occurs. The backup also should be used to restore the database upon completion of each chapter to prepare for the next chapter.

Using Your Own Computer

To use your own computer to work through the chapter examples and to complete the Hands-on Assignments and Case Projects, you will need the following:

- **Hardware:** A computer capable of using the Microsoft Windows NT, 2000 Server with Service Pack 1, 2000 Professional, or XP Professional operating system. You should have at least 256MB of RAM and between 2.75GB and 4.75GB of hard disk space available before installing the software.

- **Software:** Oracle9*i* Release 2 (9.2.0.1.0) Database. Although the examples in this textbook used the Enterprise Edition, you can also use the Personal edition. The Course Technology Kit for Oracle9*i* Software contains the database software necessary to perform all the tasks shown in this textbook. Detailed installation, configuration, and logon information for the software in this kit are provided at *www.course.com/cdkit*. Look for this book's title and front cover, and click on the link to access the information specific to this book.

When you install the Oracle9*i* software, you will be prompted to change the password for certain default administrative user accounts. Make certain that you record the names and passwords of the accounts because you may need to log in to the database with one of these administrative accounts in later chapters. After you install Oracle9*i*, you will be required to enter a user name and password to access the software. If you have installed the Personal Edition of Oracle9*i*, you will not need to enter a Connect String during the log in process.

- **Data files:** You will not be able to use your own computer to work through all the chapter examples and complete the projects in this book unless you have the data files. You can get the data files from your instructor, or you can obtain the data files electronically by accessing the Course Technology Web site at *www.course.com* and then searching for this book's title.

Each chapter includes a section entitled Set Up Your Computer for the Chapter. In most chapters, you will need to create a backup of the database files before beginning the chapter examples. These backups are crucial to restoring the database in the event an unrecoverable error occurs. To simplify the process, you should create a folder on your hard drive to store the backup copy of the database.

In addition, you should be aware that completion of examples in certain chapters can require up to an hour of time, depending on the processing speed of your computer and the size of the database. In particular, when performing the incomplete recovery procedures in Chapters 5 and 9, make certain you have enough time allocated to complete all the steps. Otherwise, you will need to restore the backup version of the database to continue with the material presented in subsequent chapters.

The database referenced throughout the book is named TESTDB. It was created using the default configuration available through the Oracle Database Configuration Assistant. When performing the examples, you will need access to a system administrator account. In a default installation, the user name for the administrator account is SYSTEM. The password for this account is specified when the database is created. If you are creating the database on your home computer, make certain you remember the password specified for the account. Otherwise, you will need to recreate the database.

The following procedures should be used when logging into the database:

- When logging into SQL*Plus, enter the user name **SYSTEM** and then the appropriate password into the dialog box. After you have logged in, at the SQL> prompt you will need to type **connect / as sysdba** and press **Enter** to enable your SYSDBA privileges.

- When you are completing examples using the Enterprise Manager Console, you will need to enter the SYSTEM user name and the appropriate password in the appropriate text boxes of the dialog box. However, you will also need to click the drop-down list beneath the password and select SYSDBA to connect to the database with the SYSDBA privileges enabled.

Practice Test Software

We have created a powerful practice test engine, a demo of which is located on a CD in the back of the book, to help you prepare for Oracle Certification Exam 1Z0-032, Oracle9i: DBA Fundamentals II. The extended demo of this assessment tool includes 50 interactive questions; they are much like the ones you would find on Oracle exams. After you have completed the practice test, it will inform you of what questions you got right and wrong with just the click of a button. For those questions answered incorrectly, the Test Engine includes references to the page(s) in this book where you can find and review the topic(s).

If you would like to obtain additional questions, you may upgrade from the demo to a more robust version of the tool for a fee. This engine offers approximately 150 additional exam practice questions. To upgrade, visit the Boson Web site at *www.boson.com.*

Visit Our World Wide Web Site

Additional materials designed especially for you might be available on the World Wide Web. Go to *www.course.com* periodically and search this site for more details.

TO THE INSTRUCTOR

In this book, students will be required to change the configuration and perform shutdown/startup procedures. If students share a single database, then only one student will be able to complete these operations at a time. Therefore, it is highly recommended that each user have access to his/her own database. The database can reside on the individual machine or be accessed across a network. However, if the client machine has both a local database and can access a remote database, TNS errors may occur after shutting down the database. If the client machine must be able to access both types of database, a simply solution is to create two Tnsnames.ora files and have the students access the appropriate file based on the situation. In addition, students will need to have access to a SYSDBA privileged account to perform the tasks presented in this book.

Each chapter includes a section entitled Set Up Your Computer for the Chapter. In most chapters, students will need to create a backup of the database files before beginning the chapter examples. These backups are crucial to restoring the database in the event an unrecoverable error occurs. To simplify the process, provide each student with a folder, either on their local machines or on a server, that can be used to store the backup files.

Some examples using the GUI environment require access to an Oracle9*i* Management Server (OMS). In a typical configuration, a workstation would have either Microsoft Windows XP Professional or Microsoft Windows 2000 Professional, while a server would have the Microsoft 2000 Server operating system installed. Because the OMS is normally run with server software, in chapters where both the command-line and GUI approaches are demonstrated, the Microsoft Windows 2000 Server is used for the GUI examples and the command-line interface is presented using the Microsoft Windows XP Professional operating system. However, all examples can be performed using either operating system environment.

To complete some of the chapters in this book, your users must have access to a set of data files. These files are included in the Instructor's Resources. They may also be obtained electronically through the Course Technology Web site at *www.course.com*.

The chapters and projects in this book were tested using the Microsoft Windows 2000 Server operating system with Service Pack 1 and Oracle9*i* Release 2 (9.2.0.1.0.) Enterprise Edition and the Microsoft Windows XP Professional operating system with Oracle9*i* Release 2 (9.2.0.1.0) Personal Edition.

 Depending on the processing speed of the computers and the size of the database, the completion of examples in certain chapters can require up to an hour of time. In particular, when performing the incomplete recovery procedures in Chapters 5 and 9, make certain enough lab time has been allocated to complete all the steps. Otherwise, the students will need to restore the backup version of the database to continue with the material presented in subsequent chapters.

Course Technology Data Files

You are granted a license to copy the data files to any computer or computer network used by individuals who have purchased this book.

1

BACKUP AND RECOVERY OVERVIEW

> **After completing this chapter, you should be able to do the following:**
> - Identify the files associated with an Oracle9*i* database
> - List the logical structure of an Oracle9*i* database
> - Identify the components of an Oracle9*i* instance
> - Identify the purpose of each background process of an Oracle9*i* instance
> - Identify the purpose of each component of the system global area (SGA)
> - Identify the purpose of the Init.ora file and the type available with the Oracle9*i* database
> - Define mean-time-between-failures (MTBF) and mean-time-to-recover (MTTR)
> - Identify the parameters that can be used to speed up instance recovery
> - Identify the various types of failures that can occur in an Oracle9*i* database
> - Use the Enterprise Manager Console to view the names and locations of the control file

The purpose of this textbook is to familiarize you with various tasks associated with the backup and recovery of an Oracle9*i* **database** and with accessing the Oracle server via a network. Regardless of how well a database management system (DBMS) is designed, developed, and implemented, there is always the possibility that hardware failure, malicious attackers, or a natural disaster can render the database inaccessible or completely useless. It is the job of the database administrator to ensure the availability of the database, and when that availability is compromised, to restore the database back to a useable state with minimal data loss in the shortest amount of time possible.

Before examining backup and recovery operations, you should first understand the architecture of an Oracle9i database. This includes the physical files stored through the operating system, as well as the logical units referenced by the DBMS. Because a database can only be accessed when an **instance** is running, you will also need to understand the various components of an Oracle9i instance. Then you will be prepared to discuss the nuances of backup and recovery.

THE CURRENT CHALLENGE IN THE JANICE CREDIT UNION DATABASE

The database used as the running case in this book is a database used by the Janice Credit Union. The credit union is a relatively small organization consisting of 45 employees at two locations. All financial transactions are stored in an Oracle9i database. Of course, the credit union has several databases. The largest is the financial database that stores all customer account information, including checking and saving accounts, certificates of deposit (CDs), loans, etc.

As you work through this chapter and the remainder of the textbook, you will learn how Carlos, who is the database administrator (DBA), backs up the database, recovers the database when a failure occurs, and ensures that the financial database is accessible to all employees at both locations. Note that in the remaining sections of this chapter, you will learn the various components of the Janice Credit Union database.

ORACLE9i DATABASE

An Oracle9i database has both a physical and a logical structure. The database is composed of three types of physical files. The physical files can be renamed, copied, moved, or deleted through the operating system. These files contain the data structure, as well as the actual data stored in the database. However, the DBMS itself traditionally stores and references data units through a logical structure. Both the physical and logical structures of an Oracle9i database are discussed in the following sections.

Physical Structure

The physical structure of the Oracle9i database refers to the actual files that are stored by the operating system. There are three types of physical files used by the Oracle9i database:

- **Control files:** A control file is a small binary file that contains the name of the database, as well as the names and locations of all the data files and redo log files. Control files always have the file extension of .ctl. The control file is a critical component of the Oracle9i database because the database is not able to locate the data files or redo log files without it. A database must have at least one control file and can have up to a maximum of eight multiplexed control files. Oracle9i recommends that there be at least two control files for every database. However, in a default installation of an Oracle9i database, there are three mirrored copies of a control file, and they are stored in the same physical location. After the installation process is completed, at least one copy of the control should be moved to another drive. Doing so allows the database administrator to have access to a current copy of the control file to recover the database in the event of a hard drive failure.

- **Redo log files:** The redo log files contain the changes that have been made to the database as a result of committing transactions. Only after the changes have been recorded in the redo log files can those changes then be written to the data files. Oracle9*i* requires a minimum of two redo log files for each database. When one of the files is filled with changed data, a log switch occurs and then any other data that is committed is written to the other redo log file. After the second (or any subsequent) log file is filled, the first redo log file is then overwritten. For example, if the first redo log file is named Redo1, and the second is named Redo2, Oracle9*i* begins writing to Redo2 after Redo1 is filled. After Redo2 is filled with committed data, the contents of Redo1 are overwritten, as illustrated in Figure 1-1. In other words, the files are written to in a circular fashion.

- **Data files:** The actual data contained within the database is physically stored in the data files that can be manipulated by the operating system. Data files have the file extension of .dbf. Do not confuse the concept of a data file by trying to think of it in terms of a specific database object such as a table named PRODUCTS. The data entered into a table might be contained in one data file, or it might be spread across several data files. In other words, if you do create a table name PRODUCTS, do not expect to find a file on the hard drive named Products.dbf. In addition, data files are used to store indexes, data dictionary information, and any other data that needs to be referenced by the database.

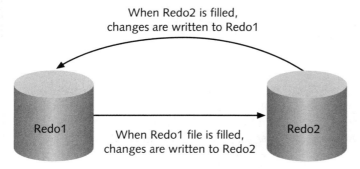

Figure 1-1   Redo log file writing sequence

 Additional files that may be referenced by the Oracle9*i* DBMS include initialization files and password files. However, these files are not technically a part of the Oracle9*i* database.

The names and locations of the files in the preceding bulleted list can easily be found through the Enterprise Manager Console. As shown in Figure 1-2, when the CONTROLFILE object is selected in the left pane of the window, the name and location of each control file is displayed in the right pane. The database selected in Figure 1-2 was created using the default installation process and has three control files. Notice that all the control files are stored on the same drive. As mentioned, this is not an ideal situation from a backup and recovery perspective.

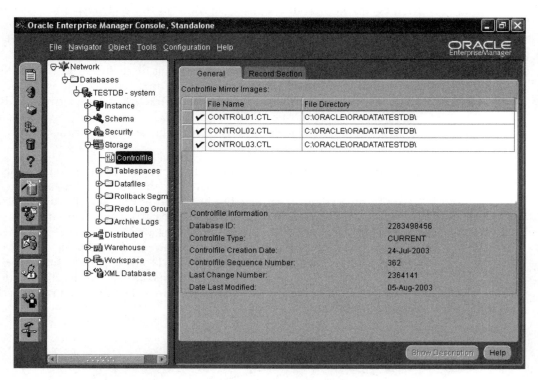

Figure 1-2   Name and location of the database control files

 Where appropriate, both the Enterprise Manager Console and a command-line environment such as SQL*Plus are used to demonstrate various backup and recovery tasks throughout the textbook. The Enterprise Manager Console provides a graphical user interface (GUI) that can be useful to novice DBAs. A command-line approach is more commonly used by DBAs with several years of administration experience.

Logical Structure

Although the operating system is used to copy, rename, and delete physical data files, Oracle9*i* usually references the actual data contained within the files through a logical structure. The largest logical structure, of course, is the database. Oracle9*i* divides the database into smaller structures, or units, that are used to manage data. The logical units consist of the following:

- **Tablespaces:** A database can be divided into tablespaces. Each tablespace is a grouping of related objects, such as indexes, table data, and so on. A tablespace consists of one or more segments. Each tablespace can belong only to one database, although a database can contain numerous tablespaces. Each tablespace is given a name so that it can be referenced by the database or by a user. All Oracle9*i* databases have a tablespace with the default name SYSTEM. With the exception of the SYSTEM tablespace, a tablespace can be taken offline or changed to read-only or read-write status. As shown in Figure 1-3, the name and size of the tablespaces can be displayed through the Enterprise Console Manager.

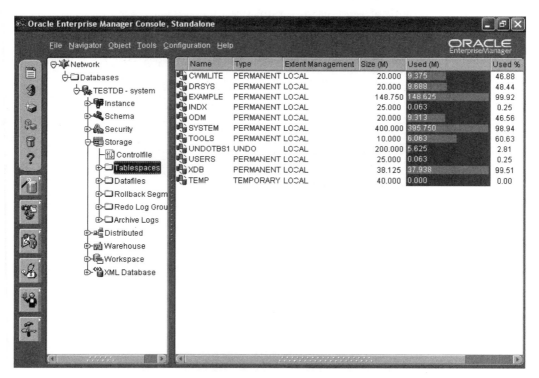

Figure 1-3   Name and size of tablespaces

- **Segments:** A segment is a logical storage structure that consists of one or more extents (see the next bullet for further discussion). There are four types of segments: data segments, index segments, undo segments, and temporary segments. **Data segments** are used to store data for the database tables. An **index segment** is used to store data regarding indexes. **Undo segments** store the data necessary to roll back changes, while **temporary segments** are created to perform sort operations or to complete the execution of certain types of SQL statements. Each segment can belong to only one tablespace.

- **Extents:** An extent consists of a set of contiguous Oracle9i data blocks. To increase the size of a segment, the database administrator or the Oracle9i DBMS can add extents to the segment. Each extent can only belong to one segment.

- **Data blocks:** A data block is the smallest storage unit that can be referenced by Oracle9i. The actual size of a data block can be specified when the database is first created with the DB_BLOCK_SIZE parameter; however, to increase read-write efficiency, it should be a multiple of an operating system block. For example, if the block size of the operating system is 2KB, then the DBA can set the Oracle9i block size as 2KB, 4KB, 8KB, and so on. After the block size for a database has been set, it cannot be changed without recreating the database. The maximum size of an Oracle9i data block is operating-system dependent.

Relationships Between the Physical and Logical Structures

Figure 1-4 depicts the relationship between the physical and logical structures within an Oracle9i database. The database is physically divided into three types of files: control files, data files, and redo log files. The physical data files are used to store the data contained in a logical tablespace. However, Oracle9i divides the tablespace into smaller logical units that can be referenced internally by the DBMS. Some backup and recovery procedures require the DBA to take a tablespace offline before a data file can be backed up, renamed, or moved through the operating system.

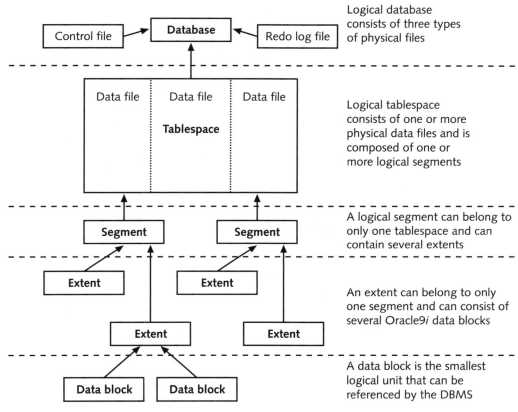

Figure 1-4   Oracle9*i* structures

ORACLE 9*i* INSTANCES

An Oracle9*i* instance consists of both background processes and memory structures. The background processes are used to perform certain tasks, which will be described in this section. The memory structures are areas of memory that have been allocated for a specific purpose, such as data dictionary information, SQL commands, table data, and so on. The background processes and memory structures are also discussed in this part of the chapter.

Background Processes

When an instance is started, there are five background processes that are started by default: **system monitor (SMON)**, **process monitor (PMON)**, **log writer (LGWR)**, **database writer (DBW*n*)**, and **checkpoint (CKPT)**. In addition, optional background processes, such as **archiver (ARC*n*)** can also be started. The following sections discuss the purpose of each process.

System Monitor (SMON)

One of the responsibilities of the SMON background process is to perform instance recovery at database startup by applying the contents of the online redo log files to the database. If the operating system fails, or if there is a power loss, any changed data that is being stored in memory (that is, has not be written to the disk) is lost.

When data changes are committed, the changes are first written to files known as redo logs before the actual data are changed. If changes have been recorded in the redo logs but not in the data files, SMON is responsible for "rolling forward" the changes and ensuring that the changes are recorded in the data files.

Process Monitor (PMON)

The PMON background process is responsible for "cleaning up" failed user processes. This includes rolling back any uncommitted transactions and releasing locks and other resources held by the failed user process.

Log Writer (LGWR)

It is the responsibility of the LGWR background process to write all changed data blocks residing in the memory structure of the Oracle9i instance to the online redo log files. The LGWR can write to multiplexed copies of the online redo log files. Even if one of the log files is unavailable, the LGWR continues to write the changes to the available copies.

Database Writer (DBWn)

The purpose of the DBWn is to write the changed data blocks from memory to the actual data files. A total of ten DBWns can be started to perform the write operations for highly dynamic databases. Each DBWn is assigned a numeric value ranging from zero to nine. By default, only one database writer (DBW0) is started when the instance is started. However, the DBA can start additional database writers during peak times of the system's operation.

Checkpoint (CKPT)

When an instance recovery is needed, there must be some way to determine the transactions that should be applied to each data file to put the database in a consistent state (for example, the entire transaction has been updated to the file(s) and not just a portion of the transaction). The CKPT background process triggers the writing of modified data and then stamps the headers in the control file and data files with the system change number (SCN). It stamps the number to identify the sequence of data that was written by the DBWn to the data files.

The CKPT can indirectly determine how long it takes to perform an instance recovery by how often a checkpoint, or the triggering event, occurs. If checkpoints take too long to occur, there will be more modified data that has to be written to the data files if instance recovery is required. However, if the checkpoints occur too often, there is a possibility of disk contention occurring (that is, the same drive needs to perform several write operations at the same time). A checkpoint can occur after a duration of time, by default every three seconds, or after a specific event, such as when a redo log file has filled and the LGWR begins writing to another log file, which is known as a log switch.

Archiver (ARC*n*)

Crucial to the ability to recover a database without data loss after a disk failure is the optional ARC*n* background process. When a log switch occurs, the contents of the filled online redo log file are copied to another file. This copy of the online redo log file is often referred to as the "archived" redo log file or "offline" redo log file. In the event that a data file is unavailable due to media failure, the archived log file can be used to recover the database without any data loss. In addition, if the archived copies of the redo log files are stored onto a different disk than the online redo log files, then the backup copy of the files is available if a disk failure occurs and it makes all the data files—including the online redo log files—unavailable.

Memory Structures

There are two memory areas used by Oracle9i: **program global area (PGA)** and **system global area (SGA)**. The contents of the PGA vary based on the configuration of the instance. The SGA consists of several structures. Of particular interest to the DBA are the shared pool, database buffer cache, and redo log buffer structures within the SGA. The PGA is allocated only when a server process is started and deallocated after the process is finished; the SGA is allocated when the instance is started.

The following list discusses the shared pool, database buffer cache, and redo log buffers within the SGA.

- **Shared pool:** The shared pool is a portion of the Oracle9i instance that contains the library cache and the data dictionary cache. The **library cache** is used to hold the most recently used SQL and PL/SQL statements, while the **data dictionary cache** holds information regarding database objects such as users, privileges, database tables, indexes, and so on.

- **Database buffer cache:** The database buffer cache contains the most recently used blocks. The data blocks remain in the database buffer cache until a server process needs space to store newly read data blocks and no empty space is available. At that point, one or more of the least recently used blocks are written to the disk. Then the blocks in the cache can be overwritten. When data requests are made, the database buffer cache is always searched first to determine whether the data is already available in memory. Such searching reduces the number of reads from the physical storage device (that is, the hard drive). If the data is not found, then the required data block is read into memory from the data file. In addition, if data blocks are modified, the changed blocks are held in the database buffer cache until they are written to the data files by the DBW*n*.

- **Redo log buffer:** The redo log buffer stores changes made by INSERT, UPDATE, DELETE, CREATE, ALTER, or DROP operations. Basically, these are changes that have been made to the data contained in the data files. Eventually the contents of the redo log buffer are written to the online redo log files to be used in the event of database failure.

INITIALIZATION FILE (INIT.ORA)

Although the initialization file technically is not a part of the database, it is the file that is referenced when the instance is started. Recall that the control file specifies the location of the data files and redo log files for the database. However, how does the database know where to find the control file itself (plus any copies of the control file)? The DBMS references the initialization file, which is stored in the same directory as the actual DBMS software.

The initialization file is generically referred to as the Init.ora file. The initialization file contains the name of the database, the location of all control files, as well as the values assigned to numerous parameters which are used to configure the database. If a parameter is not explicitly stated in the Init.ora file, then the parameter's default value is assumed during the start-up process.

Beginning with Oracle9i, there are two types of initialization files. The first type is commonly referred to as a **pfile**. It is a text file that can be edited by a DBA through any text editor, such as Notepad. The second type, referred to as an **spfile**, was introduced with the first release of Oracle9i. It is a binary file created by the Oracle server. When a database is created, the spfile is automatically created. Because the spfile is not a text file, Carlos, the database administrator from Janice Credit Union, would not be able to edit the file with Notepad. Instead, he would need to either use the Enterprise Manager Console to make the changes on his behalf or have the Oracle9i server create a pfile that can be edited through a text editor. If a pfile and an spfile file both exist for one database, by default, the spfile is referenced during instance start-up.

Because the contents of a pfile are viewable outside the Oracle9i DBMS environment, Carlos always keeps an updated copy of the pfile available for reference. In the following steps, you will create a pfile from an existing spfile through SQL*Plus. In order to create the pfile, you will need to be logged into Oracle9i with the SYSDBA privilege enabled. (For instructions on how to do this, consult your instructor.)

1. Log into SQL*Plus with a valid user account and enable the SYSDBA privilege.

2. Type **CREATE PFILE FROM SPFILE;** at the SQL> prompt and press **Enter**, as shown in Figure 1-5.

3. From the operating system, open the Init.ora file created in Step 2. (Note that the file name may vary depending on your system.) By default, it is located in a folder called Database in the home directory of your database software installation.

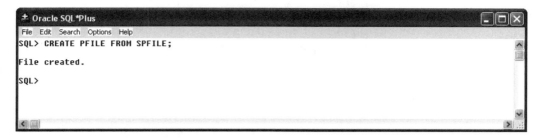

Figure 1-5   Command to create a pfile

An example of an opened pfile is shown in Figure 1-6. Notice the unusable symbols inserted into the text file. This is the tell-all sign indicating that the file was created from an spfile file. If the pfile had been created by the DBA through a text editor, it would normally have documentation and a formatted appearance.

```
INITtestdb - Notepad
File  Edit  Format  Help
*.aq_tm_processes=1■*.background_dump_dest='C:\oracle\admin\TESTDB\bdump'■
*.compatible='9.2.0.0.0'■
*.control_files='C:\oracle\oradata\TESTDB\CONTROL01.CTL','C:\oracle\oradata\TESTDB\CONTROL02.CT
L','C:\oracle\oradata\TESTDB\CONTROL03.CTL'■*.core_dump_dest='C:\oracle\admin\TESTDB\cdump'■
*.db_block_size=8192■*.db_cache_size=25165824■*.db_domain=''■
*.db_file_multiblock_read_count=16■*.db_name='TESTDB'■*.dispatchers='(PROTOCOL=TCP)
(SERVICE=TESTDBXDB)'■*.fast_start_mttr_target=300■*.hash_join_enabled=TRUE■
*.instance_name='TESTDB'■*.java_pool_size=33554432■*.job_queue_processes=10■
*.large_pool_size=8388608■*.log_archive_start=TRUE■*.open_cursors=300■
*.pga_aggregate_target=25165824■*.processes=150■*.query_rewrite_enabled='FALSE'■
*.remote_login_passwordfile='EXCLUSIVE'■*.shared_pool_size=50331648■*.sort_area_size=524288■
*.star_transformation_enabled='FALSE'■*.timed_statistics=TRUE■*.undo_management='AUTO'■
*.undo_retention=10800■*.undo_tablespace='UNDOTBS1'■
*.user_dump_dest='C:\oracle\admin\TESTDB\udump'■
```

Figure 1-6   pfile generated from an spfile

Notice the third and fourth lines of the pfile in Figure 1-6. They identify the CONTROL_ FILES parameter that specifies the name and location of all copies of the control files. As other parameters are presented in the remainder of the chapter, attempt to identify them in the pfile you created or in the example provided in Figure 1-6.

BACKUP AND RECOVERY

In some organizations, the unavailability of a database that is used for day-to-day activities, such as processing customer orders, can result in the loss of thousands or even millions of dollars in revenues. In some cases, the unavailability of a database could even result in personal injury. For example, a hospital may store physician orders and medication schedules in a database. In an extreme case, if the database is lost or the data becomes corrupted, treatment may be delayed while nurses search for the written instructions provided by the attending physician. In a real case involving one hospital, an entire day's worth of insurance claims were lost because of a database failure and the lack of proper backup procedures. When it became apparent that it would not be possible to accurately reconstruct the missing information, even manually, the result was a loss of over $100,000 dollars in insurance claims alone.

In the case of the Janice Credit Union, there are over 15,000 customers who, on average, make a deposit every 1.5 weeks. In addition, the average checking account has a balance of $2,873 with 23 checks being posted every 30 days. It has been estimated that every 10 minutes of downtime costs the credit union approximately $1,800 in lost ATM revenues, idle teller time, and so on. Because this is a relatively small credit union, a loss of even a few thousand dollars can have a critical impact on the bottom line.

Key Elements in Backup and Recovery

The term **backup** refers to having valid copies of the database files that can be used to restore the database back to its original state. There are various approaches that can be used to make copies of the database files, including using the operating system to perform user backups or through the Oracle9*i* Recovery Manager (RMAN). This textbook covers both approaches.

Recovery refers to returning the database to a desired state. In some cases, this may require experiencing some data loss. For example, if a user accidentally loaded a large amount of data into the wrong table and there was no simple way to identify and remove the data, the DBA would need to reset, or recover, the database to a previous point in time when the data is believed to have been correct.

Regarding recovery policy, the goals of the database administrator are to increase the **mean-time-between-failures (MTBF)** and decrease the **mean-time-to-recover (MTTR)**. You can increase the MTBF by examining the resource requirements or error messages contained in the alert log and trace files to anticipate problems that could result in the database being made unavailable to the users, as well as performing preventive maintenance for the hardware.

To decrease the MTTR, you can make frequent backups of the database and prepare for a variety of database failures. If the entire database is frequently backed up, fewer transactions need to be applied from the archived redo log files and the recovery process will be quicker. In addition, preparations such as having employees practice database recovery under different failure scenarios, ensuring additional equipment availability, and so on, can reduce the amount of time spent during the recovery process.

Tuning Instance Recovery

There are two initialization parameters that can be used to limit the amount of time required to perform instance recovery: FAST_START_MTTR_TARGET and FAST_START_PARALLEL_ROLLBACK.

The FAST_START_MTTR_TARGET parameter causes the DBW0 to write faster thereby reducing the number of changed data blocks, called dirty blocks, contained in memory, while the FAST_START_PARALLEL_ROLLBACK parameter can speed up the process of rolling back uncommitted changes. Because the recovery process is faster, the MTTR is decreased. These parameters are discussed in the following sections.

FAST_START_MTTR_TARGET Parameter

The FAST_START_MTTR_TARGET parameter specifies the maximum number of seconds that data should be held in memory before the DBW*n* begins writing changes to the data files. The portion of the initialization file displayed in Figure 1-7 has this parameter set at 300 seconds. If instance recovery is required, then it should take no longer than 300 seconds to write the changes necessary to put the database in a consistent state. The value assigned to the FAST_START_MTTR_TARGET parameter can range between 0 and 3600 seconds.

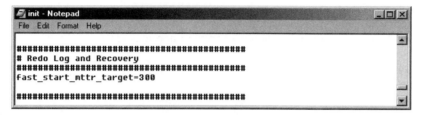

Figure 1-7   Portion of parameter file containing the FAST_START_MTTR_TARGET value

FAST_START_PARALLEL_ROLLBACK Parameter

The second initialization parameter that can be set to speed up instance recovery is the FAST_START_PARALLEL_ROLLBACK parameter. This parameter is used to determine whether SMON should use parallel processes to perform instance recovery. The parameter can be assigned the value of FALSE, HIGH, or LOW. A value of FALSE indicates that parallel processes should not be used. A value of HIGH indicates that parallel processes should be used to perform rollback operations if there are more than 100 transactions affected. A value of LOW means that the parallel process should be used even if there are less than 100 transactions affected by rollback redo operations.

In Figure 1-8, the values assigned to both the FAST_START_MTTR_TARGET and FAST_START_PARALLEL_ROLLBACK parameters are displayed by SQL*Plus through the use of the SHOW PARAMETER command.

```
± Oracle SQL*Plus                                                    [-][□][X]
File  Edit  Search  Options  Help
SQL> SHOW PARAMETER fast_start_mttr_target

NAME                                     TYPE          VALUE
---------------------------------------- ------------- -----------------------
fast_start_mttr_target                   integer       300
SQL>
SQL>
SQL> SHOW PARAMETER fast_start_parallel_rollback

NAME                                     TYPE          VALUE
---------------------------------------- ------------- -----------------------
fast_start_parallel_rollback             string        LOW
SQL>
```

Figure 1-8   Parameter settings

 If you receive the error message "ORA-00942: table or view does not exist" when attempting to execute a SHOW PARAMETER command, then have your instructor or the system administrator assign the necessary privileges, or the DBA role, to your Oracle9*i* user account and reissue the command.

Types of Failure

There are two broad categories of failures that can occur: **non-media failure** and **media failure**. Most types of non-media failure are automatically recoverable by Oracle9*i* background processes, while media failure requires the DBA to determine the necessary recovery process.

Non-media failures include process failures, instance failures, statement failures, and user errors. A **process failure** occurs when the application being used to interface with the Oracle9*i* server generates an internal program error or if the user does not log out of the Oracle9*i* server correctly, that is, performs an abnormal termination. In this case, the PMON background process is activated and correctly terminates the process.

An **instance failure** occurs when a power outage, a problem with the computer housing the Oracle9*i* server, or an issued command shuts down the database with the abort option causing the instance to shut down without updating and closing the database files. When the instance is restarted, SMON applies the contents of the redo log files to the data files and brings the database to a consistent state.

A **statement failure** occurs when there is a syntax error in an SQL or PL/SQL statement that has been submitted to the Oracle9*i* server for processing. The statement failure can simply be corrected by resubmitting the statement without the syntax error. A **user error** occurs when data or tables are incorrectly altered. A user error may require the intervention of the database administrator to correct the problem.

Media failure usually results from the loss or corruption of a file, or the failure of a controller card. When a controller card fails, the computer cannot read or write to the database files accessed through that drive. Media failures are much more critical than non-media failures because the database can be rendered inaccessible to all users for an extended period of time. In all cases, the database administrator needs to perform recovery procedures to make the database available again. The recovery procedures necessary to recover from media failures and user errors will be covered in subsequent chapters of this textbook.

Backup and Recovery Strategy Factors

There are many questions that Carlos has to address when developing backup and recovery strategies for Janice Credit Union. In terms of backing up the database, he should ask and answer the following questions so that he is prepared for any type of disaster that can affect database availability:

1. Does Janice Credit Union possess the resources to perform certain types of backups? For example, is there an individual who is trained in performing backups through the operating system or through RMAN? Is enough money and hardware available to create a duplicate system or are the database files simply copied to a storage medium?

2. Can Janice Credit Union afford to have the database offline to perform a cold backup or must the database be available 24 hours a day, seven days a week (24/7), which would require that hot backups be performed?

3. Is the value of the data sufficient to warrant the costs associated with a particular backup strategy?

4. How regularly are the backups tested to ensure that they are valid and useable if recovery becomes necessary?

5. Are the backup copies of the database secure from natural disasters or the malicious intention of others, yet still accessible in a timely fashion when needed?

Carlos should ask and answer the following questions to determine the most appropriate course of action to recover a database:

1. What type of failure occurred?

2. How quickly must the database be recovered?

3. Is a complete recovery required or should only certain data be recovered?

4. Are the required backups available and valid?

Testing a Strategy

Throughout this textbook, several backup procedures and recovery options will be presented. A DBA such as Carlos is required to weigh the advantages and disadvantages of various options to determine the most appropriate strategy for his organization. The strategy to be used in different situations should be clearly thought out, documented, and tested. Every possible scenario, such as loss of a data file, corrupted tablespace, and hard drive failure, should be simulated on a regular basis to ensure that all appropriate personnel are familiar with, and able to perform, the correct recovery procedure.

There have been many occasions in which personnel have received the training necessary to perform a certain type of recovery task, but when the actual event occurs in a real-world situation, the individuals have forgotten how to perform the task. When database failure occurs, nerves can be rattled—especially when users begin calling and demanding to know what has happened and when it will be fixed. This is not the best time for everyone to be scratching their heads in bewilderment trying to figure how to fix the problem. This is the time when the salaries commanded by Carlos and his staff are earned.

Although it is preferable that database failures never occur, this is an unrealistic expectation and everyone should be prepared to recover a database. As you progress through this textbook, you will learn different backup procedures, as well as the steps necessary to perform various types of recovery operations. The appropriateness of each procedure in a given situation depends upon the factors previously listed. For each backup and recovery procedure discussed, the advantages and disadvantages will also be presented so that you can determine the best approach for your particular situation when the need arises in a real-world setting.

CHAPTER SUMMARY

- An Oracle9i database is composed of three types of physical files: control files, data files, and redo log files.

- The Oracle9i database consists of both a physical structure (operating system files) and a logical structure.

- The Oracle9i server consists of a database and an instance. An instance consists of the background processes and memory structures.

- By default, five background processes are started during instance start-up: SMON, PMON, LGWR, DBWn, and CKPT. An optional background process that should always be used in production systems is ARCn.

- The two primary memory areas are the system global area (SGA) and program global area (PGA).

- The Init.ora file consists of the initialization parameters to be used during instance and database startup.

- There are two categories of failures that can occur in an Oracle9i database: non-media failure and media failure.

- There are a variety of backup and recovery procedures. Testing should be done to ensure the validity of the backup copies and to practice recovery methods for various scenarios.

SYNTAX GUIDE

Command	Description	Example
CREATE PFILE	Used to create the text-based pfile from the binary spfile	`CREATE PFILE FROM SPFILE;`

Parameter	Description	Example
DB_BLOCK_SIZE	Specifies the size of an Oracle9i data block; should be a multiple of an operating system block. Specified in terms of bytes	`DB_BLOCK_SIZE = 8192`
FAST_START_ MTTR_TARGET	Specifies the maximum number of seconds that data should be held in memory before the DBWn begins writing changes to the data files; valid values range between 0 to 3600 seconds	`FAST_START_MTTR_TARGET = 300`
FAST_START_ PARALLEL_ROLLBACK	Specifies whether the SMON should use parallel processes to perform instance recovery; valid values are FALSE, HIGH, or LOW	`FAST_START_PARALLEL_ ROLLBACK= FALSE`

REVIEW QUESTIONS

1. What is an Oracle9i instance?
2. Of the listed logical structures, which is the smallest?
   a. tablespace
   b. data block
   c. extent
   d. segment
3. How can a DBA decrease MTTR?
4. The largest value that can be assigned to the FAST_START_MTTR_TARGET is _____ .

5. Which of the following are components of an Oracle9*i* database?

   a. background processes and memory structures

   b. memory structures and an Oracle9*i* instance

   c. control, data, and redo log files

   d. all of the above

6. Which of the following are components of an Oracle9*i* instance?

   a. background processes and memory structures

   b. memory structures and a database

   c. control, data, and redo log files

   d. all of the above

7. Which of the following is not a valid type of segment?

   a. data segment

   b. redo log segment

   c. temporary segment

   d. index segment

8. A data file can belong to no more than _____ tablespace(s).

9. A maximum of _____ control files can be supported in one database.

10. Which of the following background processes is responsible for instance recovery?

    a. PMON

    b. LGWR

    c. ARC*n*

    d. SMON

11. Which of the following is an accurate statement?

    a. To increase the size of a data block, simply change the operating system.

    b. To increase the size of an extent, simply add more segments to the extent.

    c. To increase the size of a tablespace, simply add more redo log files to the database.

    d. To increase the size of a segment, simply add more extents to the segment.

12. What is the maximum number of DBW*n* processes allowed in a database?

    a. one

    b. two

    c. eight

    d. ten

13. What is the minimum number of control files required for a database?

14. A database must have at least _____ redo log files.

15. One goal of a database administrator should be to increase the _____.
    a. time to backup a database
    b. MTTR
    c. MTBF
    d. number of people with high-level privileges

16. Which of the following is written to in a circular fashion?
    a. data files
    b. control files
    c. redo log files
    d. data dictionary cache

17. Which of the following parameters can be used to set the maximum data that would need to be written by the DBW*n* background process in the event of an instance recovery?
    a. FAST_START_MTTR_IO
    b. FAST_START_MTTR_TARGET
    c. FAST_START_MTBF_TARGET
    d. FAST_START_PARALLEL_ROLLBACK

18. Which of the following types of failures would most likely require the intervention of the database administrator?
    a. user error
    b. process failure
    c. statement failure
    d. instance failure

19. What are the valid values that can be assigned to the FAST_START_MTTR_TARGET parameter?

20. What is the purpose of the ARC*n* background process and how does it relate to the recovery process?

HANDS-ON ASSIGNMENTS

Assignment 1-1 Determine the Size of a Tablespace

In this assignment, you will determine the size of a tablespace.

1. Start the Enterprise Manager Console.
2. Log into the database.

3. In the configuration list provided on the left side of the Enterprise Manager Console, click the + (plus sign) next to Storage.

4. After the available objects are displayed, click the + (plus sign) next to Tablespaces.

5. Using the information provided in the right pane of the Enterprise Manager Console, determine the current size of the SYSTEM tablespace.

6. Exit the Enterprise Manager Console.

Assignment 1-2 Determine the Location of Files

In this assignment, you will determine the location of data files.

1. Start the Enterprise Manager Console.

2. Log into the database.

3. In the configuration list provided on the left side of the Enterprise Manager Console, click the + (plus sign) next to Storage.

4. After the available objects are displayed, click the + (plus sign) next to Datafiles.

5. Using the information provided in the right pane of the Enterprise Manager Console, determine the names and locations of all the data files.

6. Exit the Enterprise Manager Console.

Assignment 1-3 Work with Control Files

In this assignment, you will work with control files.

1. Start the Enterprise Manager Console.

2. Log into the database.

3. Determine how many control files exist in your database and the location of those files.

4. Exit the Enterprise Manager Console.

Assignment 1-4 Work with the SHOW PARAMETERS Command

In this assignment, you will use a command to find the value of a parameter.

1. Log into SQL*Plus.

2. Use the SHOW PARAMETERS command to determine the current value assigned to the FAST_START_PARALLEL_ROLLBACK parameter.

3. Exit SQL*Plus.

Assignment 1-5 Find the Value of a Parameter

In this assignment, you will use a command to find the value of a parameter.

1. Log into SQL*Plus.
2. Use the SHOW PARAMETERS command to determine the current value assigned to the FAST_START_MTTR_TARGET parameter.
3. Exit SQL*Plus

Assignment 1-6 Change a Parameter

In this assignment, you will modify a parameter.

1. Open the **Ch1init.ora** file from the Chapter01 folder using Notepad.
2. Determine the location of the control files.
3. Change the necessary parameter to specify that, in the event of database recovery, 1800 seconds is the maximum amount of time it should take to record dirty buffers to the data files.
4. Save the changed file as **Ch1Project.ora** in your Chapter01 folder.

CASE PROJECTS

Case 1-1 Create a pfile

Create a pfile from the existing spfile for your database. Open the pfile and determine the following (if possible):

1. The size of the Oracle9i data blocks. Use the conversion rate of 1024 bytes = 1 kilobyte to determine the number of kilobytes specified in the file.
2. The number and location of all control files
3. The current setting for the FAST_START_MTTR_TARGET parameter
4. The name of the database

Case 1-2 Format a pfile

Take the pfile created in Case 1-1 and format the file to resemble the example shown in Figure 1-7. When you encounter a parameter contained in the file that was discussed in this chapter, add documentation to the file indicating the purpose of the parameter. To indicate that an entry is documentation rather than a parameter and should be ignored by Oracle9i, precede the line with a # (number sign).

2

ARCHIVING

**After completing this chapter,
you should be able to do the following:**

- Distinguish between ARCHIVELOG and NOARCHIVELOG modes
- Describe the implications of operating a database in ARCHIVELOG mode
- Describe the implications of operating a database in NOARCHIVELOG mode
- Change a database to ARCHIVELOG mode
- Perform manual archiving
- Enable automatic redo log file archiving
- Specify the destination for archived redo log files
- Explain the different archive destination parameters available in Oracle9*i*
- Specify a format for the file name of archived redo log files

In this chapter, the advantages and disadvantages of NOARCHIVELOG and ARCHIVELOG modes and various Init.ora parameters related to the archiving process are presented. In addition, steps to place a database in a specific mode and enable the archiving process are demonstrated.

The archiving mode of the database will be changed several times within the chapter. If you are unable to complete the chapter material during one session, you should place the database in NOARCHIVELOG mode at the end of each session. The necessary steps are listed in the section entitled "Changing to NOARCHIVELOG Mode in SQL*Plus" later in this chapter.

THE CURRENT CHALLENGE IN THE JANICE CREDIT UNION DATABASE

Carlos has created a new database to store physical asset information for the Janice Credit Union. One question he must address is whether the database should be operated in ARCHIVELOG or NOARCHIVELOG mode.

After media failure, an essential ingredient for restoring the contents of a database without data loss is the archived redo log files. Recall that the online redo log files contain changed data. These files are written to in a circular fashion. That is, after all the redo log files are filled, the log writer (LGWR) background process begins overwriting the contents of the first redo log file with new data changes. Unless the files have been archived, or copied, the overwritten data changes are not available if the database needs to be restored.

To specify that the redo log files must be archived before they are overwritten, the database must be in ARCHIVELOG mode. When a database is operated in ARCHIVELOG mode, the LGWR is prevented from overwriting unarchived redo log files. However, they can be overwritten without archiving when the database is operated in NOARCHIVELOG mode.

NOTES ABOUT DUAL COVERAGE WITHIN THIS CHAPTER

In this chapter, procedures are presented using both SQL*Plus and the Enterprise Manager Console. Each section heading will identify which interface will be used in that particular section. When using the Enterprise Manager Console, always remember to log in with the SYSDBA privileges enabled.

NOARCHIVELOG MODE

By default, a database is in NOARCHIVELOG mode when it is created. When changes are made in a database (e.g., data modification, new objects creation, and so forth), these changes are recorded in the online redo log files by the LGWR. As shown in Figure 2-1, after the first redo log file is filled, the LGWR then writes to the next available redo log file. After all the redo log files are full, the LGWR begins overwriting the contents of the first file. This cycle continues throughout the duration of the Oracle9i instance.

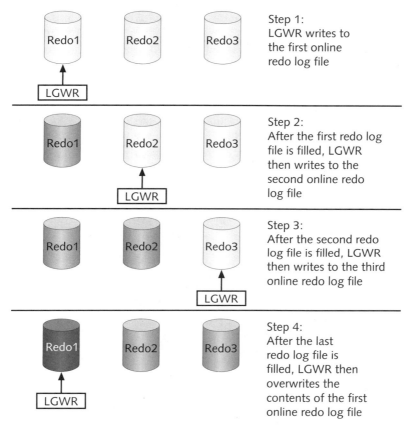

Figure 2-1   Log writing process in NOARCHIVELOG mode

Advantages and Disadvantages of NOARCHIVELOG Mode

There are advantages to operating a database in NOARCHIVELOG mode. First, Carlos would not have to be concerned with the additional storage space that would be required to store the copied files. An inexperienced database administrator tends to forget that the archived files are consuming space. If a database is in ARCHIVELOG mode and the storage location for the archived files becomes full, the archiving process cannot be continued. This eventually results in the database hanging when it attempts to overwrite an unarchived redo log file. Because NOARCHIVELOG mode does not require the files to be archived before they are overwritten, there is no danger in running out of storage space and ultimately preventing further data changes from being made within the database.

This leads to the second advantage of operating a database in NOARCHIVELOG mode—no archiving administrative tasks are required. Carlos would not be concerned about whether the redo log files are actually being archived or if the size of the redo log files are too small. If the redo log files are too small, then the LGWR may fill the redo log files faster than they are being archived. He would then need to decide whether the files should be resized or if additional archiving processes (ARC*n*) should be started. Furthermore, if automatic archiving has not been enabled, then Carlos is required to manually initiate the archiving process.

Although the initial reaction to someone unfamiliar with the Oracle9*i* database environment might be to operate in NOARCHIVELOG mode, there are disadvantages to this approach. The most significant disadvantage from a backup and recovery perspective is that in the event of media failure, any changes made after the last database backup are lost. For an organization that does not have a high volume of transactions, this may be acceptable. However, in today's highly dynamic environment, an organization usually cannot afford to lose data. This is not only due to the potential direct effect on revenue, but also the possible impact on customer and vendor relations and overall operations within the organization.

Another disadvantage to operating a database in NOARCHIVELOG mode relates to the backup process. The only option available when a database is in NOARCHIVELOG mode is a **cold backup**. A cold backup consists of shutting down the database and then copying the physical files through the operating system. The problem lies with shutting down the database. When the database is shut down, it is unavailable to the users. In an organization that is required to have the database available 24 hours per day, 7 days per week, this downtime would be unacceptable.

With the disadvantages of potential data loss and database unavailability, one might wonder why a database would ever be operated in NOARCHIVELOG mode. Most organizations operate only their test database in this mode. An organization typically has two databases: a production database and a test database. The production database is the actual database that is used to store the daily transactions of an organization, while the test database is used for training purposes or to test the effect of application program and database modifications.

 Many organizations have other databases that are used to store historical data for data mining purposes or for development purposes.

Verifying Archive Mode

The first thing Carlos should do is to verify the current archive mode of the TESTDB database. When a database is initially created, it is in NOARCHIVELOG mode. It can be changed at any time by the DBA as long as the database is in a mounted, but not opened, state. In the following sections, the archive mode of the database is verified using SQL*Plus, then again through the Enterprise Manager Console.

Verification Through SQL*Plus

There are two common approaches using SQL*Plus to determine the archiving mode of a database. The first is to retrieve the current mode from the V$DATABASE view. The second is to execute the ARCHIVE LOG LIST command. The following steps demonstrate both approaches.

> Remember to always log into the Oracle9*i* database using the SYSTEM account with the SYSDBA privileges enabled, unless instructions specify otherwise.

To determine the archiving mode of a database:

1. Log into SQL*Plus, type **SELECT log_mode FROM v$database;** at the SQL> prompt, and press **Enter**. As shown in Figure 2-2, the current logging mode is displayed after the command is executed.

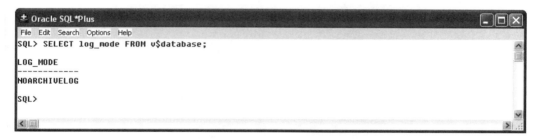

Figure 2-2   Log mode from the V$DATABASE view

2. Type **ARCHIVE LOG LIST** at the SQL> prompt, and press **Enter**. As shown in Figure 2-3, the first item listed in the results is the current logging mode of the database.

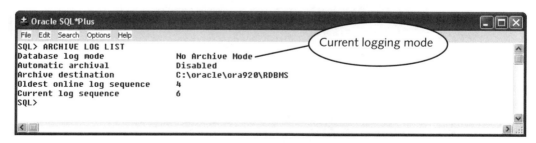

Figure 2-3   Results displayed from execution of the ARCHIVE LOG LIST command

3. Type **EXIT** at the SQL> prompt, and press **Enter** to exit SQL*Plus.

Verification Through Enterprise Manager Console

In this section, the verification process is repeated to familiarize you with some of the functionality of the Enterprise Manager Console.

 Remember to always log into the Oracle9*i* database using the SYSTEM account with the SYSDBA privileges enabled, unless instructions specify otherwise.

To work with the Enterprise Manager Console:

1. Start the Enterprise Manager Console, and then click the **+** (plus sign) next to the database name in the left pane of the window. Provide the appropriate user information when prompted.

2. Click the **+** (plus sign) next to the Instance object located in the list that appears beneath the database name in the left pane of the window, as shown in Figure 2-4.

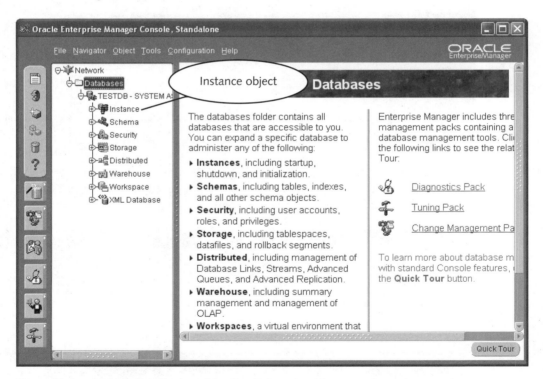

Figure 2-4   Objects list displayed by the Enterprise Manager Console

3. The Instance object appears with a list below it, as shown in Figure 2-5. Click either the word **Configuration** or the icon to the left of the word.

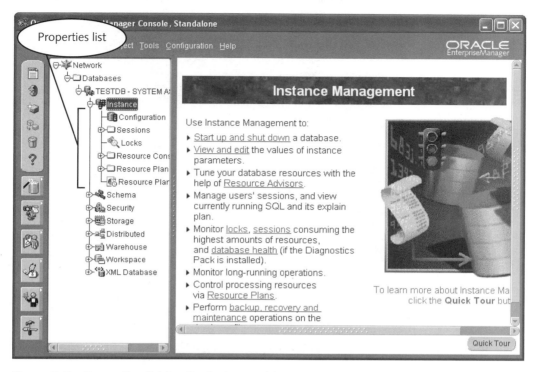

Figure 2-5   Properties list for the Instance object

4. When the general configuration for the database is displayed in the right pane of the window, locate the Archive Log Mode value displayed in the Database and Instance Information section, as shown in Figure 2-6. If the database is currently in NOARCHIVELOG mode, NOARCHIVELOG is displayed.

5. Click **File** on the menu bar, and click **Exit** to exit the Enterprise Manager Console.

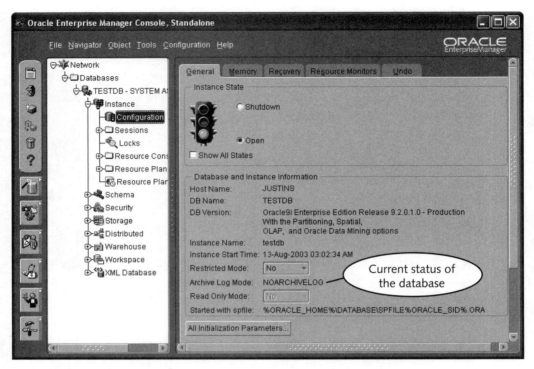

Figure 2-6   Archive Log Mode value for the TESTDB database

ARCHIVELOG MODE

When a database is operated in ARCHIVELOG mode, the redo log files are not over-written until they have been copied, or archived. As shown in Figure 2-7, the LGWR records changes in the first redo log file. When the first redo log file is full, the LGWR begins to record subsequent changes in the second redo log file. When the LGWR switches files, known as a **log switch**, the ARC*n* background process begins creating an archive of the filled redo log file by copying the contents to another file. After the second redo log file is filled, the LGWR begins recording changes into the next available redo log file. If archiving the first redo log file has been completed, the ARC*n* begins copying the second file.

2

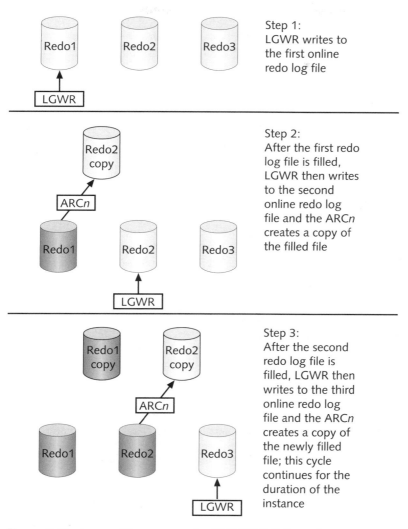

Step 1:
LGWR writes to
the first online
redo log file

Step 2:
After the first redo
log file is filled,
LGWR then writes
to the second
online redo log
file and the ARCn
creates a copy of
the filled file

Step 3:
After the second
redo log file is
filled, LGWR then
writes to the third
online redo log
file and the ARCn
creates a copy of
the newly filled
file; this cycle
continues for the
duration of the
instance

Figure 2-7   Log writing process in ARCHIVELOG mode

A DBA can force a log switch to occur by issuing the `ALTER SYSTEM SWITCH LOGFILE;` command when the database is open. A DBA might force a log switch when renaming or dropping redo log files, or to ensure that the archiving process is working.

When a database has only two redo log files, the LGWR begins overwriting the contents of the first redo log file if the archiving process for that file has been completed. If a large number of transactions occur in a short period of time or a few transactions make a large number of changes to the database, it is possible that the second redo log file is filled before the archiving process for the first redo log file has been completed. If the first file has not been archived, the LGWR cannot switch to that file and record changes. The result is that the database "hangs" until the archiving process has been completed and new changes can be recorded in the redo log file.

Note that the term "hang" basically means that the database does not allow further operations to be performed. This possible scenario is why Oracle recommends that a database should have at least three redo log files. In the event that the second redo log file becomes full while the first file is still being copied, the LGWR can begin storing changes into a third redo log file.

How Many Redo Log Files Do You Need?

Oracle9*i* allows a maximum of ten redo log files. The contents of the first redo log file can be overwritten only after the file has been archived and all other available redo log files have been filled. There is no set number of redo log files that every organization should have for a database. For example, the new database for the physical assets information is not updated as frequently as the financial database that stores customers' deposits and withdrawals.

Of course, Carlos would need to monitor how frequently the LGWR is unable to perform a log switch. If he notices that log switches are being delayed, additional redo log files can be added to the database or additional archiving (ARC*n*) processes can be started. Oracle9*i* allows a maximum of ten redo log files per database and a maximum of ten archiving processes per Oracle9*i* instance.

A database can also have automatic archiving enabled. When automatic archiving is enabled, the ARC*n* process archives a redo log file without DBA intervention. When automatic archiving is disabled, the DBA must enter the appropriate command to trigger the archiving process.

Advantages and Disadvantages of ARCHIVELOG Mode

Basically, the advantages and disadvantages to operating a database in ARCHIVELOG mode are the reverse of NOARCHIVELOG mode. Advantages include:

- The database can be recovered without data loss in the event of a media failure.
- A hot backup can be performed allowing users access to the database during the backup process.

If a media failure occurs, the database is recreated by retrieving files from the backup copy of the database. However, changes made to the database after the last backup was performed are not included in those files. Luckily, those changes are stored in the archived redo log files. Carlos would simply need to apply the contents of the archived files to the restored database resulting in recovering the database up to the point that the media failure occurred. In other words, no data will have been lost. However, this scenario is based upon one major assumption—the archived files and the database backup are not stored on the same drive that had the media failure.

The second advantage is that a **hot backup** can be performed on a database that is operating in ARCHIVELOG mode. During a hot backup, the users are able to access the database. After the tablespace has been backed up, changes made during the backup process and the current system change number (SCN) are then applied to the tablespace to bring it up to date. This provides flexibility to a DBA such as Carlos because the database does not have to be shut down at regular intervals to perform a backup.

However, there are drawbacks to operating a database in ARCHIVELOG mode. These include additional storage space requirements and increased administrative and maintenance tasks for the DBA such as Carlos. If left unchecked, the archived files can consume a tremendous amount of storage space. If the database cannot archive the redo log files, the database hangs when the LWGR attempts to overwrite an unarchived file. In most organizations, only one or two days worth of archived log files are kept on a hard drive. Earlier archived files that were created before the last cold backup would not be necessary for data recovery and simply can be deleted. This, of course, increases the maintenance tasks that Carlos would need to be perform.

Changing Archive Mode

A database can be placed in ARCHIVELOG mode using the ALTER DATABASE command with the ARCHIVELOG logfile clause. This command must be executed while the database is mounted, but not opened. In addition, the DBA issuing the command must have the ALTER DATABASE system privilege. After the command is executed, the database is opened using the ALTER DATABASE command with the OPEN startup clause so that users can access the database. The following section will demonstrate how to change the archiving mode of the TESTDB database using SQL*Plus and the Enterprise Manager Console.

Changing to ARCHIVELOG mode with SQL*Plus

 Remember to always log into the Oracle9*i* database using the SYSTEM account with the SYSDBA privileges enabled, unless instructions specify otherwise.

1. After logging into SQL*Plus, type **SHUTDOWN** at the SQL> prompt, and press **Enter** to shut down the database, as shown in Figure 2-8.

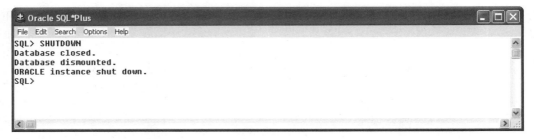

Figure 2-8   Shutting down the database in SQL*Plus

2. Type **STARTUP MOUNT** at the SQL> prompt, and press **Enter** to mount the database, as shown in Figure 2-9.

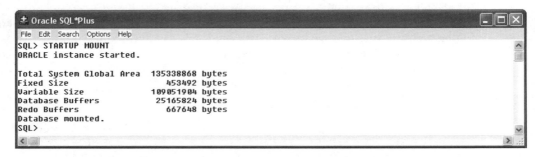

Figure 2-9   Mounting the database

3. Type **ALTER DATABASE ARCHIVELOG;** at the SQL> prompt, and press **Enter** to place the database in ARCHIVELOG mode. After the command is executed, the message shown in Figure 2-10 is displayed.

Figure 2-10   Execution of command to place database in ARCHIVELOG mode

4. Type **ALTER DATABASE OPEN;** at the SQL> prompt, and press **Enter** to open the database and make it accessible to the users. As shown in Figure 2-11, the SQL> prompt is displayed after the database has been opened.

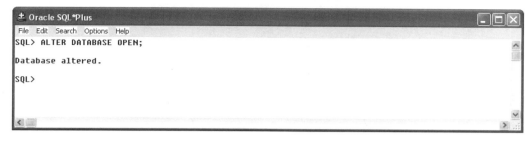

Figure 2-11   Opening the database after it was placed in ARCHIVELOG mode

5. Type ARCHIVE LOG LIST at the SQL> prompt, and press Enter to verify that the database is now in ARCHIVELOG mode. The DATABASE LOG mode displays the value of "Archive Mode" as shown in Figure 2-12.

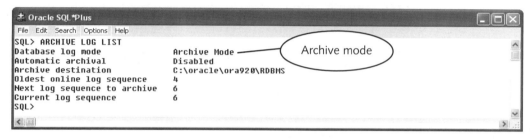

Figure 2-12   Verifying archive mode

Changing to NOARCHIVELOG Mode in SQL*Plus

Before demonstrating how to change the archiving mode through the Enterprise Manager Console, the database needs to be placed back into NOARCHIVELOG mode. The following steps switch the database back to NOARCHIVELOG mode.

Remember to always log into the Oracle9i database using the SYSTEM account with the SYSDBA privileges enabled, unless instructions specify otherwise.

1. Type SHUTDOWN at the SQL> prompt, and press Enter to shut down the database, as shown in Figure 2-13.

Figure 2-13 Shutting down the database

2. Type **STARTUP MOUNT** at the SQL> prompt to mount the database. After the database has been mounted, the statistics shown in Figure 2-14 are displayed.

3. Type **ALTER DATABASE NOARCHIVELOG;** at the SQL> prompt, and press **Enter** to switch the database to NOARCHIVELOG mode, as shown in Figure 2-15.

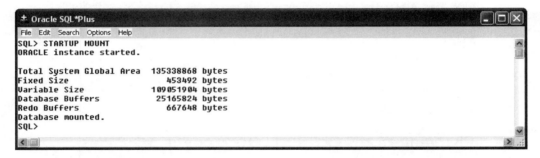

Figure 2-14 Mounting the database

Figure 2-15 Switching the database to NOARCHIVELOG mode

4. Type **ALTER DATABASE OPEN;** at the SQL> prompt to open the database, as shown in Figure 2-16.

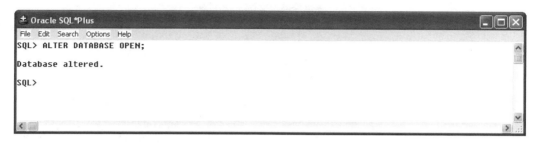

Figure 2-16   Opening the database after switching to NOARCHIVELOG mode

5. Type **EXIT** at the SQL> prompt, and press **Enter** to exit SQL*Plus before proceeding to the Enterprise Manager Console.

 Although a user can be logged into both SQL*Plus and the Enterprise Manager Console at the same time, doing so can quickly cause the maximum number of allocated processes to be reached, preventing other users from accessing the database.

Changing Archive Mode with the Enterprise Manager Console

This section will demonstrate how to perform the task of placing a database in ARCHIVELOG mode through the Enterprise Manager Console.

1. After starting and logging into the Enterprise Manager Console, click the **+** (plus sign) next to the Instance object for the database.

2. Click **Configuration** from the Instance properties list. The right pane of the window displays the General configuration for the database, as shown in Figure 2-17.

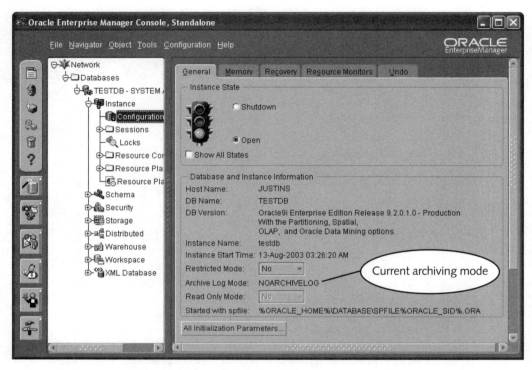

Figure 2-17   General configuration of the database

3. Verify that the current Archive Log Mode of the database is NOARCHIVELOG, as shown in Figure 2-17.

4. Click the **Recovery** tab at the top of the Configuration window to display Recovery information, as shown in Figure 2-18.

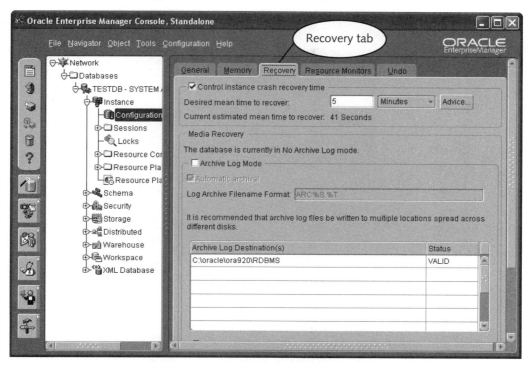

Figure 2-18   Recovery properties for the database

5. Check the **Archive Log Mode** check box in the Media Recovery portion of the window. After checking the check box, a checkmark appears, as shown in Figure 2-19.

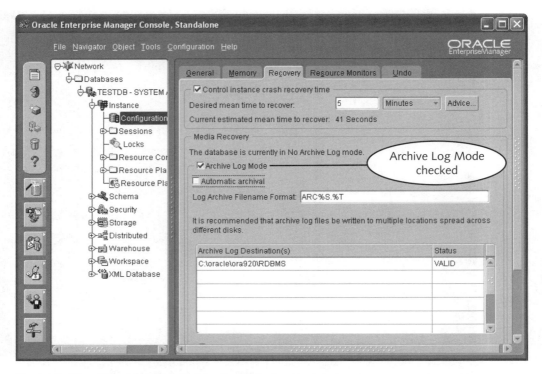

Figure 2-19   Checking the Archive Log Mode check box

6. If a checkmark also appears in the Automatic archival check box beneath the
Archive Log Mode check box, uncheck the check box to indicate that manual
archiving mode is desired.

7. Click the **Apply** button at the bottom of the screen to have Oracle9*i* switch
the database to ARCHIVELOG mode. If the button is not visible, scroll
down using the window's scroll bar shown in Figure 2-20 until the button
appears. Then click it.

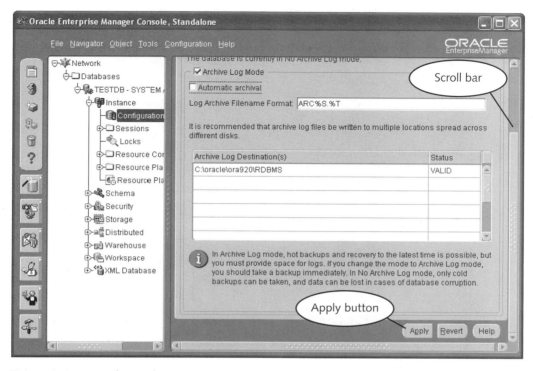

Figure 2-20   Applying changes

8. The Shutdown Options window shown in Figure 2-21 appears next. Click **OK** to bounce (i.e., shutdown and restart) the database.

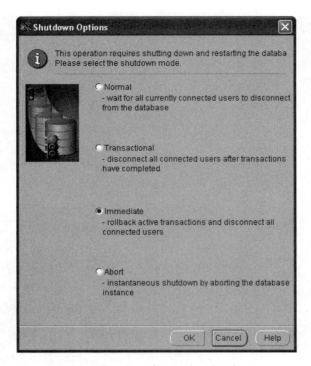

Figure 2-21   Shutting down the database

9. After the database has been reopened, click **Close**, as shown in Figure 2–22, to close the dialog box and return to the Enterprise Manager Console.

Figure 2-22   Shutting down and restarting the database completed

10. After the General Configuration of the database is visible, verify that the Archive Log Mode is now listed as ARCHIVELOG, as shown in Figure 2-23.

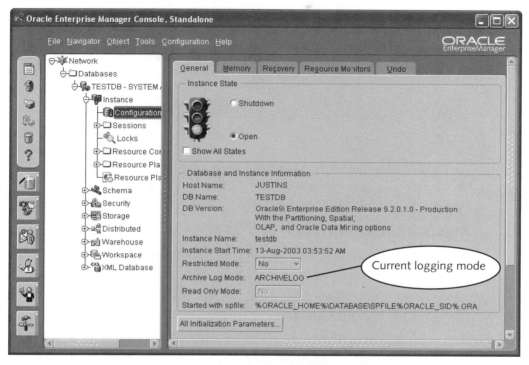

Figure 2-23   Database restarted in ARCHIVELOG mode

11. Click **File** and then **Exit** to exit the Enterprise Manager Console.

Manual Archiving

Notice back in Figure 2-20 that automatic archiving was disabled even after execution of `ALTER DATABASE ARCHIVELOG;`. To archive any filled redo log files, Carlos must issue the `ALTER SYSTEM` command with the appropriate ARCHIVE_LOG clause. Some of the keywords that can be used with the `ALTER SYSTEM` command are provided in Table 2-1.

Table 2-1   ALTER SYSTEM command keywords

Keyword	Description
ALL	Manually archives all filled online redo log files
NEXT	Manually archives the next filled online redo log file
START	Enables automatic archiving
STOP	Disables automatic archiving

To issue the ALTER SYSTEM command with an ARCHIVE_LOG clause, the SYSDBA or SYSOPER privilege must be enabled.

In the following sections, manual archiving of the redo log files will be demonstrated. To have the database respond as if all the redo log files have been filled and to cause the database to hang, a script file named Ch02Script.sql needs to be executed.

Manual Archiving with SQL*Plus

To perform manual archiving with SQL*Plus, complete the following steps.

To manually archive:

1. Log into SQL*Plus and type **start <*path*> ch02script.sql**, where <*path*> is the location of the script file, as shown in Figure 2-24, to generate log switches, which force the redo log files to be archived. Press **Enter**.

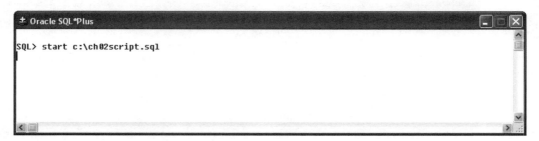

Figure 2-24   Execution of script file

The database hangs when the script file is executed because the online redo log files need to be archived. To manually archive the redo log files, the DBA must create another session with the Oracle9*i* server to execute the necessary command.

2. Open SQL*Plus again in another window, and log into the database. Do not close the SQL*Plus window used in Step 1.

3. When the SQL> prompt appears in the second SQL*Plus window, type **ARCHIVE LOG ALL**, and press **Enter** to archive all filled redo log files. As shown in Figure 2-25, there are two redo log files that need to be archived.

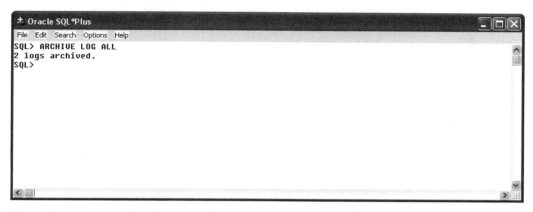

Figure 2-25   Manual archiving of filled redo log files

4. After the message appears that the redo log files have been archived, type **EXIT** at the SQL> prompt, and press **Enter** to close the second SQL*Plus window. After the first SQL*Plus window is visible again, the SQL> prompt is displayed, as shown in Figure 2-26.

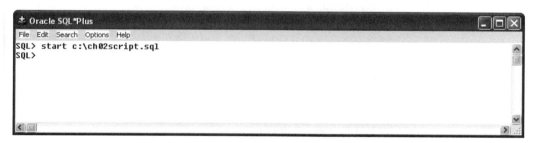

Figure 2-26   Database is again available after manual archiving

5. Type **ARCHIVE LOG ALL** at the SQL> prompt, and press **Enter**. Because none of the redo log files need archiving, the error message shown in Figure 2-27 is displayed. This message simply indicates that the procedure is not necessary at this time, not that a problem exists with the database.

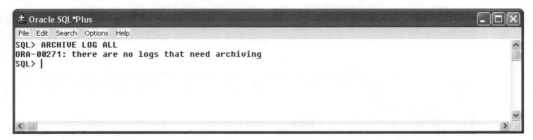

Figure 2-27   Error message from attempting to unnecessarily archive the redo log files

 If the second SQL*Plus window appears to "freeze" and become inaccessible, you can use the operating system's Task Manager to end the process.

 Because database configurations vary, you may need to repeat Step 5 before the error message is displayed.

6. Type **EXIT** at the SQL> prompt, and press **Enter** to exit SQL*Plus.

Manual Archiving with the Enterprise Manager Console

Next, the manual archiving process for the filled redo log files will be performed through the Enterprise Manager Console. The Ch02script.SQL script file will be used again to cause the database to hang creating the need to manually archive the files.

To archive the files:

1. Start the Enterprise Manager Console and log into the database, and then click the **Database Applications** button on the left edge of the window. After the application buttons have appeared as shown in Figure 2-28, click the **SQL*Plus Worksheet** button.

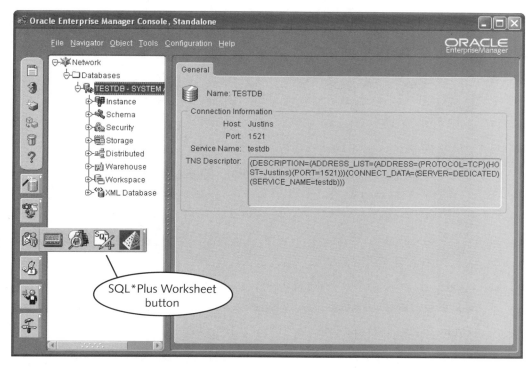

Figure 2-28   SQL*Plus Worksheet button

2. After the SQL*Plus Worksheet window appears, type **START**
   <path>\ch02script.sql>, where *<path>* is the location of the file in the
   Chapter02 folder. Click the **Execute** button, as shown in Figure 2-29, to execute the command.

3. Click the **Oracle Enterprise Manager** button on the Windows XP taskbar
   to make the Enterprise Manager Console visible. After it is visible, click the **+**
   (plus sign) next to Instance, and then click **Configuration**.

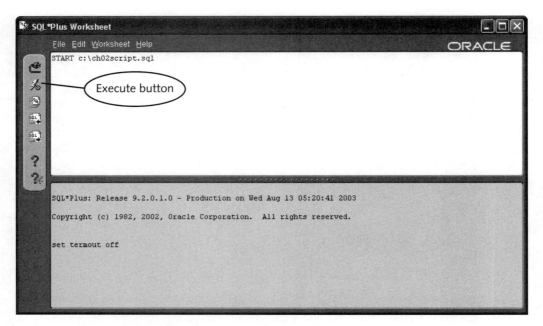

Figure 2-29   Execution of script file SQL*Plus Worksheet

4. From the Object menu bar in the Enterprise Manager Console, point to Manually Archive. After the submenu appears, click the **All** option to archive all filled redo log files, as shown in Figure 2–30.

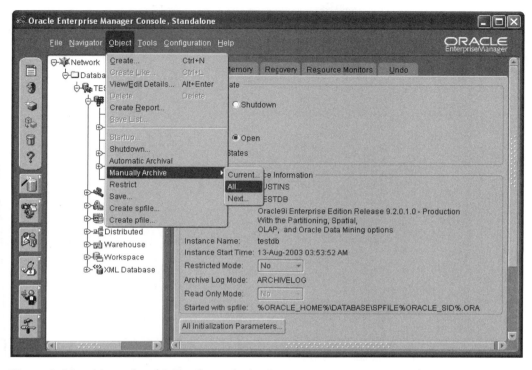

Figure 2-30   Manual archiving through the Enterprise Manager Console

5. A dialog box appears prompting for the location to store the archived redo log files. Click **OK** to accept the Destination directory already displayed in the text box, as shown in Figure 2-31.

Figure 2-31   Dialog box to specify location for the archived redo log files

6. After the filled archive redo log files have been archived, attempt to manually archive the redo log files again by repeating Steps 4 and 5. If no files need archiving, the error message displayed in Figure 2-32 appears. Click **OK** to close the dialog box.

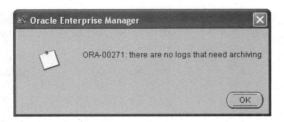

Figure 2-32 Error message

7. Return to the SQL*Plus Worksheet and click **File**, then **Exit** to exit the application, as shown in Figure 2-33.

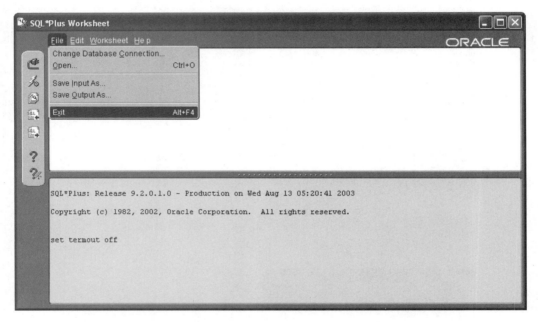

Figure 2-33 Exiting the SQL*Plus Worksheet

8. After the SQL*Plus Worksheet window has closed, exit the Enterprise Manager Console.

Automatic Archiving

Rather than having a DBA such as Carlos initiate the archiving process every time the redo log files become full, automatic archiving can be enabled. There are two methods to enable automatic archiving. The first method is to issue the ALTER SYSTEM command with the ARCHIVE LOG START clause. However, this change is not permanent because if the database is restarted, the value assigned in the initialization file is used to determine whether automatic archiving is enabled.

The second method is to actually change the LOG_ARCHIVE_START parameter of the Init.ora file. This ensures the change is permanent, regardless of how many times the database is shut down and restarted. By default, the parameter is assigned the Boolean value of FALSE. To enable automatic archiving, simply edit the pfile and change the assigned value to TRUE. Because the ALTER SYSTEM command was demonstrated in the previous section, the following example changes the LOG_ARCHIVE_START parameter to enable automatic archiving.

Enabling Automatic Archiving with SQL*Plus

To enable automatic archiving with SQL*Plus, execute the following steps.

To enable automatic archiving:

1. Create a pfile from the current spfile using the CREATE PFILE FROM SPFILE command.

2. Use the Find feature of your text edit or to locate the LOG_ARCHIVE_START parameter. If it already exists in the pfile, change the value to **FALSE**. Then save the file.

3. If the Find feature from Step 2 did not find the LOG_ARCHIVE_START parameter, move the text editor's cursor to the end of the file and type ***.log_archive_start = TRUE**, as shown in Figure 2-34, to add the LOG_ARCHIVE_START parameter and enable automatic archiving.

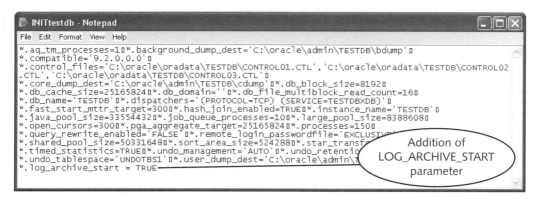

Figure 2-34 LOG_ARCHIVE_START parameter added to the Init.ora file

Because the database is started by default with the parameters contained in the spfile, the database must be restarted with the modified contents of the pfile. At this point, there are two options: recreate the spfile from the pfile or issue a startup command that references the pfile. In the following steps, the latter approach is used.

4. Log into SQL*Plus and then shut down the database using the SHUTDOWN command.

5. At the SQL> prompt, type **startup pfile = '<*path\filename*>'** where *path* represents the location of the file and the *filename* is the name of the file, as shown in Figure 2-35.

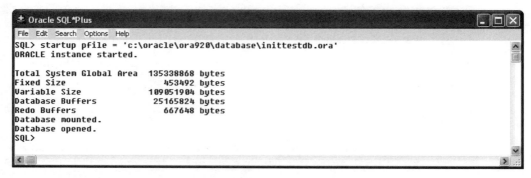

Figure 2-35   Restarting the database with the pfile

6. After the database has been restarted, type **SHOW PARAMETER LOG_ARCHIVE_START** at the SQL> prompt, as shown in Figure 2-36, and press **Enter** to display the value of this parameter.

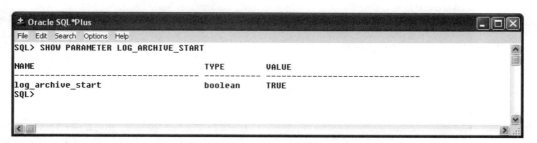

Figure 2-36   Value of the LOG_ARCHIVE_START parameter

7. Exit SQL*Plus by typing **EXIT** at the SQL> prompt, and pressing **Enter**.

Note that an alternative to the SHOW PARAMETERS command shown in Step 5 is to execute the ARCHIVE LOG LIST command. The results of this command are shown in Figure 2-37.

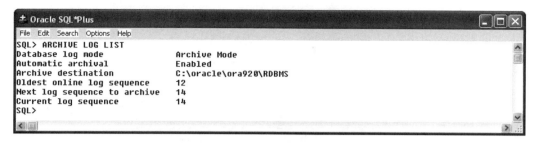

Figure 2-37   Status of automatic archiving

Before practicing the steps necessary to enable automatic archiving through the Enterprise Manager Console, the pfile must be edited and either the LOG_ARCHIVE_START parameter set to FALSE or the parameter deleted. Because the database was started with the pfile, the spfile was never updated. Therefore, the pfile can simply be changed without having to recreate the spfile.

To edit the pfile:

1. Open the pfile in Notepad and if the LOG_ARCHIVE_START parameter was added in Step 2 of the preceding step sequence, delete the text addition. Otherwise, simply change the parameter value to FALSE.

2. After making the necessary change to the pfile, save the modified file and exit Notepad.

Enabling Automatic Archiving with the Enterprise Manager Console

To enable automatic archiving with the Enterprise Manager Console, execute the following steps.

To enable automatic archiving:

1. After starting and logging into the Enterprise Manager Console, display the Configuration properties for the database instance.

2. Click the **All Initialization Parameters** button in the right pane of the window. The button is shown in Figure 2-38.

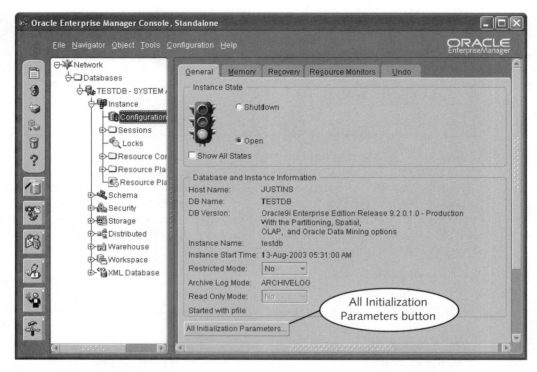

Figure 2-38   General configuration properties

3. When the Initialization Parameters window appears, scroll down to the LOG_ARCHIVE_START parameter, as shown in Figure 2-39.

Parameter Name	Value	Default	Dynamic	Category
log_archive_start	FALSE			Archive
log_archive_trace	0	✔	✔	Archive
log_buffer	524288	✔		Redo Log and Recovery

Figure 2-39   Initialization parameters (partial display)

4. Click the **value** column of the LOG_ARCHIVE_START parameter, and click **TRUE** from the drop-down list that appears.

5. Click the **Apply** button shown in the lower-right corner of Figure 2-40 to reset the parameter value.

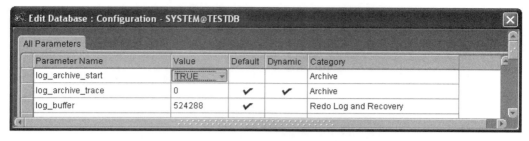

Parameter Name	Value	Default	Dynamic	Category
log_archive_start	TRUE			Archive
log_archive_trace	0	✔	✔	Archive
log_buffer	524288	✔		Redo Log and Recovery

Figure 2-40   New value selected for the LOG_ARCHIVE_START parameter (partial display)

If a window appears suggesting modifications for the pfile, click the **Close** button.

6. When the Shutdown Options window appears, as shown in Figure 2–41, click **OK** to shut down and restart the database.

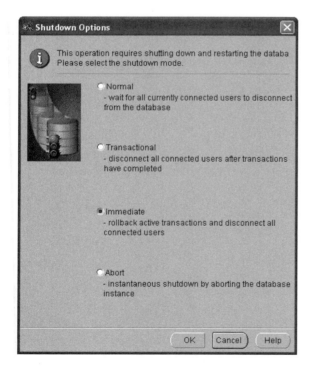

Figure 2-41   Shutting down the database

7. After the database has been shut down and restarted, click **Close** to close the Bouncing database window shown in Figure 2-42.

Figure 2-42   Shutdown and restart of database completed

8. Click the **Recovery** tab at the top of the right pane of the Enterprise Manager Console window and verify that the check box for Automatic archival contains a checkmark, as shown in Figure 2-43.

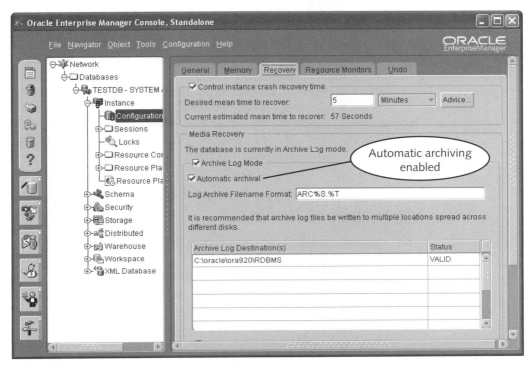

Figure 2-43   Current setting for automatic archival

When attempting to enable automatic archiving for the first time through the Enterprise Manager Console, most novices try clicking the Automatic archival check box rather than changing the LOG_ARCHIVE_START parameter. However, this approach enables automatic archiving only for the current instance. To make the change permanent, the initialization parameter must be changed to TRUE.

ARCHIVING PARAMETERS

Different parameters can be included in the Init.ora file to specify the number of copies that should be made of the archived redo log files, where they should be stored, and even the format of the file name for the archived files. In addition, Carlos can specify how many archiving processes should be used to create the archived files.

Destination Parameters

By default, the archived redo log files are stored in a folder named RDBMS. However, there are two sets of parameters that can be used to specify the storage location of the archived redo log files. These parameters can also be used to ensure that multiple copies of the archived files are created. Why have multiple copies of the same file? If only one copy of the archived files exists, and it becomes inaccessible (e.g., media or hard drive failure), then the subsequent archived files are useless. Therefore, if there are multiple copies, and each copy is stored on a different drive, at least one copy should always be available if database recovery becomes necessary because of the failure of a single hard drive. It seems that redundancy is not always a bad thing in the world of databases.

LOG_ARCHIVE_DEST and LOG_ARCHIVE_DUPLEX_DEST Parameters

The first set of parameters consists of the LOG_ARCHIVE_DEST and LOG_ARCHIVE_DUPLEX_DEST parameters. The parameter(s) can simply be typed into the Init.ora file, followed by the desired archive location, as shown in Figure 2-44. Together, these parameters allow Carlos to specify a maximum of two locations for the archived redo log files.

Note that some DBAs prefer to have more than two copies of the files. Also, larger organizations have the flexibility of storing files on a file server located on a network. Although this scenario cannot be supported by the LOG_ARCHIVE_DEST and LOG_ARCHIVE_DUPLEX_DEST parameters, it is supported by the LOG_ARCHIVE_DEST_N parameter.

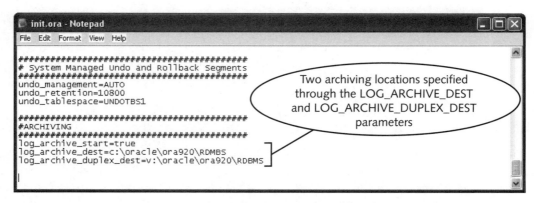

Figure 2-44   Example Init.ora file with LOG_ARCHIVE_DEST and LOG_ARCHIVE_DUPLEX_DEST parameters included (partial file shown)

LOG_ARCHIVE_DEST_*N* Parameter

The LOG_ARCHIVE_DEST_*N* parameter, where *N* is assigned a value from 1 to 10, allows Carlos to specify up to ten locations for storing the archived files. In addition, one of these locations can be on a different computer, such as a file server or a remote location for an alternate computer site, to be used in the event of a natural or man-made disaster at the primary computer center. When LOG_ARCHIVE_DEST_*N* parameters are listed in the Init.ora file, the DBA cannot include the LOG_ARCHIVE_DEST and LOG_ARCHIVE_DUPLEX_DEST parameters. In other words, these parameters are mutually exclusive.

An example Init.ora file containing the LOG_ARCHIVE_DEST_*N* parameter is shown in Figure 2-45. Notice the syntax for specifying the archiving destinations. First, the destinations are listed within double quotation marks. Second, if the destination for the archived file is located on the same computer as the Oracle9*i* server, then the LOCATION keyword is used. However, to reference a remote location such as an archival database on another computer, the SERVICE keyword is required.

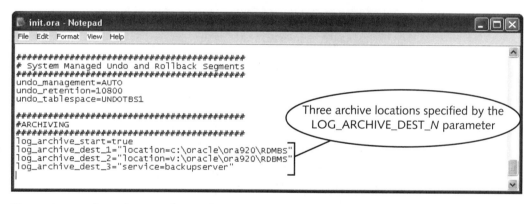

Figure 2-45 Specifying archiving locations with the LOG_ARCHIVE_DEST_*N* parameter (partial file shown)

Destination Availability

Suppose the database attempts to archive a redo log file and one of the destinations specified by a destination parameter is unavailable. Do all copies of the filled redo log need to be made before database operations can resume, or just one or two? The LOG_ARCHIVE_MIN_SUCCEED_DEST parameter can be included in the Init.ora file to indicate the number of "local" archived copies of the redo log file that must be made before the file is overwritten. In addition, Carlos can specify exactly which destination(s) must successfully receive a copy of the file by including the MANDATORY keyword.

Note that the keyword OPTIONAL can be used to indicate that archiving to a specified destination is not required, as long as archiving to the destinations that have been specified as mandatory is successful. Optional archiving locations are specified to provide additional copies of the archived redo log files but are not necessarily the first copies that would be retrieved by a DBA in the event of media failure.

In the example given in Figure 2-46, there are three archive destinations included in the Init.ora file. The destination specified by the LOG_ARCHIVE_DEST_1 parameter contains the keyword MANDATORY to indicate that at a minimum, this location must receive a copy of the archived log file. The destination specified by LOG_ARCHIVE_DEST_2, however, is denoted as OPTIONAL.

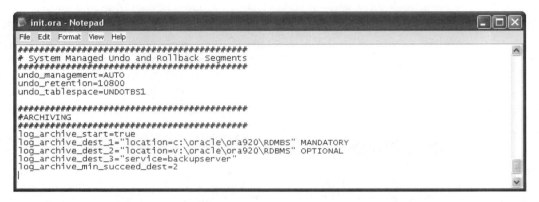

Figure 2-46   Destinations that must receive a copy of the archived redo log files (partial file shown)

The LOG_ARCHIVE_MIN_SUCCEED_DEST parameter has also been added to the Init.ora file shown in Figure 2-46 to indicate that two locations must receive a copy of the archived file. Although only one of the LOG_ARCHIVE_DEST_*N* parameters is specified as MANDATORY, the value of two assigned to the LOG_ARCHIVE_MIN_SUCCEED _DEST parameter requires that both local destinations receive a copy of the archived file before the online redo log files are overwritten. In other words, if the value of the LOG_ARCHIVE_MIN_SUCCEED _DEST is greater than the number of local destinations that are designated as MANDATORY, that value overrides any OPTIONAL keywords until the required number of destinations receive a copy of the archived file. When locations are specified by the LOG_ARCHIVE_DEST and LOG_ARCHIVE_DUPLEX_DEST parameters, the location specified by the first parameter is required (i.e., MANDATORY by default) while the second location is optional.

Filename Parameter

To make the names of archived log files more informative, a format can be specified for the names assigned to the archived files. The format for the file name is indicated using the LOG_ARCHIVE_FORMAT parameter in the Init.ora file. The symbols that can be used in the file name and/or extension of an archived log file are given in Table 2-2.

Table 2-2 File name options

Option	Description
%s	Includes the log sequence number as part of the file name
%S	Includes the log sequence number padded to the left with zeros as part of the file name
%t	Includes the thread number as part of the file name
%T	Includes the thread number padded to the left with zeros as part of the file name

In the example shown in Figure 2-47, the LOG_ARCHIVE_FORMAT parameter has been added to the Init.ora file with the format model of ARC%S.%T to be used for the file name of the archive files.

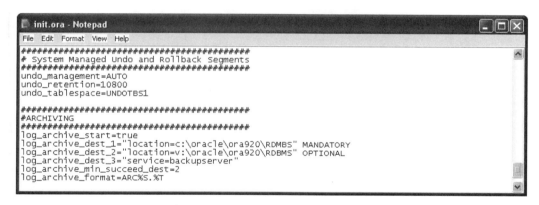

Figure 2-47 LOG_ARCHIVE_FORMAT parameter (partial file shown)

The format model requires that the file name given to each copy of the archived redo log file begin with the letters ARC followed by the log sequence number. The %T included in the format model indicates that the thread number is to be used as the extension for the file. When used, "S" and "T" are capitalized. Note the use of zeros (for padding) in Figure 2-48.

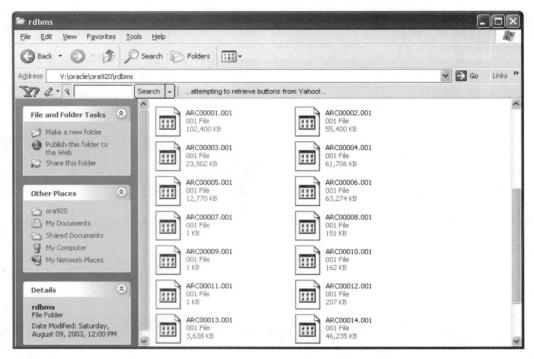

Figure 2-48   Filenames of archived redo log files

If you are not completing the hands-on assignments at this time, the database should be placed back in NOARCHIVELOG mode after completing the steps demonstrated in this chapter.

CHAPTER SUMMARY

❏ By default, databases are created in NOARCHIVELOG mode unless specified otherwise. Unless an organization can afford potential data loss, databases are normally operated in ARCHIVELOG mode.

❏ In NOARCHIVELOG mode, the redo log files are eventually overwritten after they are filled. In ARCHIVELOG mode, the redo log files are copied, or archived, before they are overwritten.

❏ Data loss can result when media failure occurs if the database is in NOARCHIVELOG mode.

❏ A database in NOARCHIVELOG mode can be backed up only by using the cold backup approach, which requires the database to be shut down during the backup process.

2

❐ A database in ARCHIVELOG mode can be backed up while the database is still in use using the hot backup approach. This allows the database to be recovered to a specific point in time without any data loss.

❐ To enable automatic archiving, set the LOG_ARCHIVE_START parameter in the Init.ora file to TRUE. Setting the parameter to FALSE requires the intervention of the DBA to manually archive the log files, which is not the typical implementation.

❐ The V$DATABASE view can be queried to determine whether the database is in ARCHIVELOG mode.

❐ To change the archive mode of a database, the user must be connected to the database using an account with the SYSDBA role.

❐ The database must be mounted, but not opened, to issue the ALTER DATABASE command to change the archiving mode of the database

❐ The ARCHIVE LOG LIST command can be used to determine whether automatic archiving is enabled and whether archiving is enabled.

❐ Filled redo log files can be archived manually by executing the ARCHIVE LOG command.

❐ Manual archiving options are available from the Objects menu located on the menu bar in the Enterprise Manager Console.

❐ Two sets of mutually exclusive parameters are available to indicate how many copies of the archived redo log files should exist and what the paths are to their locations. The parameters are stored in the Init.ora file.

❐ The LOG_ARCHIVE_MIN_SUCCEED_DEST parameter can be used to indicate how many copies of the archived log files must successfully reach their destination before the redo log files can be overwritten.

❐ A format can be applied to the archived redo log files to make it easier to identify files when needed and to make the names more meaningful.

SYNTAX GUIDE

Command	Description	Example
ALTER DATABASE ARCHIVELOG	Places the database in ARCHIVELOG mode; can be executed only when the database is mounted, but not opened	`ALTER DATABASE ARCHIVELOG;`
ALTER DATABASE NOARCHIVELOG	Places the database in NOARCHIVELOG mode; can be executed only when the database is mounted, but not opened	`ALTER DATABASE NOARCHIVELOG;`

Parameter	Description	Example
ALTER SYSTEM SWITCH LOGFILE	Forces a switch operation for the redo log files and archives the files if the database is in ARCHIVELOG mode	`ALTER SYSTEM SWITCH LOGFILE;`
ARCHIVE LOG ALL	Manually archives all filled redo log files	`ALTER LOG ALL`
ARCHIVE LOG LIST	Displays the current archive settings for the database	`ARCHIVE LOG LIST`
ARCHIVE LOG NEXT	Manually archives the next filled redo log file	`ARCHIVE LOG NEXT`
LOG_ARCHIVE _START	Used to enable/disable automatic archiving. An assigned value of TRUE enables automatic archiving; a value of FALSE requires manual archiving	`LOG_ARCHIVE_START = TRUE`
LOG_ARCHIVE _DEST	Used to specify the location of the archived redo log files	`LOG_ARCHIVE_DEST = c:\oracle\ ora920\rdbms`
LOG_ARCHIVE _DEST_N	Used to specify the location of the archived redo log files; N represents a number (1-10) to specify a maximum of ten locations for storing the archived files	`LOG_ARCHIVE_DEST_1 = "location=v:\ oracle\ ora920\rdbms"`
LOG_ARCHIVE_ DUPLEX_DEST	Used in conjunction with the LOG_ARCHIVE_DEST parameter to specify a second location for the archived redo log files	`LOG_ARCHIVE_ DUPLEX_DEST = v:\oracle\ ora920\rdbms`
LOG_ARCHIVE_ MIN_SUCCEED _DEST	Specifies how many of the specified destinations for the archived redo log files must succeed	`LOG_ARCHIVE_MIN_ SUCCEED_DEST = 2`
LOG_ARCHIVE_ FORMAT	Specifies the format for the file name assigned to an archived file	`LOG_ARCHIVE_FORMAT = ARC%S.%T`

REVIEW QUESTIONS

1. Which of the following archiving modes requires the database to be shut down to back up the database?

   a. NOARCHIVELOG mode

   b. ARCHIVELOG mode with automatic archiving enabled

   c. ARCHIVELOG mode with automatic archiving disabled

   d. all of the above

2

2. In what state must the database be when changing a database into NOARCHIVELOG mode?

   a. started

   b. allocated

   c. mounted

   d. opened

3. Which type of backup requires that the database be unavailable to users during the backup process?

   a. cold

   b. lukewarm

   c. warm

   d. hot

4. What is the maximum number of locations that can be specified by the LOG_ARCHIVE_DEST_N parameter?

   a. 25

   b. 10

   c. 5

   d. 2

5. What is one disadvantage to operating a database in NOARCHIVELOG mode?

6. The LOG_ARCHIVE_START parameter must be assigned which value to specify that manual archiving is required?

   a. ENABLE

   b. TRUE

   c. DISABLE

   d. FALSE

7. If a database that is in NOARCHIVELOG mode experiences a media failure, the database has to be re-created from scratch. True or False?

8. Which of the following commands can be used to archive filled redo log files?

   a. ALTER DATABASE ARCHIVEFILES;

   b. ALTER DATABASE ARCHIVE;

   c. ARCHIVE ALL LOGS

   d. ARCHIVE LOG ALL

9. Which of the following is an accurate statement?

   a. Archived log files should be stored on the same hard drive as the Oracle9*i* data files.

   b. Archived log files should be stored on the same hard drive as the Oracle9*i* program files.

   c. Archived log files should be stored on a different partition of the same hard drive as the Oracle9*i* redo log files.

   d. none of the above

10. What are some of the advantages to operating a database in ARCHIVELOG mode?

11. A tablespace cannot be taken offline immediately if the database is in which archive mode?

    a. NOARCHIVELOG mode

    b. ARCHIVELOG mode with automatic archiving enabled

    c. ARCHIVELOG mode with automatic archiving disabled

    d. all of the above

12. Which of the following commands can be used to display the current archive mode and archive file destination for a database?

    a. ARCHIVE LOG ALL

    b. SHOW LOG LIST

    c. SHOW LOGGING PARAMETER

    d. SHOW PARAMETERS

13. By default, the archived redo log files are stored in the _____ directory.

14. Which of the following keywords is used to denote an archival database in a remote location in the LOG_ARCHIVE_DEST_*N* parameter?

    a. LOCATION

    b. REMOTE

    c. SERVICE

    d. GOTO

15. Which of the following parameters can be used to indicate the number of archive destinations that must successfully be reached before the online redo log files can be overwritten?

    a. LOG_ARCHIVE_DEST_*N*

    b. LOG_ARCHIVE_MIN_SUCCEED_DEST

    c. LOG_ARCHIVE_SUCCEED_MIN

    d. LOG_ARCHIVE_MIN_SUCCEED

16. To indicate that a specific archive destination must be reached before the online redo log files can be overwritten, the word _____ should be included in the value assigned to the LOG_ARCHIVE_DEST_N parameter.

    a. MANDATORY

    b. OPTIONAL

    c. DEFAULT

    d. SUCCEED

17. Which of the archiving modes is more appropriate for an organization that needs the database available 24/7 (24 hours a day/seven days a week) and why?

18. Which of the following format options includes the log sequence number in the file name of the archived log files?

    a. %T

    b. %S

    c. %t

    d. %L

19. A database must be mounted before it can be changed from ARCHIVELOG to NOARCHIVELOG mode. True or False?

20. Which of the following options allows the database to continue operation without hanging even if the specified archive log destination does not receive a copy of the archived file?

    a. MANDATORY

    b. DEFAULT

    c. OPTIONAL

    d. NOSUCCEED

HANDS-ON ASSIGNMENTS

Assignment 2-1 Performing Manual Archiving with SQL*Plus

In this assignment, you learn about manual archiving with SQL*Plus.

1. Open SQL*Plus and log into your database.

2. Determine the current archiving mode of the database.

3. Perform the necessary steps to change the database to ARCHIVELOG mode with automatic archiving disabled.

4. Verify that the database is in ARCHIVELOG mode with automatic archiving disabled.

5. Execute the Ch02Script.sql file in SQL*Plus.

6. Initiate a manual archiving of all filled redo log files.

7. Exit SQL*Plus.

Assignment 2-2 Enabling Automatic Archiving Through SQL*Plus

In this assignment, you learn how to use automatic archiving in SQL*Plus.

1. Open SQL*Plus and log into your database.

2. Determine the current archiving mode of the database.

3. Perform the necessary steps to change the database to ARCHIVELOG mode with automatic archiving enabled.

4. Verify that the database is in ARCHIVELOG mode with automatic archiving enabled.

5. Exit SQL*Plus.

Assignment 2-3 Placing a Database in NOARCHIVELOG Mode Through SQL*Plus

In this assignment, you learn how to place a database in NOARCHIVELOG mode.

1. Open SQL*Plus and log into your database.

2. Determine the archiving mode of the database.

3. Perform the necessary steps to change the database to NOARCHIVELOG mode.

4. Verify that the database is in NOARCHIVELOG mode.

5. Execute the Ch02Script.sql file in SQL*Plus and determine its effect upon the database after the redo log files are filled.

6. Exit SQL*Plus.

Assignment 2-4 Placing a Database in ARCHIVELOG Mode Using the Enterprise Manager Console

In this assignment, you learn how to place a database in ARCHIVELOG mode.

1. Open the Enterprise Manager Console and log into your database.

2. Determine the current archiving mode of the database.

3. Perform the necessary steps to change the database to ARCHIVELOG mode with automatic archiving disabled.

4. Verify that the database is in ARCHIVELOG mode with automatic archiving disabled.

5. Execute the Ch02Script.sql file.

6. Manually archive any filled redo log files.

7. Exit the Enterprise Manager Console.

2

Assignment 2-5 Enabling Automatic Archiving Using the Enterprise Manager Console

You learn how to enable automatic archiving in this assignment.

1. Open the Enterprise Manager Console and log into your database.
2. Determine the current archiving mode of the database.
3. Perform the necessary steps to change the database to ARCHIVELOG mode with automatic archiving enabled.
4. Verify that the database is in ARCHIVELOG mode with automatic archiving enabled.
5. Exit the Enterprise Manager Console.

Assignment 2-6 Changing a Database to NOARCHIVELOG Mode Using the Enterprise Manager Console

In this assignment, you learn how to change a database to NOARCHIVELOG mode.

1. Open the Enterprise Manager Console and log into your database.
2. Determine the current archiving mode of the database.
3. Perform the necessary steps to change the database to NOARCHIVELOG mode.
4. Verify that the database is in NOARCHIVELOG mode.
5. Exit the Enterprise Manager Console.

Assignment 2-7 Enabling Automatic Archiving for the Current Instance Using the Enterprise Manager Console

In this assignment, you learn the steps to enable automatic archiving.

1. Open the Enterprise Manager Console and log into your database.
2. Determine the current archiving mode of the database.
3. Perform the necessary steps to change the database to ARCHIVELOG mode with automatic archiving enabled only for the current instance.
4. Verify that the database is in ARCHIVELOG mode with automatic archiving enabled.
5. Exit the Enterprise Manager Console.

Assignment 2-8 Modifying the Initialization File

Modifying the initialization file is what you learn how to do in this assignment.

1. Create a copy of the pfile for your current database.
2. Open the current copy of the pfile in Notepad.
3. Save the pfile with the file name Ch2practice.

4. Edit the file and define four archiving destinations. Make certain one of these destinations is an archival database on a remote computer.

5. Add a parameter to specify that at least two of the archiving destinations must successfully receive a copy of the archived file before the contents of the online redo log file can be overwritten.

6. Add a parameter to specify that the format of the name of the archived file consists of the word ARCHIVE followed by the unpadded log sequence number. Do not indicate an extension for the file.

7. Save the revised copy of the pfile as Ch02hp8.

8. Close any open windows.

CASE PROJECTS

Case 2-1 Determining Which Mode to Use

Carlos needs to determine whether the database for the physical asset information should be operated in ARCHIVELOG or NOARCHIVELOG mode. When the database is first created, information regarding buildings, furniture, equipment, and so forth owned by the Janice Credit Union will be entered into the database. This includes data such as the date of purchase, depreciation rate, current value, and property asset number. After all the information for the current assets have been entered, the database will change only when new assets are purchased and at the end of the year when the amount of depreciation needs to be revised.

You have been asked for your opinion regarding the operating mode for the database. Decide whether the database should be operated in ARCHIVELOG mode with automatic archiving enabled, ARCHIVELOG mode with automatic archiving disabled, or NOARCHIVELOG mode. Create a memo that states your opinion and your rationale.

Case 2-2 Adding Steps to the Manual

The procedure manual needs to have information to which employees can refer when performing certain tasks through the Enterprise Manager Console. The manual should contain the steps necessary to complete each of the following tasks:

❑ Placing a database in ARCHIVELOG mode

❑ Placing a database in NOARCHIVELOG mode

❑ Enabling automatic archiving for the current instance

❑ Enabling automatic archiving for the current and future instances

❑ Enabling manual archiving

❑ Specifying destinations for archived redo log files

Create two lists of all the necessary steps to perform the indicated tasks. The first list should identify how the tasks are performed using SQL*Plus, while the second list should identify the steps required when using the Enterprise Manager Console.

3

USER-MANAGED BACKUPS

**After completing this chapter,
you should be able to do the following:**

♦ Perform a cold backup of a closed database

♦ Create a hot backup of tablespaces in an open database

♦ Describe the alternatives for backing up a control file

♦ Identify the purpose of DBVERIFY

A key to database recovery is availability of the data necessary to reconstruct the contents of the database. Normally, a DBMS is configured to store its files in multiple locations to prevent a single point of failure. If multiple copies of the data exist, a problem with one computer should not result in the database becoming unavailable. The DBA, or in some cases the software itself, simply identifies a new location for referencing the database files, thus allowing database operations to resume. As an added precaution, backup, or offline, copies of the database files are created in the event the online copies become unavailable due to media failure, user errors, and so forth.

Recall from Chapter 1 that more frequent backups of the whole database can decrease the mean-time-to-recovery (MTTR). In Chapter 2, you learned how to generate backup (archived) copies of the online redo log files. However, the database is also composed of data files and control files. To create a backup of the whole database, any initialization and password files should also be copied. There are various methods for creating backup copies of these files. Database files can be backed up through the operating system, Recovery Manager (RMAN), or third-party software. In this chapter, you will learn how to perform user-managed backups with operating system commands issued through SQL*Plus.

THE CURRENT CHALLENGE IN THE JANICE CREDIT UNION DATABASE

Imagine what would happen to Carlos if a hard drive crash occurred—such as the read-write head physically touching the disk and all the data files regarding Janice Credit Union's customer accounts becoming lost—and there were no other electronic copies of the data. Not only would Carlos be shown the nearest exit, but the credit union could become subject to several lawsuits.

In the information age, organizations rely heavily on the data stored in databases. At a minimum, the data depicts all the economic events that have occurred within an organization. If this data becomes unavailable, operations can come to a complete halt. In the case of Janice Credit Union, there would be no way to determine the current balance of customer accounts, payment status on loans, and so forth. Basically, there would be complete chaos while the credit union attempted to re-create the data manually.

Of course, this scenario is just for illustrative purposes. Organizations avoid these types of situations by developing disaster recovery plans. A **disaster recovery plan** identifies the appropriate action to take to minimize the impact of threats, both natural and man-made, against an organization. The plan includes a section directly related to information technology, including the DBMS and the data it manages.

For example, Carlos may identify flooding and hard drive failure as two potential threats to the department. The impact of a flood could be offset by ensuring that a duplicate of the database exists at another location—such as the second branch of the credit union. To minimize the effect of a hard drive failure, Carlos could protect the database through the use of a **redundant array of independent disks (RAID)**. In essence, RAID is a method of configuring multiple hard drives by writing copies of data to more than one disk. If there is a hard disk failure, a copy of the data is still available on another disk.

SET UP YOUR COMPUTER FOR THE CHAPTER

Before performing the steps in this chapter, make certain your database is in NOARCHIVELOG mode. If you need help putting the database into NOARCHIVELOG mode, refer to the "Changing to NOARCHIVELOG mode in SQL*Plus" section in Chapter 2.

NOTES ABOUT DUAL COVERAGE WITHIN THIS CHAPTER

This chapter will focus only on the command-line approach to performing backup tasks. The Backup Wizard accessed through the Enterprise Manager Console is based on Recovery Manager. Therefore, the Enterprise Manager Console is not discussed in this chapter.

PERFORMING A COLD BACKUP OF A CLOSED DATABASE

Carlos knows that a cold backup, also called an offline backup, is required when the database is operated in NOARCHIVELOG mode. Such a backup can be performed only when the database is not open, which means no users can access the database while the backup is being performed.

To perform a cold backup, the database should be shut down with an option that ensures the database is in a consistent state. When the database is shut down with the ABORT option, the database requires instance recovery before the database can be re-started. Therefore, the database should be shut down only with the NORMAL (default), TRANSACTIONAL, or IMMEDIATE options. When a database is shut down with one of these options, any data contained in the memory structure is automatically recorded in the appropriate data file. However, because there are no archived redo log files, the database can be recovered only to the point of the last valid cold backup.

The following sections will guide you through locating the data files for the database and performing a cold backup. When performing these steps, make certain you are logged into SQL*Plus with the SYSDBA privilege enabled. (See your instructor if you are unsure about how to do this.)

The examples presented in this section assume that you have a folder available on the C: drive named Backuparea to store the backup copies of the data files. Your instructor will inform you of any necessary changes that should be made to the examples based upon the configuration of your computer and/or network.

 In a real-world environment, the backup copies of database files are never placed on the same drive as the actual database files. However, to simplify the examples in this chapter, the backup is performed to a different folder on the same hard drive.

Locating the Database Data Files

Before performing a cold backup, the data files associated with the database must be located. The simplest way to find this information is to retrieve the NAME column from the V$DATAFILE view through SQL*Plus.

To identify the name and location of each data file:

1. Log into SQL*Plus, and enable the SYSDBA privilege, if necessary.

2. Type **SELECT name FROM v$datafile;**.

   Your screen should resemble, but not be identical to, Figure 3–1, which provides a list of the data files associated with the TESTDB. Of course, your database may include data files not shown in this display, depending on the configuration of your database.

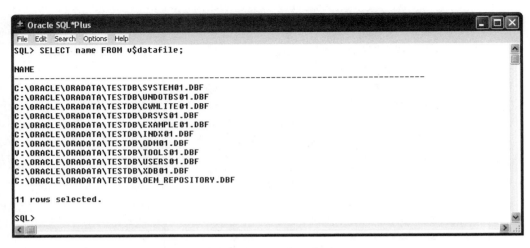

Figure 3-1   Name and location of the TESTDB data files

In Figure 3-1, the results included 11 data files that need to be copied. Notice that each data file is assigned the extension .DBF. All data files created by Oracle9i are automatically given this extension by default. However, user–created tablespaces include this file extension only if it is included with the data file name in the CREATE TABLESPACE command.

Of course, this view only identifies the location of the data files. The location of other database files can be located by viewing the database parameters.

Performing a Cold Backup

To perform a cold backup, the database needs to be shut down. The appropriate operating system command is used to copy the specified files. Then the database is restarted. The following steps demonstrate how to make a backup copy of the database data files. To create a backup of the whole database, then the control file, initialization file, and any other files referenced by the database must be copied.

To make a backup copy:

1. At the SQL> prompt, type **SHUTDOWN** to shut down the database with the normal option, as shown in Figure 3-2. This causes the data files to be updated with the changed contents of the memory structures, and the data file headers and control file stamped with the current SCN.

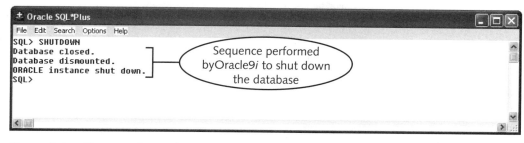

Figure 3-2 Shutting down the database

2. At the SQL> prompt, type **$ COPY *source destination***, as shown in Figure 3-3. Substitute the current location of the files where *source* is indicated and the location where the backup copies should reside where *destination* is indicated.

Figure 3-3 Command to copy identified data files

To create a copy of the whole database, the control file and initialization files should also be copied. For a database in ARCHIVELOG mode, the archived redo log files would also be included. Simply re-execute Step 2 for each file that needs to be copied.

3. After control has returned to SQL*Plus, type **STARTUP** at the SQL> prompt. After the database has been opened, it is available to the users.

The $ (dollar sign) in the command shown in Figure 3-3 is used to indicate that the subsequent command is to be executed by the host environment (i.e., the operating system). The dollar sign is followed by the actual command to be passed to the operating system. In this case, the COPY command is used to instruct Microsoft Windows 2000 to make a copy of the indicated files and store them in the specified destination. Because all the data files in this example were stored in one location and they all have the same extension, one command was sufficient.

The *.DBF in the source portion of the COPY command instructs the operating system to copy all files that have the extension of .DBF, regardless of the file name. The file name is indicated by the * (asterisk), a wildcard character that is recognized by the Windows operating system. In the destination parameter for the COPY command, the *.* indicates that the copied versions of the files should keep the same file name and extension as the original files. This portion is optional.

When the command is executed, a window appears displaying the files being copied by the operating system, as shown in Figure 3-4. After the files have been copied, the window closes and control returns to SQL*Plus.

To verify that the data files have been backed up:

1. Click the **Start** button in the lower-left corner of the task bar.

2. From the Start menu, click **My Computer**.

Figure 3-4 Files copied by the operating system

3. After the window identifying the available drives for your computer appears, click the **drive icon** for the drive referenced previously in the COPY command. Click the appropriate **folder icon** from each subsequent window until the final window displaying the backup copies of the data files is displayed.

4. In Figure 3-5, note that an examination of the Backuparea folder reveals that a copy of each data file has been made and is stored in the correct location.

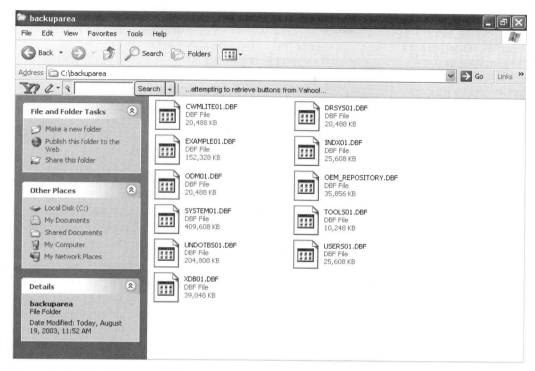

Figure 3-5   Verification of file copies in Windows

HOT BACKUP OF AN OPEN DATABASE

A hot backup, also called an online backup, is a backup that is performed while the database is still open. A hot backup can be performed only for a database that is in ARCHIVELOG mode. Carlos knows that the procedure is similar to a cold backup, in that the operating system is used to create a copy of the database files. However, because the database is still open, the tablespace associated with each data file being copied must be placed in backup mode during the backup process.

In the following sections, a tablespace will be placed in backup mode. Then a hot backup of the associated data files is performed. After the backup is complete, the tablespace is taken out of backup mode.

A cold backup can be performed for both ARCHIVELOG mode and NOARCHIVELOG mode databases. However, a hot backup can be made of a database in ARCHIVELOG mode only.

Special Considerations for Backing Up a Tablespace

A tablespace consists of one or more data files. A tablespace is a logical structure that can be referenced by Oracle9*i*. However, the data is physically stored by the operating system in data files. When you are backing up the data stored in an Oracle9*i* database, you are basically making copies of the data files. In terms of backup and recovery, there are two factors you need to consider when working with tablespaces: backup mode and read/write modes. The following sections will discuss each of these factors.

Backup Mode for Tablespaces

If a backup is being performed while the database is open, a tablespace must be placed in backup mode before its data files can be backed up through the operating system. A tablespace is placed in backup mode with the `ALTER TABLESPACE tablespacename BEGIN BACKUP` command, where *tablespacename* is the name of the tablespace being copied. When this command is executed, data changes are not written to the associated data file.

After the tablespace is placed in backup mode, the required operating system command can be issued to make a copy of the appropriate data file(s). After a copy has been made, the tablespace can be made available to the users again by issuing the command `ALTER TABLESPACE tablespacename END BACKUP`. When the tablespace is taken out of backup mode, the header of each corresponding data file is updated to the current checkpoint for the database. This procedure must be repeated for every tablespace in the database, until all tablespaces have been copied.

In a hectic environment where the phone is constantly ringing or management is demanding immediate response to a request, the DBA might forget to take a tablespace out of back up mode. If a tablespace is in backup mode when a database is shut down, it must be taken out of backup mode to start the database again. Problems can also occur if you try to back up other data files or if database recovery becomes necessary. Most DBAs, such as Carlos, create a script file that contains all the commands necessary to back up the database, including the ALTER TABLESPACE commands. The script file is then executed at regularly scheduled times to perform the backup process.

To determine whether the data file associated with a tablespace is in backup mode, you can view the contents of the V$BACKUP view with a SELECT statement. The V$BACKUP view displays the backup status of all online data files. To display the contents of the V$BACKUP view, you type the following code:

```
SELECT * FROM V$BACKUP;
```

If a data file is denoted as "NOT ACTIVE," then it is not in backup mode, as shown in Figure 3-6 (note that the display on your computer will differ). However, if the status of a file is "ACTIVE," then the data file is in backup mode.

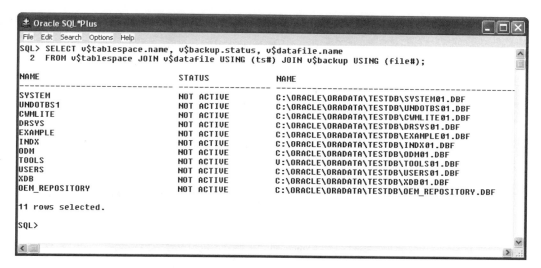

Figure 3-6   Query to view backup status

Read-write and Read-only Modes

During backup and recovery procedures, it is also important to determine whether a tablespace is in read-only or read-write mode. If a tablespace is read-only, changes cannot be made to the data contained in that tablespace. If the tablespace was read-only during the last backup, and the status has not been changed to read-write mode since that backup, then another copy of the tablespace is not required because its contents have not been modified. In other words, you only need one backup copy of a read-only tablespace. This can save time when performing backup operations. The downside, of course, is remembering to include the tablespace in the backup routine if the status is ever changed.

The read/write status of tablespaces can be determined by issuing this code:
`SELECT enabled, name FROM v$datafile;`

Performing a Hot Backup

The following sections will guide you through performing a hot backup. Before performing these steps, make certain that the database is in ARCHIVELOG mode (as discussed in Chapter 2) and that you are logged into the database with the SYSDBA privilege enabled. Database archiving mode can be changed by shutting down the database, mounting the database, issuing the ALTER DATABASE command, and then opening the database.

Identifying Tablespaces and Their Corresponding Data Files

Your first step is to determine which data file is associated with which tablespace:

1. At the SQL> prompt, type the following code:

   ```
   SELECT v$tablespace.name, v$datafile.name
   FROM v$tablespace JOIN v$datafile USING (ts#);
   ```

2. Wait while the query is executed. The results display the name of each tablespace contained in the database in the first column and the name of the data file to be backed up in the second column, as shown in Figure 3-7.

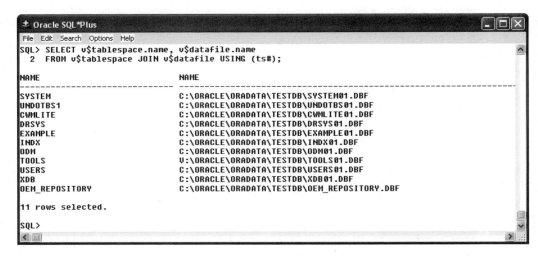

Figure 3-7 Query results

 If the results displayed by the query issued in Figure 3-7 wrap to two lines, you need to reset the linesize variable for your display to a larger value, using the command SET LINESIZE <*integer*>, where <*integer*> is the number of characters to be included on each line displayed.

Performing a Hot Backup of a Tablespace

Rather than placing all the tablespaces in backup mode and then copying the data files, you should always back up one tablespace at a time during a hot backup. The process of copying all the data files could take a very long time and it is possible that the redo log files would become filled, causing the database to hang. Therefore, you should always alter one tablespace and back up the appropriate data file(s) individually. The following steps allow you to perform a hot backup of the SYSTEM tablespace. Remember to substitute the appropriate location and file names for your database when copying the data files.

To perform a hot backup:

1. Type **ALTER TABLESPACE system BEGIN BACKUP;** at the SQL> prompt. Figure 3-8 shows the message returned by Oracle9*i* after the command has been executed.

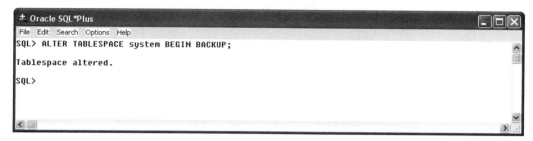

Figure 3-8   Executed ALTER TABLESPACE command

2. Type **$ COPY <source location>\ORADATA\PRACTICE\SYSTEM01.DBF <destination location>** at the SQL> prompt where *<source location>* is the current location of the file and *<destination location>* is the location to which the file is being copied. While the command is being executed by the operating system, a command-line window appears. After the data file has been copied, control returns to SQL*Plus, and the SQL> prompt appears as shown in Figure 3-9.

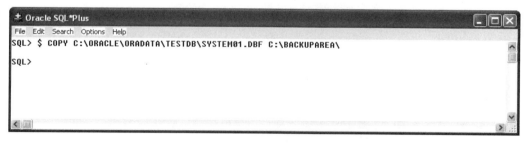

Figure 3-9   Executed operating system COPY command

3. Type **ALTER TABLESPACE system END BACKUP;** at the SQL> prompt. After the command is executed, the tablespace is accessible by the database, and the message "Tablespace altered." appears, as shown in Figure 3-10.

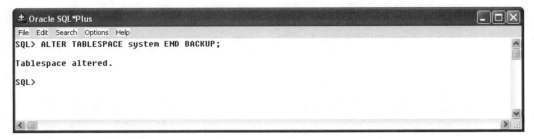

Figure 3-10 Removing a tablespace from backup mode

4. Type **SELECT v$tablespace.name, v$backup.status, v$datafile.name FROM v$tablespace JOIN v$datafile USING (ts#) JOIN v$backup USING (file#);** to make certain that no tablespace is currently in backup mode. Your results should resemble those displayed in Figure 3-11.

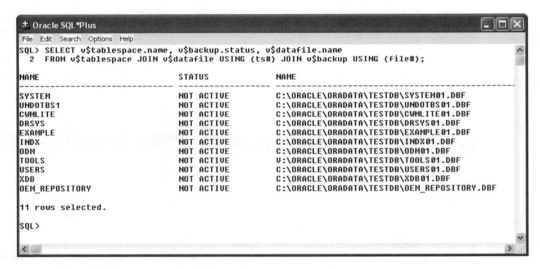

Figure 3-11 Querying the status of tablespaces

5. Repeat Steps 1 through 3 for each tablespace and data file until all the data files for the database have been copied.

After completing these steps, place the database back in NOARCHIVELOG mode.

CONTROL FILE BACKUPS

Carlos knows that the physical structure of the database is stored in the control file. If the control file is lost due to media failure or it becomes corrupted, the database cannot be opened. Therefore, he always has more than one copy of a control file. Although he can have up to eight mirrored copies of the control file for a database, Carlos still makes backup copies of the control file to ensure that he has a backup of the entire Janice Credit Union database, not just of the data files.

There are three approaches for backing up the control file:

- The first approach is to simply make a copy of the control file through the operating system at the same time you are backing up other files in the database. This allows you to have a backup of the whole database. However, in the event there is a problem with the contents of the control file at the time of the backup, or if a structural change is made to the database before the scheduled time of a regular backup, it is advisable to also make a copy of the control file using one of the alternate approaches as a safety precaution.

- The second approach is to create a binary copy of the control file. If the original control file becomes unusable and no mirrored copies are available, the binary copy can be used in place of the original control file.

- The third approach generates a text file containing the necessary commands to recreate the lost control file through Oracle9*i*. The generated text file is commonly referred to as a **trace file**. There are various types of trace files available in Oracle9*i*. Trace files can be generated to log or audit the activity of users, or they can be generated by background processes to assist the DBA in identifying errors.

To create a binary copy of a control file, simply enter the command **ALTER DATABASE BACKUP CONTROLFILE TO 'location';** where *location* specifies the destination and file name for the backup copy of the control file. After the command is successfully executed, Oracle9*i* returns the message "Database altered." as shown in Figure 3-12.

Figure 3-12 Creation of a binary copy of a control file

The code **ALTER DATABASE BACKUP CONTROLFILE TO TRACE;** is used to create a trace file containing the command(s) necessary to recreate the original control file. After the command shown in Figure 3-13 is executed, the trace file is stored in the location specified by the USER_DUMP_DEST parameter in the Init.ora file. By default, this location is in the \UDUMP subdirectory or folder within the Oracle9*i* home directory.

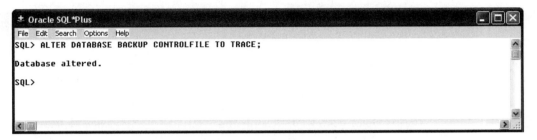

Figure 3-13 Creating a trace file containing commands to recreate a control file

After the file has been created, you can view the contents of the trace file with any text editor. In Figure 3-14, the trace file has been opened in Notepad. The # (pound sign) shown at the beginning of various lines within the file indicates that the line is a comment to provide documentation to the user.

Figure 3-14 Portion of the text backup copy of a control file

The text contains several options that can be taken based on the desired configuration of the database. The documentation contained within the trace file is designed to explain the various commands presented. Figure 3-15 shows the end of the trace file where the actual command is located that can be used to create a new control file for the database. The values assigned for various clauses in the CREATE CONTROLFILE command shown in Figure 3-15 are based upon the values assigned to the database at the time the trace file was created. If any of these values are changed, the control file should be backed up immediately to ensure that a valid control file can be recreated if necessary.

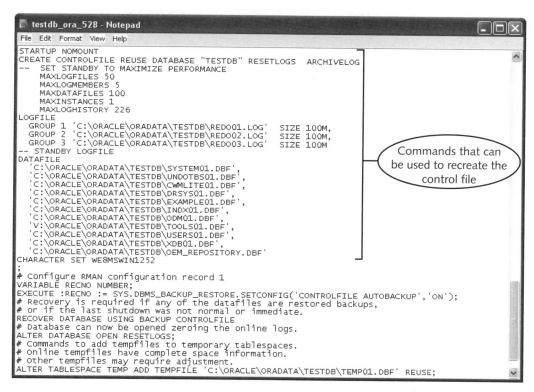

Figure 3-15   Latter portion of the trace file

USING **DBVERIFY** TO ANALYZE DATA BLOCKS

A recurring nightmare of many database administrators (Carlos and his team included) is to have a database failure and then realize that the backup copies of the database are corrupted and cannot be used to recover the database. The time and effort spent creating the backup is wasted if the copies are useless. In the case of Janice Credit Union, Carlos verifies the validity of the backup copies of the database by ensuring that all necessary files have been backed up and that the individual files are not corrupt.

Carlos regularly schedules backup and recovery practice sessions for his employees. This is to ensure that each employee is competent to perform a variety of tasks, even if it is not a part of the individual's regular job. In different sessions, an employee is required to perform either a backup of a database or to recover a database. When practicing recovery of a database, a copy of the current backup files is used. This allows Carlos to verify that all required files are actually being included in the backup procedure.

There are several utilities available to determine whether individual files are corrupt. As previously mentioned, one problem that can result from a hot backup is if the DBA forgets to place the tablespace in backup mode before copying the data file(s). If Oracle9*i* attempts to write to the data file while it is being copied, data blocks can become fractured or incomplete. Therefore, it is always a good idea to double-check whether the data blocks are corrupt. The blocks can be verified using the DBVERIFY utility. It is a package that is included in every Oracle9*i* installation.

This utility is run from the command prompt available within the operating system. The utility is invoked by typing **dbv** at the command-line prompt. In addition, you need to indicate the name and location of the data file to be verified using the FILE parameter. Unless the block size used by the database is the same size as an operating system block, the block size should also be specified. If you are uncertain of the correct block size, the simplest solution is to attempt to verify a data file without specifying the appropriate block size. If the block size of the data file does not match the block size of the operating system, an error message is returned indicating the correct block size. Then you can simply reissue the command with the correct block size.

To verify the data blocks in the copy of the SYSTEM tablespace you previously created:

1. Click the **Start** button on the task bar.

2. Click **Run** from the Start menu.

3. Type **cmd** to access the operating system command line. A window similar to the one shown in Figure 3-16 appears.

4. At the system prompt, type **dbv file=<*location of file*>\ system01.dbf blocksize=8192**, and press **Enter**. The FILE parameter is used to indicate which file is to be analyzed while the BLOCKSIZE parameter specifies the block size to be used during the verification process.

5. If an error message is returned indicating that the block size is incorrect, reissue the command substituting the correct block size contained in the error message. After verification of the file's data blocks has been completed, the statistics of the process are displayed, as shown in Figure 3-17.

In the output shown in Figure 3-17, the integrity of all the data blocks within the file was verified as indicated by the zero values assigned to the Total Pages Failing entries. Had any of the Failing or Corrupt entries displayed a non-zero value, the DBA would need to double-check the data blocks of the original data file. If the problem existed in the original file, the data blocks would need to be marked as corrupted so that the database would not attempt to write to those data blocks. The blocks can be marked as corrupt using a package called DBMS_REPAIR. If the problem does not exist in the original data file, this indicates that the problem is with the backup copy of the data file. In this case, the DBA can simply make another copy of the data file.

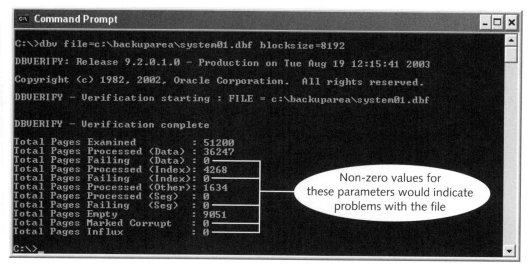

Figure 3-16   Command-line access window

Figure 3-17   Statistics returned by the DBVERIFY utility

Refer to the *Oracle9i Database Administrator's Guide* for the DBMS_REPAIR procedures and command syntax required based on the type of error returned by DBVERIFY.

If you are not completing the Hands-on Assignments at this time, the database should be placed back in NOARCHIVELOG mode after completing the steps demonstrated in this chapter.

CHAPTER SUMMARY

- A cold backup is required if the database is in NOARCHIVELOG mode.
- A cold backup requires the database to be shut down during the backup process.
- A hot backup can be performed only on databases that operate in ARCHIVELOG mode.
- A tablespace must be placed in backup mode before the associated data files are backed up through the operating system.
- If a tablespace is left in backup mode, it can cause problems during database startup or recovery.
- A read-only tablespace needs to be backed up only once.
- The control file can be backed up through the operating system or by Oracle9*i*. Oracle9*i* can be used to generate a binary file or a trace file containing the commands that can be used to recreate a control file.
- The integrity of the data blocks of the backup copies of the data files can be verified using the DBVERIFY utility.

SYNTAX GUIDE

Command	Description	Example
ALTER TABLESPACE *tablespace* BEGIN BACKUP	Places a tablespace in backup mode for a hot backup of its associated data file(s)	`ALTER TABLESPACE SYSTEM BEGIN BACKUP;`
ALTER TABLESPACE *tablespace* END BACKUP	Takes a tablespace out of backup mode	`ALTER TABLESPACE SYSTEM END BACKUP;`
ALTER DATABASE BACKUP CONTROLFILE TO '*location*'	Creates a binary copy of the control file; the file is stored in the location specified by the USER_DUMP_DEST parameter in the Init.ora file	`ALTER DATABASE BACKUP CONTROLFILE TO 'c:\backup area\ control backup';`
ALTER DATABASE BACKUP CONTROLFILE TO TRACE	Creates a trace file containing the necessary commands to re-create the control file	`ALTER DATABASE BACKUP CONTROLFILE TO TRACE;`

Parameter	Description	Example
USER_DUMP_DEST	Specifies the location to store user-generated dump files	`user_dump_dest = 'c:\oracle\ora92\ udump\'`

View	Description	Example
V$BACKUP	Displays the backup status of all online data files	`SELECT * FROM V$BACKUP;`
V$DATAFILE	Displays information contained in the control file regarding data files; includes name and location of the files	`SELECT enabled, name FROM V$DATAFILE;`

3

REVIEW QUESTIONS

1. Which of the following types of backups require a tablespace to be in backup mode before the associated data file is copied?

   a. cold backup

   b. hot backup

   c. offline backup

   d. all of the above

2. What type of backup can be performed for a database in NOARCHIVELOG mode?

3. Which of the following Init.ora parameters specifies the location of a trace copy of the database control file?

   a. USER_DUMP_DEST

   b. USER_DEST_DUMP

   c. DEST_DUMP_USER

   d. none of the above

4. A read-only tablespace does not ever need to be backed up. True or False?

5. What is the status of a tablespace in the V$BACKUP view if the tablespace is in backup mode?

6. To perform an offline backup, a database cannot be shut down using which of the following options?

   a. NORMAL

   b. IMMEDIATE

   c. TRANSACTIONAL

   d. ABORT

7. Which of the following lines of code place a tablespace named USERS in backup mode?

   a. `ALTER TABLESPACE users BEGIN BACKUP;`

   b. `ALTER DATABASE users BEGIN BACKUP;`

   c. `ALTER DATAFILE users BEGIN BACKUP;`

   d. `ALTER TABLE users BEGIN BACKUP;`

8. Which of the following views can be used to determine whether a tablespace is in backup mode?

   a. V$DATAFILE

   b. V$TABLESPACE

   c. V$BACKUP

   d. DBA_TABLESPACE

9. If a tablespace is in backup mode, the status of the tablespace is displayed as which of the following?

   a. STALE

   b. BACKUP

   c. IN PROGRESS

   d. none of the above

10. Why is it necessary to back up the control file as well as the data files?

11. Which of the following statements is correct?

    a. Backup copies of the data files should be stored on the same drive as the original data files.

    b. Backup copies of the data files should be stored in the same directory as the original data files.

    c. Backup copies of the data files should not be stored in the same folder as the original data files.

    d. Backup copies of the data files should not be stored on the same drive as the data files.

12. A hot backup can be performed only when which of the following is true?

    a. A second drive is available for the backup copies of the files.

    b. No users are accessing the database.

    c. The database is in ARCHIVELOG mode.

    d. The database is in NOARCHIVELOG mode.

13. A trace file can be created that contains the necessary commands to recreate any lost data files. True or False?

14. Which of the following views can be joined to the V$TABLESPACE view to determine the name and location of each data file that is associated with each tablespace contained in the database?

    a. V$BACKUP

    b. V$DATAFILE

    c. V$DATAFILES

    d. all of the above

15. Which of the following is used in SQL*Plus in the Microsoft Windows 2000 environment to indicate that the subsequent command should be executed by the operating system?

    a. !

    b. $

    c. #

    d. –

16. Which of the following is used to denote a comment in a trace file generated by backing up a control file?

    a. !

    b. $

    c. #

    d. –

17. If a backup of a data file is created, and it exists in the folder designated by the COPY command, then the DBA can assume the file is not corrupt and can be used for recovery purposes. True or False?

18. Which of the following lines of code takes a tablespace named TOOLS out of backup mode?

    a. `ALTER TABLESPACE tools BEGIN BACKUP;`

    b. `ALTER DATABASE tools END BACKUP;`

    c. `ALTER TABLESPACE tools END BACKUP;`

    d. `ALTER TABLESPACE tools BEGIN BACKUP MODE;`

19. Which of the following parameters is used to specify the file that is to be analyzed by the DBVERIFY utility?

    a. SPFILE

    b. PFILE

    c. FILE_ANALYZE

    d. FILE

20. If a tablespace is not in backup mode, the status of the tablespace in the V$BACKUP view is displayed as _____?

a. COMPLETED

b. BACKUP

c. IN PROGRESS

d. none of the above

Hands-on Assignments

Assignment 3-1 Determining Backup Mode

In this assignment, you determine the backup mode in which you are operating.

1. Log into SQL*Plus, and enable the SYSDBA privilege.
2. Query the V$BACKUP view to determine whether any tablespaces in your database are in backup mode.
3. Exit SQL*Plus.

Assignment 3-2 Locating Database Files

In this assignment, you locate database files.

1. Log into SQL*Plus, and enable the SYSDBA privilege.
2. Query the V$DATAFILE view to determine the location of all data files.
3. Type **SHOW PARAMETER FILE** at the SQL> prompt to determine the location of all copies of the control file and the initialization file (spfile).
4. Exit SQL*Plus.

Assignment 3-3 Creating a Cold Backup of a Closed Database

You create a cold backup of a closed database in this assignment.

1. Log into SQL*Plus, and enable the SYSDBA privileges.
2. Shut down the database.
3. Perform a cold backup of all data files, any archived redo log files, any control files, and any initialization file used by the database.
4. Start up the database.
5. Exit SQL*Plus.

Assignment 3-4 Creating a Script File to Perform a Cold Backup of a Whole Database

You create a script file for use in backups in this assignment.

1. Click the **Start** button from the task bar of the operating system.
2. Click **All Programs** from the Start menu.
3. Click **Accessories** from the Programs submenu.
4. Click **Notepad** from the Accessories submenu.
5. Enter the commands, in the correct sequence, required to perform Steps 2 through 4 of Assignment 3-3. The overall sequence of the commands should be as follows: shut down the database, copy data files, copy any archived redo log files, copy control files, copy the initialization file (if appropriate), and start up the database. Remember to include semicolons at the end of the ALTER DATABASE commands.
6. Click **File** from the Notepad menu bar.
7. Click **Save As** from the File menu.
8. Type **CH03ColdBackup.sql** for the file name.
9. Click the Save as type drop-down list, and then click **All Files**.
10. Click **Save** to save the file.
11. Close all open windows.

Assignment 3-5 Executing a Script File to Perform a Cold Backup of a Whole Database

In this assignment, you execute a script file to perform a database backup. This assignment assumes that you have completed Assignment 3-4.

1. Log into SQL* Plus, and enable the SYSDBA privilege.
2. Type **START *location*\CH03ColdBackup.sql** at the SQL> prompt to execute the script file created in Assignment 3-4. Substitute the appropriate drive letter and folder name(s) in the command.
3. After execution of the script, use the operating system to verify that the files were stored in the correct location.
4. Exit SQL*Plus.

Assignment 3-6 Identifying Tablespaces and Associated Data Files

In this assignment, you learn how to identify tablespaces and their associated data files.

1. Log into SQL*Plus, and enable the SYSDBA privilege.

2. Query the V$DATAFILE view, and identify the data file(s) associated with the TOOLS tablespace.

3. Exit SQL*Plus.

Assignment 3-7 Backing Up a Tablespace From an Open Database

Backing up a tablespace is an important skill. You learn how to back up a tablespace in this assignment.

1. Log into the database, and enable the SYSDBA privilege.

2. Shut down the database.

3. Place the database in ARCHIVELOG mode so that you can perform a hot backup.

4. Open the database.

5. Place the TOOLS tablespace in backup mode.

6. Create a backup of the data file(s) associated with the TOOLS tablespace.

7. Take the TOOLS tablespace out of backup mode.

8. Place the database back in NOARCHIVELOG mode.

9. Close all open windows.

Assignment 3-8 Verifying Data Block Integrity with DBVERIFY

In this assignment, you learn how to verify data block integrity.

1. Click the **Start** button on the taskbar of the operating system.

2. Click **Run** from the Start menu.

3. Type **cmd** to access the command-line window.

4. Verify the integrity of the data blocks for the backup copy of the data file(s) created in Assignment 3-3.

5. Close all open windows.

CASE PROJECTS

Case 3-1 Scripts Used to Back Up Databases

When the employee who normally backs up the database is on vacation or out sick, another employee is assigned the task of backing up the database. Carlos is concerned about the possibility of certain files being omitted because this task is not part of the substitute employee's normal job routine. Therefore, Carlos has asked that you create a script file that can be used to perform a hot backup of the entire database.

Of course, the script needs to include placing each data file in backup mode, as well as take them out of backup mode after each data file has been copied. The script should also include the necessary command to create a backup copy of the control file, and so forth. Because a hot backup can be performed only on a database that is in ARCHIVELOG mode, the script also needs to copy all archived redo log files. Save the script as Ch03HotBackup.sql.

Case 3-2 Procedures Used with Backups

Carlos has assigned you the task of updating the procedure manual with the procedures for performing cold and hot backups. The new sections to be added to the procedure manual must specify the steps necessary to perform the following tasks:

- How to determine the name and location of all database files
- How to perform a cold backup
- How to perform a hot backup
- How to create a binary backup of the control file
- How to create a trace file of the control file
- How to verify the integrity of the data blocks in the backup files

4

USER-MANAGED COMPLETE RECOVERY

**After completing this chapter,
you should be able to do the following:**

◆ Identify the difference between a complete and incomplete recovery

◆ Perform a complete recovery for a database in NOARCHIVELOG mode

◆ Recover the SYSTEM tablespace for a database in ARCHIVELOG mode

◆ Open an ARCHIVELOG mode database before recovery of a non-SYSTEM tablespace

◆ Rename a data file

◆ Recover a missing or corrupted control file

◆ Explain the implications of recovering a read-only tablespace for NOARCHIVELOG and ARCHIVELOG databases

---

A reccurring nightmare of Carlos and many IT employees is the dreaded call in the middle of the night regarding some catastrophic event that has rendered the database and all backups useless. Of course, this would only be a dream because no DBA worth his or her salary would ever let a situation occur where no backup of the database were available—at least not one who had ever suffered through that situation before. As you progress through this chapter, you will realize why it is so important to ensure that backup copies of the database files are always available. When proper backup procedures have been followed, the database recovery process is a dream. However, when backup procedures are flawed because the files are unavailable (because of media failure, no one remembering where they are, etc.), database recovery can become an absolute nightmare.

THE CURRENT CHALLENGE IN THE JANICE CREDIT UNION DATABASE

In this chapter, Carlos is faced with several types of database failure problems. In one case, a data file for a database in NOARCHIVELOG mode becomes unavailable and the database must be recovered using a cold backup, and data loss occurs. In other instances, the data files for an ARCHIVELOG mode database become inaccessible and the database is recovered from a hot backup. Finally, situations resulting in the loss of a control file require Carlos to recreate the control file or copy one of the images of the control file as a replacement.

SET UP YOUR COMPUTER FOR THE CHAPTER

Certain steps included in this chapter require an error to occur when attempting to access a database file. Before attempting any task presented in this chapter, perform a cold backup of the whole database in the event recovery becomes necessary. In the event the database becomes unrecoverable, this cold backup can be used to restore and recover the original database. Store this backup in a location separate from any backups generated while performing the steps presented in this chapter or when completing the end-of-chapter assignments.

NOTES ABOUT DUAL COVERAGE IN THIS CHAPTER

This chapter will focus only on the command-line approach to performing backup tasks using SQL*Plus. The Recovery Wizard accessed through the Enterprise Manager Console is based on Recovery Manager. Therefore, the Enterprise Manager Console will be included in later chapters that are dedicated to using Recovery Manager to restore and recover a database.

DATABASE RECOVERY

There are two types of recovery that can be performed in an Oracle9i environment: complete and incomplete. With a complete recovery, all data, including data contained in the archived redo log files, is restored. If the database is operated in NOARCHIVELOG mode, there may be some data loss unless the media failure occurred immediately after the backup process was performed and before any changes were made to the database. However, if all data contained within the backup files are restored, then it is considered a **complete recovery**. On the other hand, an **incomplete recovery** occurs when only a portion of the data in the archived redo log files is restored. This chapter addresses performing user-managed complete database recovery in various situations.

The term **database recovery** refers to resetting a database to a point immediately prior to a database failure. A **database failure** is when the database becomes inaccessible to the user, regardless of whether the user is an application program or a human. A failure can be due to an error, such as dropping the wrong table, which normally requires an incomplete recovery, or due to a media failure.

The term **media** refers to the actual storage material, such as magnetic tape or disk, used to permanently hold the database files. Data is written to or read from the media using a type of hardware device known as a drive. A **media failure** occurs when a file cannot be accessed by the drive. The failure can be hardware related, such as when the drive stops working completely. Of course, a media failure could also simply be a case of a data file being physically moved to another location and the database has not been updated with the correct reference. This would still be considered a media failure by the database because the file cannot be accessed.

The general phrase used to refer to a media failure is that the file is "unavailable." When a file becomes unavailable, **media recovery** is required. Either the database needs to be updated with the correct location of database file or, in the case of a hardware problem, the problem needs to be corrected. Then a backup of the file is restored if the original cannot be retrieved, and its contents are updated so that the file is consistent with the other database files.

The steps to perform media recovery vary depending on the situation. In some cases, the operating system is used to simply overwrite the database files with previous backed-up copies and open the database with the DBMS software. Sometimes commands may need to be executed in Oracle9*i* to indicate the new location of a file or to apply the contents of archived redo log files. More extreme problems may require the DBA to recreate control files.

NOARCHIVELOG Recovery

The simplest database to recover is a database that is operated in NOARCHIVELOG mode. Why? Well, you do not have to be concerned with any archived redo log files and nothing has to be updated in the database. The recovery process simply requires the DBA to copy files from the last valid database backup over the current files and then start up the database.

Data Loss

Of course, there is the problem with data loss. Always keep in mind the caveat "pay me now, or pay me later" when a database is in NOARCHIVELOG mode. Carlos, for instance, could operate the financial database in NOARCHIVELOG mode, but if a database failure occurs, and there is no backup, then the credit union would have no way of identifying the current balance of each customer's account, which customers have made their loan payments, and so on.

In a service environment that relies almost entirely on electronic data, the loss of even a few transactions would have a negative impact on the organization's image and ability to operate. In addition, it would be a safe bet that at least one individual would be searching for a new job, with little chance of getting a good reference from the credit union. In other words, the backup process may be faster and money may be saved by not buying additional storage space, but data loss is the ultimate price.

Restoring All Files

The key concept behind recovering a NOARCHIVELOG database is that all of the backed up database files must be restored in the recovery process, even if only one file is unavailable. If any of the backup files are excluded from the recovery process, the database is in an inconsistent state. In other words, the system change number (SCN) used to identify the transactions contained in the database for the recovered file is not the same as the SCN in the control file and the database generates an error message. The problem of data loss is avoided with ARCHIVELOG databases by applying the contents of the archived redo log files to the recovered file. However, this is not an option for NOARCHIVELOG databases.

The Janice Credit Union Database

At Janice Credit Union, the database that is used for testing structural or programming changes has an unavailable data file. Somehow the SYSTEM tablespace has become corrupt and cannot be accessed. Recall that the SYSTEM tablespace contains the data dictionary for the database. Unless the database configuration has been customized, it is also the default tablespace for the users. Therefore, any tables, indexes, and so forth created by the users are contained in this tablespace. (However, in a real-world environment, database users should be assigned a different default tablespace.) In this particular database, the default configuration was used when the database was created.

Because the database is not used in the normal operations of the credit union, the database is operated in NOARCHIVELOG mode. Luckily, a policy had been established years ago that required any user testing changes on the database to create a backup of the database before any changes were initiated, as well as a backup after the test was successfully completed.

Although there are no archived redo log files that can be used to prevent the loss of some of the database changes and test data, at least there is the option of recovering the data up to the point of the last cold backup. The following steps demonstrate how to recover a NOARCHIVELOG database, just as would be done at Janice Credit Union.

To emphasize how data loss can occur, the simulation includes the creation of a database table immediately after a cold backup is created. Of course, after the database files are restored, the table no longer exists because it is not included in the backup copy of the files and there are no archived redo log files applied during the recovery process for a database in NOARCHIVELOG mode.

To recover a NOARCHIVELOG database:

1. Log into the database and enable the SYSDBA privileges.

2. Verify that the database is in NOARCHIVELOG mode. If the database in not in NOARCHIVELOG mode, refer to Chapter 2 for more information on how to change it into this mode.

3. Shut down the database.

4. Use the COPY command to create a cold backup. Remember to copy all of the data files, all control files, the redo log files, and the initialization files referenced by your database.

5. Start up the database.

6. Type **START <*path*>ch04script1.sql**, where <*path*> is the location of the script file for Chapter 4, as shown in Figure 4-1.

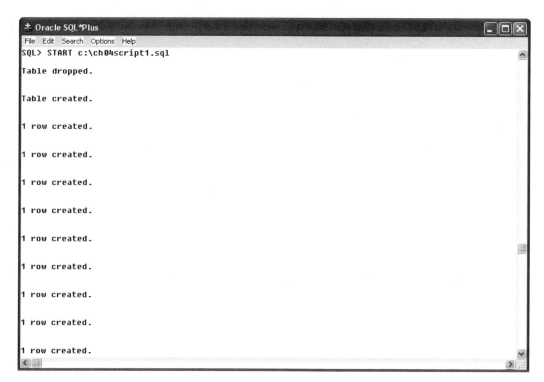

Figure 4-1   Script file execution (partial output shown)

7. Type **SELECT * FROM janicecustomers;** to verify that the table was created, as shown in Figure 4-2.

```
Oracle SQL*Plus
File  Edit  Search  Options  Help
SQL> SELECT * FROM janicecustomers;

 CUSTOMER# LASTNAME   FIRSTNAME   ADDRESS                CITY          ST ZIP
 --------- ---------- ----------  ---------------------  ------------- -- -----
      1001 MORALES    BONITA      P.O. BOX 651           BURBANK       CA 91510
      1002 THOMPSON   RYAN        P.O. BOX 9835          SANTA MONICA  CA 90404
      1003 SMITH      LEILA       P.O. BOX 66            BURBANK       CA 91510
      1004 PIERSON    THOMAS      69821 SOUTH AVENUE     BURBANK       CA 91510
      1005 GIRARD     CINDY       P.O. BOX 851           BURBANK       CA 91510
      1006 CRUZ       MESHIA      82 DIRT ROAD           BURBANK       CA 91510
      1007 GIANA      TAMMY       9153 MAIN STREET       BURBANK       CA 91508
      1008 JONES      KENNETH     P.O. BOX 137           BURBANK       CA 91508
      1009 PEREZ      JORGE       P.O. BOX 8564          BURBANK       CA 91510
      1010 LUCAS      JAKE        114 EAST SAVANNAH      SANTA MONICA  CA 90404
      1011 MCGOVERN   REESE       P.O. BOX 18            BURBANK       CA 91510
      1012 MCKENZIE   WILLIAM     P.O. BOX 971           BURBANK       CA 91508
      1013 NGUYEN     NICHOLAS    357 WHITE EAGLE AVE.   BURBANK       CA 91510
      1014 LEE        JASMINE     P.O. BOX 2947          BURBANK       CA 91508
      1015 SCHELL     STEVE       P.O. BOX 677           BURBANK       CA 91508
      1016 DAUM       MICHELL     9851231 LONG ROAD      BURBANK       CA 91508
      1017 NELSON     BECCA       P.O. BOX 563           BURBANK       CA 91508
      1018 MONTIASA   GREG        1008 GRAND AVENUE      BURBANK       CA 91510
      1019 SMITH      JENNIFER    P.O. BOX 1151          BURBANK       CA 91508
      1020 FALAH      KENNETH     P.O. BOX 335           SANTA MONICA  CA 90404

20 rows selected.

SQL>
```

Figure 4-2 Contents of the JANICECUSTOMERS table

8. Type **SHUTDOWN** to shut down the database.

9. Using My Computer from the operating system, move the System01.dbf data file to a different folder to simulate a media failure. Because the files from the cold backup in Step 1 have not been verified, as a safety precaution, it is better to simply move the data file rather than delete it.

10. Type **STARTUP** to reopen the database. When the database is unable to locate the System01.dbf file, the error message shown in Figure 4-3 is displayed.

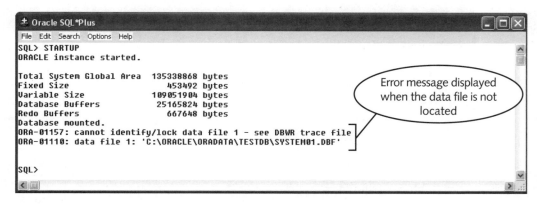

Figure 4-3 Database error

Before recovering the database by restoring *all* of the database files, Carlos decides to put the procedure to a test. Recall that when a NOARCHIVELOG database is being recovered, all of the backed up database files must be copied back to the database folder. However, Carlos wants to see what will happen if only the backup copy of one file, the System01.dbf data file, is used.

To test the procedures:

1. Type **SHUTDOWN** to shut down the database. The error message shown in Figure 4-4 is displayed, indicating that the database was never opened, but then the remaining portion of the shutdown process is completed.

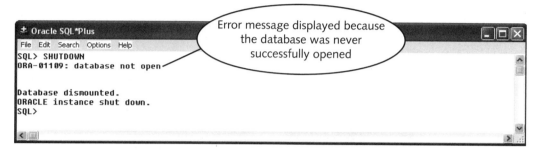

Figure 4-4   Shutting down the database

2. Restore (copy) the backup copy of the System01.dbf file to the database data folder using the operating system. This step can be performed through SQL*Plus or directly from the operating system using My Computer.

3. Type **STARTUP** to restart the database. You receive the error message shown in Figure 4-5 indicating that media recovery is necessary to place the database in a consistent state.

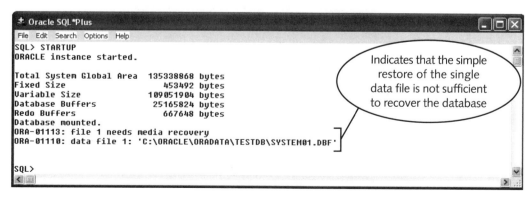

Figure 4-5   Media recovery error message

4. Type **SHUTDOWN**. Again, Oracle9*i* displays an error message that the database is not open before performing the remaining tasks required to shut down the database.

Media recovery requires the use of archived redo log files to place the restored data file in a consistent state. Because the database was operated in NOARCHIVELOG mode, there is no way to perform a media recovery for the System01.dbf file. Therefore, the only way to recover the database is to restore all the database files from the cold backup because the entire backup set contains the database when it was in a consistent state.

Carlos is satisfied that a NOARCHIVELOG database cannot be recovered by simply restoring the missing data file. You now help him perform the correct recovery procedure.

To perform the recovery procedure:

1. Use the operating system to restore the copies of all the database files created during the cold backup procedure.

2. Type **STARTUP** and the database is opened.

3. Type **SELECT * FROM janicecustomers;** to verify that the table no longer exists. When the table cannot be located, the error message shown in Figure 4-6 is displayed.

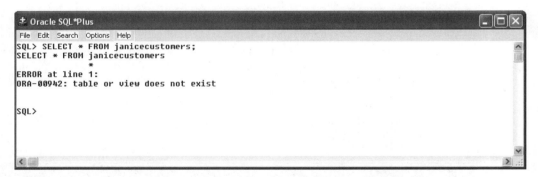

Figure 4-6   Data not contained in cold backup is missing after recovery

Restoring a NOARCHIVELOG database is an easy process as long as a valid backup of all the database files exists. However, remember that any database changes made after the cold backup was performed will be lost. The only way to recover lost data is to re-enter it. Therefore, databases are rarely operated in NOARCHIVELOG mode. Normally, only databases used for testing or training purposes are used in this mode.

ARCHIVELOG RECOVERY

Although recovery of an ARCHIVELOG database can be slightly more complex than the recovery of a NOARCHIVELOG database, the overall procedure is the same. The main difference is that the contents of the archived redo log files need to be written to the restored data file(s) to update the file with changes that have occurred since the backup was created. The media recovery process also updates the restored data file with the correct SCN after the log files have been applied so that the database will be in a consistent state. As mentioned, this option is not available with a NOARCHIVELOG database.

The basic procedure for recovering a database in ARCHIVELOG mode is to place a copy of the backed-up data file in the database data folder, then mount the database, which starts the instance and opens the control file. You then can issue the RECOVER command and reapply the transactions in the archived redo logs.

Of course, there are some additional steps that most DBAs, including Carlos, perform in an actual work environment. In the real world, many organizations cannot afford to have the entire database unavailable while recovery is being performed. Fortunately, that is one of the wonderful benefits of ARCHIVELOG mode—it can be opened while recovery is being performed. This allows users to continue working with unaffected tablespaces within the database until the recovery process is completed.

To open the database after the DBMS has detected the unavailability of a data file, the DBA simply issues the ALTER DATABASE...OFFLINE command to take the data file requiring recovery offline. Because the database is no longer concerned about that particular data file, the database still can be opened and made available to the users. After the data file has been recovered, the DBA issues the ALTER DATABASE...ONLINE command to make the recovered data file available to users.

However, there is one restriction when placing tablespaces offline: a SYSTEM tablespace cannot be taken offline. Recall that the SYSTEM tablespace contains the data dictionary. The data dictionary houses all information regarding database objects. If the SYSTEM tablespace is taken offline, then none of the database tables, indexes, and so forth can be identified. Therefore, it is impossible to open a database without the SYSTEM tablespace.

RECOVERING THE SYSTEM TABLESPACE FOR AN ARCHIVELOG DATABASE

In the following example, assume that the financial database has a corrupt or unavailable System01.dbf data file. To attempt to make the database available to users as quickly as possible, Carlos will try to take the tablespace offline and then open the database before starting the recovery process. To also demonstrate that data loss normally does not occur during media failure of an ARCHIVELOG database, the JORDERS table will be created after the hot backup of the data file has been performed and just before the media failure occurs.

To work with the database:

1. If not already logged into SQL*Plus, log in and enable the SYSDBA privilege.

2. Perform the steps necessary to change the database into ARCHIVELOG mode.

3. Type **ALTER TABLESPACE system BEGIN BACKUP;**. After the command is executed, the message shown in Figure 4-7 is displayed.

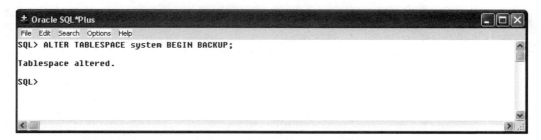

Figure 4-7   Placing the SYSTEM tablespace in backup mode

4. Type **$ COPY <*source_path*>system01.dbf <*destination_path*>**, where <*source_path*> is the current location of the data file and <*destination_path*> is the location where the copy of the file should be stored, as shown in Figure 4-8, to create a backup copy of the file.

Figure 4-8   Creating a backup copy of the data file

5. Type **ALTER TABLESPACE system END BACKUP;** as shown in Figure 4-9. This makes the data file available to users after the hot backup of the file is completed.

6. Type **START <*path*>ch04script2.sql** as shown in Figure 4-10 to execute the script file and create the JCHECKS table.

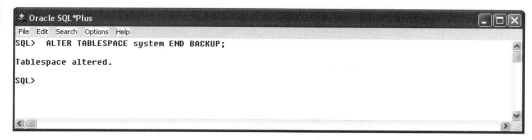

Figure 4-9   Taking the SYSTEM tablespace out of backup mode

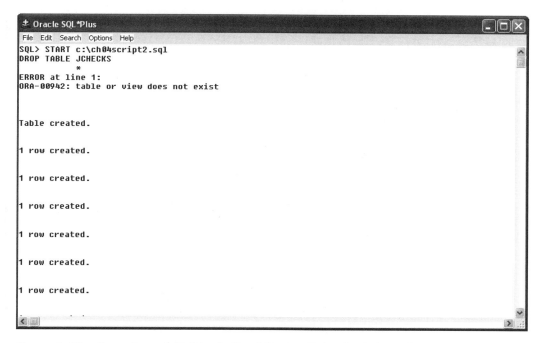

Figure 4-10   Execution of Ch04script2.sql file (partial output shown)

7. Type **SELECT * FROM jchecks;** to verify that the table was created. The contents of the table are displayed as shown in Figure 4-11.

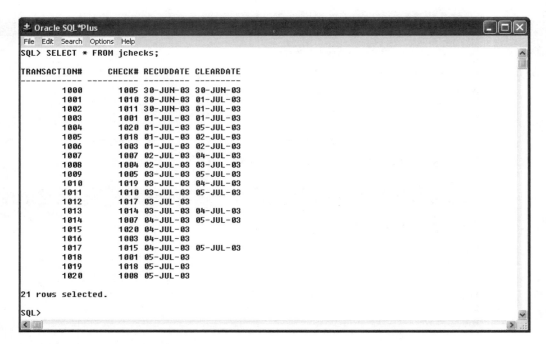

Figure 4-11 Contents of the JCHECKS table

Now that the hot backup has been performed and new data added to the data file, Carlos creates a media failure by deleting (moving) System01.dbf.

To delete System01.dbf:

1. Type **SHUTDOWN** to shut down the database.

2. Move the System01.dbf data file using My Computer through the operating system. Make certain you do not overwrite the backup copy previously made of this data file.

3. Type **STARTUP** to open the database. When Oracle9*i* is unable to find the data file, the error messages shown in Figure 4-12 are displayed.

After the media failure has occurred, Carlos performs the necessary steps to restore the data file. Ideally, Carlos would be able to open the database and allow users to access any tables not contained in the SYSTEM tablespace. However, as previously mentioned, Oracle9*i* does not allow the database to be opened if the SYSTEM tablespace is not available.

To restore the file:

1. Type **ALTER DATABASE DATAFILE '<*path*>system01.dbf' OFFLINE;**, as shown in Figure 4-13, to specify that the SYSTEM tablespace is to be taken offline.

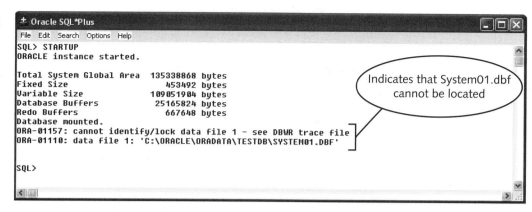

Figure 4-12   Errors generated by unavailability of System01.dbf data file

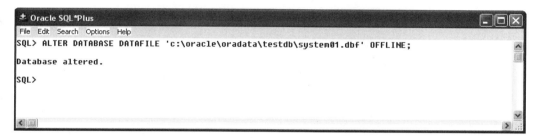

Figure 4-13   Command to take the System01.dbf data file offline

2. Type **ALTER DATABASE OPEN;** to open the database. However, as shown in Figure 4-14, because the SYSTEM tablespace is unavailable, the database is not opened.

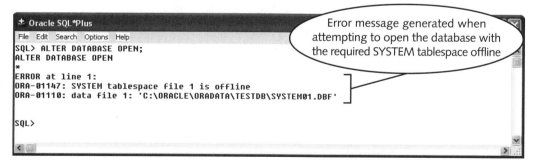

Figure 4-14   Failed attempt to open the database after media failure

3. At the SQL> prompt, type **$ COPY *<source_path>*\system01.dbf *<destination_path>*\system01.dbf** to restore the copy of the System01.dbf data file, as shown in Figure 4-15.

Figure 4-15   Restoring copy of lost data file

4. Type **RECOVER datafile '*<path>*\system01.dbf';**, where *<path>* specifies the location of the data file, to perform a media recovery of the restored data file, as shown in Figure 4-16.

Figure 4-16   Media recovery of a data file

5. Type **ALTER DATABASE DATAFILE '*<path>*\system01.dbf' ONLINE;** to place the recovered data file back online, as shown in Figure 4-17.

6. Type **ALTER DATABASE OPEN;** to open the database.

```
Oracle SQL*Plus
File  Edit  Search  Options  Help
SQL> ALTER DATABASE DATAFILE 'c:\oracle\oradata\testdb\system01.dbf' ONLINE;

Database altered.

SQL> |
```

Figure 4-17   Placing a data file online

7. Type **SELECT * FROM jcl cks;** at the SQL> prompt, and press **Enter** as shown in Figure 4-18 to v ify that the table created after the hot backup was performed has been re vered.

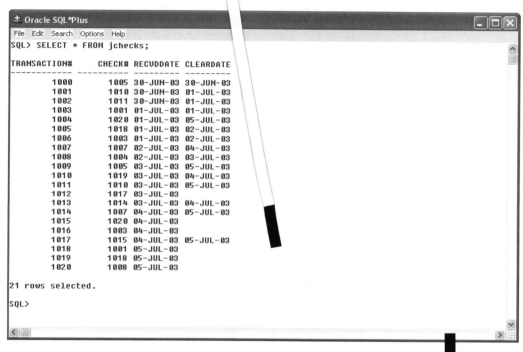

Figure 4-18   Recovered data

As shown in the preceding step sequence, even though data was e red into the database after a hot backup was performed, an ARCHIVELOG d ase can be recovered without data loss. Of course, for this procedure to work c ectly, the backup files and all redo log files archived after the backup was perform must be available. This is one of the arguments for mirroring, or duplicating, files oss several drives, as well as for keeping the backup copy of files on a different d from the original files. If a media failure results from a hard drive failure, and that rd drive also contains the backup files and the archived redo log files, then the en database is lost.

OPENING THE DATABASE BEFOR RECOVERY

At Janice Credit Union, th nancial database has experienced a media failure because someone has accidentall eted the Tools01.dbf data file. Because the majority of the daily transactions are posted t this database, the downtime for the database must be minimized.

To make the database available to the users as quickly as possible, Carlos takes the Tools01.dbf data file offline before the database is opened. By taking the data file offline, Oracle9*i* does not attempt to locate the file before opening the database. Therefore, no error occurs and the database is opened, giving users access to unaffected data files within the database. After the data file has been restored, the data file can be placed online and the entire database becomes fully operational.

To ensure that a backup copy of Tools01.dbf is available, a hot backup needs to be performed for the data file.

To create a backup of the data file:

1. Type **ALTER TABLESPACE tools BEGIN BACKUP;** to place the tablespace in backup mode.

2. Type **$ COPY *<source_path>*tools01.dbf *<destination_path>*,** where *<source_path>* is the current location of the data file and *<destination_path>* is the location where the copy of the file is to be stored.

3. Type **ALTER TABLESPACE tools END BACKUP;** to take the tablespace out of backup mode.

Now that you are certain that a backup copy of the data file exists, the following steps will create a media failure by moving the Tools01.dbf file from its current location.

To move the file:

1. Type **SHUTDOWN** to shut down the database.

2. Use My Computer from the operating system to move the file to another location. Remember not to place the file in the same location as the original cold backup because the new file causes the set of backup files to be invalid.

3. Type **STARTUP** to start the database. When the location of the data file cannot be verified, the message shown in Figure 4-19 is displayed.

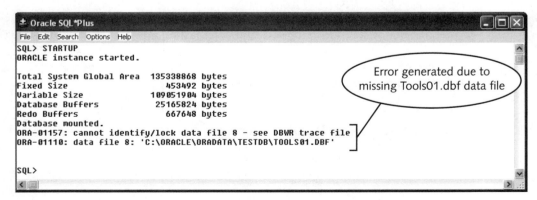

Figure 4-19   Error message due to missing data file

Because the missing data file is not associated with the SYSTEM tablespace, the data file can be taken offline and then the database can be opened. In the following steps, Carlos restores and recovers the data file after opening the database. After the data file is recovered, it can then be placed back online.

To restore and recover the data file:

1. Type **ALTER DATABASE DATAFILE '<path>tools01.dbf' OFFLINE;**, as shown in Figure 4-20, to take the data file and the associated tablespace offline.

Figure 4-20   Taking the Tools01.dbf data file offline

2. Type **ALTER DATABASE OPEN;** to open the database.

3. Type **$ COPY <source_path>tools01.dbf <destination_path>** , where *<source_path>* is the location of the backup copy of the data file and *<destination_path>* is the location where the original data file was stored, as shown in Figure 4-21.

Figure 4-21   Restoring the Tools01.dbf data file

4. Type **ALTER DATABASE DATAFILE '<path>tools01.dbf' ONLINE;**, where *<path>* is the location of the data file. As shown in Figure 4-22, an error message is returned indicating a media recovery is required to update the contents of the data file.

5. Type **RECOVER DATAFILE '<path>tools01.dbf';** to instruct Oracle9*i* to perform a media recovery for the data file.

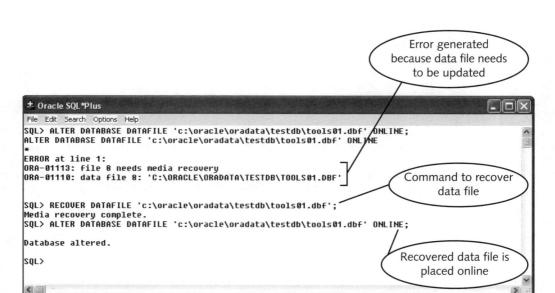

Figure 4-22   Recovering the data file and placing it online

6. After the media recovery is completed, type **ALTER DATABASE DATAFILE '*<path>*tools01.dbf' ONLINE;**, where *<path>* is the location of the data file.

After the data file has been recovered and placed online, the entire database is now available to the users.

MOVING OR RENAMING DATA FILES

The previous examples in this chapter have focused on a data file being unavailable and simply replacing the appropriate data file(s). However, what if the media failure is due to a hardware problem preventing the data file from being copied back to its former location? If the data file is placed on a different drive or in a different folder, Oracle9*i* will not know the location of the file and there will still be a media failure. If a data file needs to be moved to a different location, or even if it is simply being renamed for some reason, Oracle9*i* needs to be updated with the new location, or name, of the file. This can be accomplished by issuing the ALTER DATABASE command with the RENAME clause.

In this section of the chapter, Carlos will move the Tools01.dbf file from its current location to the TOOLS folder. Because the new information must be updated to all copies of the control file, the database must be mounted, but not open, when it is updated with the correct information. When a database is mounted, the control file is accessed to determine the location of all the physical files associated with the database. However, when the database is opened, all the data files are accessed by the database. If the database cannot locate a data file, an error message is generated and the database is not opened.

This example moves a data file from one location to another on the same drive for simplicity purposes only. In a real-world environment, the data file would normally be relocated to a different drive if the purpose of the relocation were to prevent a single point of failure.

To move the file:

1. Type **SHUTDOWN** at the SQL> prompt, and press **Enter**.

2. Use the operating system (either directly or through SQL*Plus) to move the Tools01.dbf data file to its new location.

3. Type **STARTUP MOUNT** at the SQL> prompt, and press **Enter** to mount the database.

4. Type **ALTER DATABASE RENAME FILE '<source_path>tools01.dbf'**
   TO '<destination_path>tools01.dbf';, where *<source_path>* is the current location of the data file as listed in the control file and *<destination_path>* is the new location. This is shown in Figure 4-23.

5. Type **ALTER DATABASE OPEN;** at the SQL> prompt, and press **Enter**. Because the database has been updated with the correct data file location, no media failure occurs.

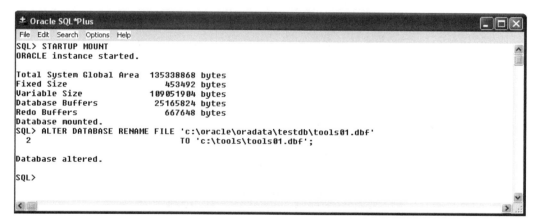

Figure 4-23   Updating the database with new data file location

Hands-on Assignment 4-1 at the end of the chapter will have you place the Tools01.dbf data file back into its original location.

RECOVERING A CONTROL FILE

The control file contains the physical structure of the database, including the name and location of all data files and redo log files. In addition, it is also accessed constantly when the database is open because the file must be updated whenever a log switch occurs, to record SCNs, and so forth. So what happens if a control file becomes inaccessible or corrupt? Basically, the database becomes unusable. In fact, the database cannot even be mounted without the control file.

If there are mirrored copies of the control file and one of the copies becomes inaccessible for the financial database at Janice Credit Union, one of the valid copies can be used as a replacement. However, if all of the copies of the control file are corrupted or lost, the control file must be recreated.

In the following sections, the necessary procedures Carlos needs to follow to ensure a copy of the control file is always available or can be recreated are presented. A copy of the control file will be used to recover the database and then the procedure to recreate the control file from the commands stored in a trace file is demonstrated.

Steps to Recover the Control File with a Mirrored Copy

The following steps will demonstrate how to recover a lost or corrupted control file using a mirrored copy of the file.

To recover the file:

1. Type **SHUTDOWN** to shut down the database.

2. Delete the file named Control01.ctl using My Computer from the operating system.

3. Type **STARTUP** at the SQL> prompt, and press **Enter** to start up the database. When all copies of the control file cannot be found, the error message shown in Figure 4-24 is displayed.

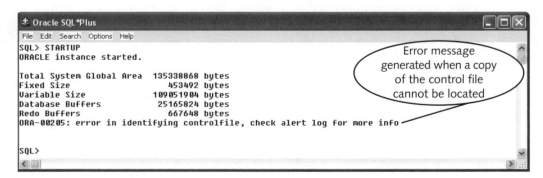

Figure 4-24   Error message caused by missing control file

4. Type **$ COPY <*path*>\control02.ctl <*path*>\control01.ctl** , where <*path*> is the location of the file, as shown in Figure 4-25, to create a new copy of the missing control file from a mirrored copy of the file.

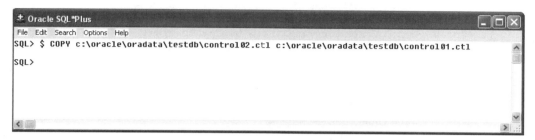

Figure 4-25   Making a new copy of missing control file

5. Type **ALTER DATABASE MOUNT;** at the SQL> prompt, and press **Enter** to mount the database. Mounting of the database indicates that the control file has been successfully restored.

6. Type **ALTER DATABASE OPEN;** at the SQL> prompt, and press **Enter** to open the database, as shown in Figure 4-26.

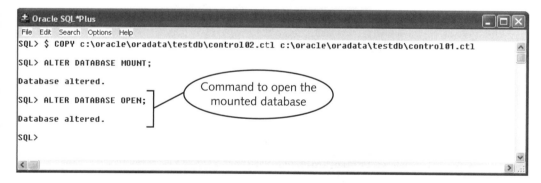

Figure 4-26   Database successfully opened

Now that you have seen how to successfully restore a control file by copying another control file, the next section will demonstrate the steps necessary to recreate a control file when all copies become unusable.

Steps to Recreate A Control File

One of the backup procedures for a control file is to create a trace file. In the following section, Carlos will create a trace file and then edit that file to create a new control file after all copies of the existing control file are lost. By default, the trace file is stored in the \oracle\admin*databasename*\udump folder.

The name of the file can be confusing because the default name of the trace file contains the name of the database followed by a series of numbers to make the file name unique. Most administrators rename the file to make it easier to identify. In this particular example, the quickest way to identify the appropriate file is to simply check the time of the computer, and then locate the file in the UDUMP folder with the corresponding time. When a control file is backed up to a trace file, Oracle9*i* places several comments in the file as documentation. Before the commands contained in the trace file can be executed to recreate the control file, the documentation either needs to be deleted from the file or changed to comments for the file to properly execute.

1. Type **ALTER DATABASE BACKUP CONTROLFILE TO TRACE;** to create the trace file.

2. Type **SHUTDOWN** to shut down the database.

3. Through the operating system, locate the trace file created in Step 1.

4. Open the trace file with a text editor, such as Notepad, as shown in Figure 4-27.

Figure 4-27   Partial contents of original trace file

5. Click **File** from the menu bar in Notepad, and then click **Save As**.

6. When prompted, type **controlscript.sql** as the file name, and change the file type to **All Files** as shown in Figure 4-28. Click **Save**.

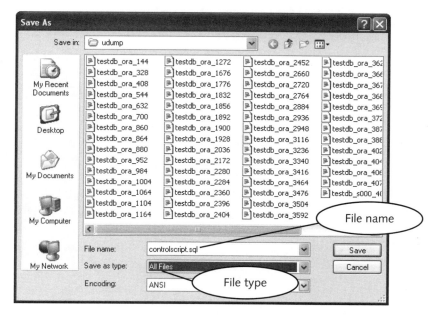

Figure 4-28   Saving copy of trace file with new file name

7. Edit the trace file by using two dashes (--) to comment out all documentation. This includes the text lines contained at the top of the file, as well as all lines throughout the document that are preceded by the number sign (#). In Figure 4-29, two dashes have been used to comment out the appropriate lines.

 The Replace command available in the Edit menu of Notepad can be used to substitute dashes wherever the number sign appears within the file.

8. Click File from the menu bar, and click Save to save the changed file.

9. Delete (or simply move) all copies of the mirrored control file through the operating system.

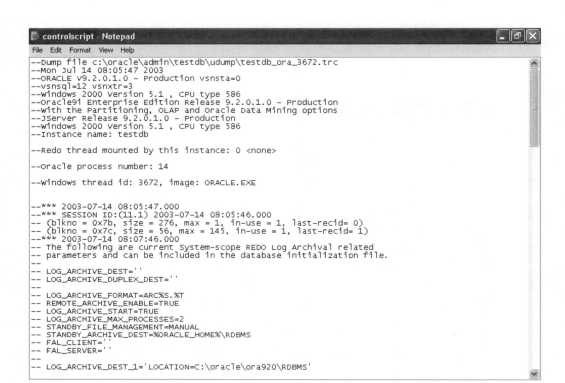

Figure 4-29   Modified trace files

10. Type **STARTUP** to start the database. Because Oracle9*i* cannot find the control file, the error message in Figure 4-30 is displayed.

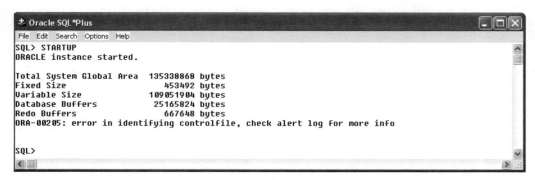

Figure 4-30   Error generated from missing control file

11. Type **SHUTDOWN** at the SQL> prompt to shut down the database. The error messages in Figure 4-31 are generated because the missing control file prevented the database from being mounted.

Figure 4-31   Error from shutting down unmounted database

12. To execute the commands contained in the trace file, type **START** **<*path*>controlscript.sql**, where *<path>* is the location of the modified trace file saved in Step 8, as shown in Figure 4-32.

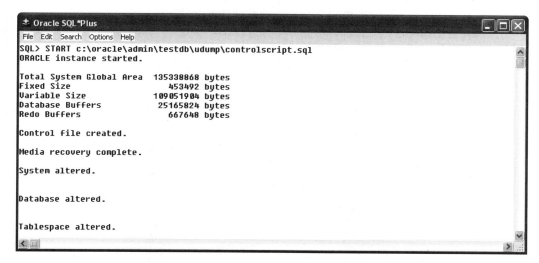

Figure 4-32   Portion of display from execution of commands within the trace file

13. Bounce the database by typing **SHUTDOWN** at the SQL> prompt and pressing **Enter** to verify that the control file has been properly created. After the database has been shut down, type **STARTUP** to restart the database.

In a real-world situation, it is more likely that there would be a gap in time between the creation of the trace file and when the control file would need to be recreated. Therefore, there would be several changes contained in the archived log files that must be updated to the control file. Oracle9i automatically prompts the user to determine which changes should be updated to the new control file. Because these prompts will be used in Chapter 5 to assist in performing incomplete recoveries, they will not be demonstrated in this chapter.

RECOVERING A READ-ONLY TABLESPACE

A tablespace can be assigned read-only or read-write status. A read-only tablespace is a tablespace that cannot be updated. Basically, it is only used for SELECT statements. However, data stored in a read-write tablespace can be updated, deleted, as well as retrieved.

Recovery of a read-only tablespace deserves special attention because the status of the tablespace can change. For example, suppose a read-only tablespace in a database is backed up. At some point after the backup is performed, the tablespace is changed to read-write. If a media failure occurs, what will be the status of the tablespace after it is recovered?

If the read-only tablespace is part of a database that is in NOARCHIVELOG mode, then the tablespace will also be in read-only status after the database is recovered. Why? Recall that when a database in NOARCHIVELOG mode is recovered, there are no archived redo log files available. Therefore, the status of the tablespace cannot be updated during the recovery process. This is true for the all the database files. However, if the database is in ARCHIVELOG mode, the status of the tablespace is updated during the recovery process. So if the status of the tablespace is changed after backing up the associated tablespace, then the change is also recorded in the archived redo log files. After the recovery process is completed, the tablespace is in the same state as when the media failure occurred.

 If you are not completing the hands-on assignments at this time, the database should be placed back in NOARCHIVELOG mode after completing the steps demonstrated in this chapter.

CHAPTER SUMMARY

- ❏ Media failure occurs when a file cannot be accessed.

- ❏ All backup files must be restored to recover a database that is in NOARCHIVELOG mode to ensure the database is in a consistent state.

- ❏ Any changes made to a NOARCHIVELOG mode database after it has been backed up are lost if the database ever needs to be recovered.

- An ARCHIVELOG database with a data file that needs media recovery can still be opened by taking the data file offline and then opening the database.

- The SYSTEM tablespace cannot be taken offline. It must be recovered before the database can be opened.

- The RECOVER command is used to update the contents of a data file with the archived redo log files.

- A data file can be renamed or moved as long as the control file is updated using the RENAME clause of the ALTER DATABASE command.

- If only one copy of a control file becomes unavailable, a valid copy of the mirrored file can be used to replace the file and open the database.

- If all copies of the control file are unavailable, the control file must be recreated before the database can be opened.

- For a NOARCHIVELOG database, the read–write status of a tablespace after recovery is the same as the status of the tablespace when it was backed up. For an ARCHIVELOG database, the status is updated when the contents of the archived redo log files are written to the recovered data files.

4

SYNTAX GUIDE

Command	Purpose	Example
ALTER DATABASE RENAME FILE	Used to specify a new name or location for a data file	`ALTER DATABASE RENAME FILE 'c:\tools01.dbf' TO c:\oracle\tools01.dbf';`
ALTER DATABASE DATAFILE... OFFLINE|ONLINE	Used to place a data file offline or online	`ALTER DATABASE DATAFILE 'c:\tools01.dbf'; OFFLINE`
RECOVER DATAFILE	Used to apply the necessary archived redo log files to the specified data file	`RECOVER DATAFILE 'c:\tools01.dbf';`
ALTER TABLESPACE...BEGIN|END BACKUP	Used to place a tablespace in backup mode or to change the status of the tablespace to read-only or read-write	`ALTER TABLESPACE system BEGIN BACKUP;`

REVIEW QUESTIONS

1. To recover a database that is in _____ mode, the initialization file, control files, redo log files, and data files must be restored, even if only one data file is unavailable.

2. Which of the following commands is used to update a restored data file with the contents of the archived redo log files?
   a. ALTER TABLESPACE
   b. ALTER DATABASE
   c. RECOVER DATAFILE
   d. RECOVER FILE

3. Which of the following is the simplest way to recover a damaged control file if a mirror image of the control file is available?
   a. Recreate the database using the Database Configuration Assistant.
   b. Make a copy of one of the remaining control files to replace the damaged control file.
   c. Execute the commands stored in the trace file created by ALTER DATABASE BACKUP CONTROL FILE TO TRACE.
   d. none of the above; a control file cannot be recovered

4. How can a database in ARCHIVELOG mode be opened if one of its data files is unavailable?

5. If a tablespace has read-only status when a hot backup is performed, what state must be recorded in the control file to recover the tablespace?
   a. the same state the tablespace was in immediately before it became unavailable
   b. read-only
   c. write-only
   d. a deleted state since a read-only tablespace cannot be recovered

6. A(n) _____ recovery is when all available data is recovered, rather than just a portion of the available data.

7. Which of the following scenarios is considered media failure?
   a. a corrupted data file
   b. a deleted data file
   c. a hard drive crash
   d. all of the above
   e. none of the above

8. Which of the following commands is used to rename a data file?

    a. RENAME DATA FILE

    b. ALTER TABLESPACE

    c. NEW NAME

    d. ALTER DATABASE

9. A complete recovery can be performed for a database in only which of the following modes?

    a. NOARCHIVELOG

    b. ARCHIVELOG

    c. READ ONLY

    d. both a and b

10. Data loss usually occurs when recovery is performed on a database in _____ mode.

11. A control file can be recreated only for a database that is mounted. True or False?

12. A database cannot be _____ if a data file is unavailable, unless the data file is taken offline.

13. Issuing the RECOVER command does not update the data files of a database that is in ARCHIVELOG mode. True or False?

14. If a database is in NOARCHIVELOG mode and the backup files are used to recover the database, which type of recovery has been performed?

    a. incomplete

    b. complete

    c. partial

    d. none of the above

15. If all the control files for a database are unavailable, the database can be started to which stage using the STARTUP command?

    a. open

    b. mount

    c. nomount

    d. none of the above

16. If all the control files for a database are unavailable, the control file must be _____ .

17. Can a database be opened if one of the data files cannot be located?

18. A tablespace that was read-only at the time of a cold backup, but that was later changed to read-write status, is _____ if the database needs recovery before the next backup occurs.

4

19. A database that is in NOARCHIVELOG mode can be started to the _____ stage of the startup process if a data file is unavailable.

20. The contents of restored data files cannot be updated during the recovery process for a database that is in NOARCHIVELOG mode because no archived redo log files exist. True or False?

HANDS-ON ASSIGNMENTS

Assignment 4-1 Changing the Location of a Tablespace

Perform the following tasks to move the Tools01.dbf back to its original location:

1. Place the database in NOARCHIVELOG mode.
2. Use My Computer from the operating system to move the **Tools01.dbf** data file back to where it was located before it was moved in the chapter example.
3. Update the database with the new location of the Tools01.dbf data file.
4. Open the database.

Assignment 4-2 Restoring a Data File for a Database in NOARCHIVELOG Mode

Perform the following tasks to restore a data file for a database in NOARCHIVELOG mode:

1. Verify that the database is in NOARCHIVELOG mode. If it is not, then perform the steps necessary to place the database in NOARCHIVELOG mode.
2. Shut down the database.
3. Perform a cold backup of the database.
4. Delete (or move) the **Indx01.dbf** data file.
5. Open the database.
6. After receiving an error message indicating that the data file cannot be identified, shut down the database.
7. Use the backup copies made in Step 3 to restore the database files.
8. Start the database.

Assignment 4-3 Recovering a Data File for a Database in ARCHIVELOG Mode

Perform the following tasks to recover the Indx01.dbf data file:

1. Place the database in ARCHIVELOG mode.
2. Perform a hot backup of the **Indx01.dbf** data file.
3. Shut down the database.

4. Delete (or move) the **Indx01.dbf** data file.

5. Start the database.

6. When the error message is displayed indicating that the data file cannot be identified, perform the necessary steps to open the remaining database files so unaffected portions of the database can be available to users.

7. Restore and recover the missing data file.

8. Make the recovered data file available to the database users.

Assignment 4-4 Recovering a Read-Only Tablespace

Perform the following tasks to recover the Indx01.dbf data file after the tablespace has been changed to read-write status:

1. Verify that the database is in ARCHIVELOG mode.

2. Change the tablespace associated with Indx01.dbf to read-only status using the command **ALTER TABLESPACE <tablespace_name> READ ONLY;**, where <tablespace_name> is the name of the tablespace associated with the Indx01.dbf data file.

3. Perform a hot backup of the Indx01.dbf data file.

4. Change the tablespace associated with **Indx01.dbf** to read-write status using the command **ALTER TABLESPACE <tablespace_name> READ WRITE;**, where <tablespace_name> is the name of the tablespace associated with the Indx01.dbf data file.

5. Shut down the database.

6. Delete (or move) the **Indx01.dbf** data file.

7. Open the database.

8. Perform the necessary steps to recover the missing data file.

9. Open the database.

10. Query the V$DATAFILE view to determine the current read-write status of the tablespace.

Assignment 4-5 Restore a Control File Using a Mirrored Copy

Perform the following tasks to restore a lost or corrupted control file:

1. Verify that the database is in ARCHIVELOG mode.

2. Shut down the database.

3. Delete the **Control02.ctl** file.

4. Open the database.

5. After the instance is started and the error message indicating that the control file cannot be identified is displayed, shut down the database.

6. Make a copy of the **Control01.ctl** file, and assign **Control02.ctl** as the file name for the copy.

7. Reopen the database.

Assignment 4-6 Recreating a Control File from a Trace File

Perform the following tasks to recreate a corrupt or lost control file when no mirrored copies of the control file are available:

1. Issue the command to create a backup of the control file to a trace file.

2. Shut down the database.

3. Delete (or move) all copies of the control file.

4. Start the database.

5. After the instance is started and the error message indicating that the control file cannot be identified is displayed, shut down the database.

6. Make the necessary modifications to the trace file created in Step 1 to create a new control file.

7. Create a new control file from the modified trace file created in Step 6.

8. After the new control file has been created, bounce the database to verify that there are no problems with the new control file.

9. Shut down the database.

Assignment 4-7 Recovering a Read-Only Tablespace for a NOARCHIVELOG Mode Database

Perform the following tasks to recover the Tools01.dbf data file after the tablespace has been changed to read-write status:

1. Change the database to NOARCHIVELOG mode.

2. Change the tablespace associated with Tools01.dbf to read-only status using the command **ALTER TABLESPACE <*tablespace_name*> READ ONLY;**, where <*tablespace_name*> is the name of the tablespace associated with the Tools01.dbf data file.

3. Perform a hot backup of the **Tools01.dbf** data file.

4. Change the tablespace associated with Tools01.dbf to read-write status using the command **ALTER TABLESPACE <*tablespace_name*> READ WRITE;**, where <*tablespace_name*> is the name of the tablespace associated with the Tools01.dbf data file.

5. Shut down the database.

6. Delete (or move) the **Tools01.dbf** data file.

7. Open the database.

8. Perform the necessary steps to recover the missing data file.

9. Open the database.

10. Query the V$DATAFILE view to determine the current read–write status of the tablespace.

Assignment 4-8 Restoring a Control File for a NOARCHIVELOG Mode Database

You can perform the following tasks to restore a control file for a NOARCHIVELOG mode database.

1. Verify that the database is in NOARCHIVELOG mode.

2. Shut down the database.

3. Create a cold backup of the database.

4. Move the **Control01.ctl** file. Do not delete the backup copy of the file.

5. Attempt to open the database. When the error message is displayed indicating that the control file cannot be located or identified, perform the necessary steps to restore the control file.

CASE PROJECTS

Case 4-1 Accessing a Database

Just as Carlos was about to leave the office on Thursday evening, he receives a frantic phone call from the payroll clerk in the Accounting Department. The payroll clerk is apparently unable to access the database that houses all the payroll information, and now cannot process the payroll checks, including Carlos' check, that need to be distributed to the employees in the morning. The database for Human Resources is managed by a junior DBA who has already left the office. Carlos needs to correct the database problem before leaving, or he will not receive a paycheck in the morning.

Carlos attempts to log into the database from his office; however, he receives an error message stating that the database is unavailable. When he issues the command to start the database, an error message indicating that the 'c:\oracle\oradata\hr\system01.dbf' data file cannot be identified is displayed. Assuming that the database is operated in ARCHIVELOG mode, identify the exact procedure Carlos needs to follow to recover the database with no data loss. Remember that Carlos is not familiar with this particular database and does not know where the backup copies of the database files are located, so the most appropriate steps to locate the files need to be included in the procedure.

Case 4-2 Modifying the Procedure Manual

Continuing with the procedure manual you began in Chapter 1, prepare lists that identify the necessary steps to accomplish the following:

❐ Perform media recovery required to restore a missing data file for a NOARCHIVELOG database.

❐ Perform media recovery required to recover a missing data file for an ARCHIVELOG database.

❐ Perform media recovery required to recover a control file for an ARCHIVELOG database when only one copy of the control file is unavailable.

❐ Perform media recovery required to recover a control file for an ARCHIVELOG database when there are no copies of the control file available.

5

USER-MANAGED
INCOMPLETE RECOVERY

**After completing this chapter,
you should be able to do the following:**

♦ Explain the difference between a complete and an incomplete
  database recovery

♦ State when an incomplete database recovery is required

♦ Identify the command to perform a database recovery

♦ Identify the options available for performing an incomplete recovery

♦ Explain the disadvantage of performing a cancel-based recovery

♦ Perform a time-based recovery

♦ Explain the advantage of performing a change-based recovery

♦ Identify the purpose of using LogMiner to analyze redo log files

♦ Use LogMiner to perform an analysis of the online redo log files

In Chapter 4, you learned to perform a complete recovery by applying the contents of the archived redo log files to the data files. However, there are times when the archived or online redo log files are not available or a complete recovery can re-introduce a problem (such as a dropped table) into the database. In these cases, the database needs to be recovered only to a certain point, known as an incomplete recovery. During an incomplete recovery, the recovery process can be stopped based on various factors, such as a specific archived redo log file, time, or a system change number (SCN). In this chapter, you will learn how to perform an incomplete recovery based on each of these factors.

To help determine the exact stopping point for a database recovery, many DBAs use the LogMiner utility, a package available in every Oracle9*i* installation. The purpose of the LogMiner utility is to analyze the contents of the archived redo log files. This allows a DBA to identify when the undesirable transaction occurred, in terms of time or SCN. The DBA can then use this information as the control criteria for the recovery process. A basic example of using LogMiner to determine the SCN for a transaction is included in this chapter.

THE CURRENT CHALLENGE IN THE JANICE CREDIT UNION DATABASE

At 9:25 a.m. Carlos receives an urgent phone call. A user has accidentally dropped a critical table used to process deposits and withdrawals for the customers of Janice Credit Union. Oops! After jotting down a note to remember to ask the senior DBA in charge of the financial database why a user would have the necessary privilege to drop such an important table, Carlos asks the caller a few questions. First, how long ago was the table dropped? Second, what was the name of the table? Finally, was another table created with the same name? Luckily, Carlos finds out the TRANSACTION table was dropped at 9:18 a.m. and the user did not create another table with the same name. At least now Carlos has a starting point for recovering the lost table.

In this type of situation, a DBA must determine whether it is more reasonable to have the user re-enter data or to recover the database to a point before the table was dropped. If the dropped table was small and there is a hardcopy of its contents, the best solution would be to rebuild the table and have the user re-enter the data.

However, if it is an extremely large table that would seriously affect business operations, then the alternative is to recover the database. The downside to the recovery solution is that to restore the database to a previous state, data will be lost because an incomplete recovery must be performed.

SET UP YOUR COMPUTER FOR THE CHAPTER

Before performing any of the tasks demonstrated in this chapter, make a cold backup of the database. Because this backup can be used to recover the database in the event an error occurs while performing the various recovery scenarios, make certain that backups created throughout the chapter do not overwrite this original cold backup. To perform an incomplete recovery, the database must be in ARCHIVELOG mode, with automatic archiving enabled. After making the cold backup, place the database in ARCHIVELOG mode, using the steps presented in Chapter 2.

NOTES ABOUT DUAL COVERAGE WITHIN THE CHAPTER

When performing an incomplete recovery through the Enterprise Manager Console, you are actually interacting with Recovery Manager. The use of Recovery Manager to perform an incomplete recovery is the topic of Chapter 9 in this textbook. Because this chapter addresses user-managed incomplete recovery, only the command-line approach will be demonstrated using SQL*Plus.

INCOMPLETE RECOVERY

A database recovery is considered to be incomplete if all the available transactions are not applied to the data files during the recovery process. When transactions are applied to data files, any DML or DDL operations that were explicitly or implicitly committed and that are not already contained in the data file are written to the data files during database recovery. This type of situation arises when an archived or online redo log file is lost or unavailable, or if the database needs to be restored to a previous state. The basic procedure is to restore the latest copy made of the necessary data file, and then stop the recovery process using the RECOVER DATABASE command. Using this command stops the contents of the redo log files from being applied shortly before the error or problem is introduced back into the database.

> If a database freezes, or hangs, because of a problem with the current redo log file, an incomplete recovery is not required. The DBA simply issues the command ALTER DATABASE CLEAR UNARCHIVED LOGFILE GROUP <#>;, where <#> is the number of the file to be overwritten or re-created if the file is unavailable. A cold backup should be performed immediately after the log file has been cleared.

Options for an Incomplete Recovery

There are three options available when performing an incomplete recovery: cancel-based, time-based, and change-based. They are briefly reviewed in this section of the chapter; each is discussed in detail later in the chapter.

In a **cancel-based recovery**, the DBA steps through the recovery process, and then stops the process just before the archived log file suspected of containing the undesired change is applied to the database. However, of the three, this approach results in the greatest amount of data loss. Why? Because a cancel-based recovery prevents all the contents of the suspect archived log file from being updated to the database, as shown in Figure 5-1. Suppose the erroneous transaction is the last transaction recorded in the file. If Carlos uses the cancel-based recovery approach, then none of the contents of that file are updated to the data files, not even the valid transactions.

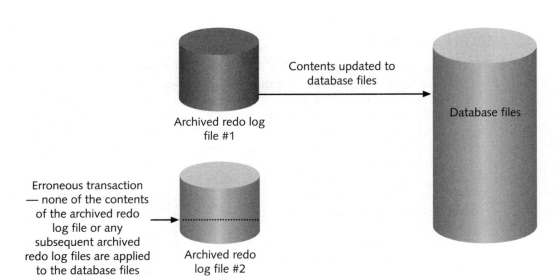

Figure 5-1   Cancel-based recovery

The second option is to perform a **time-based recovery**. When a transaction is recorded in the redo log file, it is timestamped with the exact time the transaction was completed. If the user immediately realizes that an error has occurred and calls the IT department with the approximate time of the event, then it is possible for the DBA to recover the database to a state just prior to the time specified by the user. Although the amount of data loss is less than that associated with cancel-based recovery, there are problems associated with this approach. One problem is that there may have been several transactions that occurred at that same time. In this case, all those transactions are lost, as shown in Figure 5-2. In addition, the DBA has to make the assumption that the user noted the correct time the event occurred. This is definitely not a safe assumption because the user's clock or watch may be fast. Also, the user may not have noticed the exact time the event actually occurred, but instead noticed the time at which he or she *detected* that the problem had occurred.

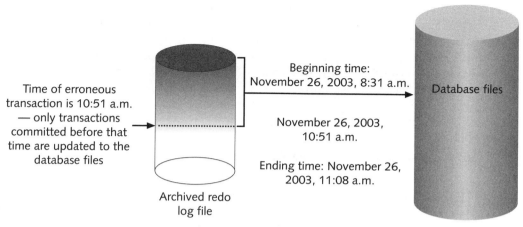

Figure 5-2   Time-based recovery

The most precise option is to perform a **change-based recovery** (see Figure 5-3), and its use hinges on the SCN. If the DBA is fortunate enough to know the exact SCN that was assigned to the transaction that caused the error to occur, then the database can be restored to the transaction immediately prior to the error. This option provides the DBA with the most control over the recovery process and the least amount of data loss. However, in an environment with several users and a high volume of transactions, it can be difficult to find the necessary SCN.

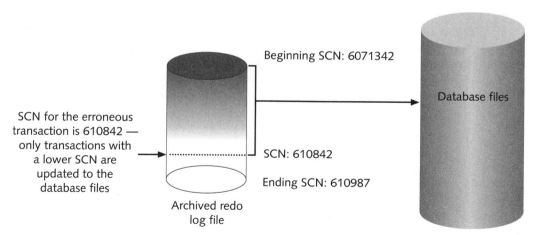

Figure 5-3   Change-based recovery

Finding the SCN

The simplest way to find the SCN for a specific transaction is to use the LogMiner utility. When a transaction is recorded in an online redo log file, and subsequently in an archived redo log file, the time the transaction is committed and its SCN are recorded. The LogMiner utility is used to analyze the transactions contained in the redo log files and allows the DBA to locate the specific command that created the problem. Although the results provided by LogMiner contain both the time and the SCN, a DBA would use the information to perform a change-based recovery rather than a time-based recovery to minimize data loss. However, always keep in mind, any transactions that occur after the specified stopping point for the incomplete recovery are not contained in the recovered database, regardless of the approach taken.

Steps During an Incomplete Recovery

As soon as Carlos determines that an incomplete recovery is necessary, the database is immediately shut down. This prevents users from adding data to the database or making other changes that will be lost when the recovery process is performed. In addition, an e-mail is sent to all users notifying them that a problem with the database has occurred and that data will be lost. Carlos also provides a timeframe for how long the database will be unavailable and how much data loss should be expected. He does this by including a statement in the e-mail that the database will be down for approximately 20 minutes and data entered after 9:15 a.m. will be lost. Luckily, Janice Credit Union has established procedures for data verification, and Carlos refers the users to those procedures so that they can determine what data must be re-entered into the database.

After the database is shut down, a cold backup is performed. This cold backup is essential if there is a problem with the recovery process. When a database is recovered, the redo log files are reset. As a result, the database may overwrite the contents of previously archived redo log files because the file names of the previous files can be reused when the database is reset. If the log files are reset when the database is opened and the DBA realizes that the error occurred earlier than originally thought, there is no way to retry the recovery process without having a copy of all database files from before the log files were reset. If the recovery needs to be retried, all the database files can be copied back from the cold backup and the recovery can be attempted again.

After the cold backup is performed, the copy of the correct data file from the last backup created *before* the error occurred should be used to replace the data file containing the error. The database is then mounted, but not opened. The RECOVER DATABASE command with the appropriate option for the desired recovery type is then executed. If a cancel-based recovery is specified, the DBA is asked to confirm each archived log file that is being applied to the data file during the recovery process. When all the valid files are applied, the DBA cancels the process, and the database is then opened.

After the database is opened, the DBA should verify that the original problem has been corrected (that the dropped table has been recovered, and so on). If the database has been recovered to the appropriate point, another cold backup should immediately be performed in the event another problem occurs before the next scheduled hot backup or in the event that the RESETLOGS option of the ALTER DATABASE command resets the log files and they no longer will be valid. If the problem is not corrected, the DBA needs to restore all files from the cold backup taken immediately before attempting the recovery and try again. Once successful, the database can be restarted and the users notified that the database is available.

CANCEL-BASED RECOVERY

When the exact time that the problem was created or the SCN for the transaction is not known, Carlos has to perform a cancel-based recovery. In addition, a cancel-based recovery is required if one of the online redo log files becomes corrupt or unavailable. With a cancel-based recovery, Carlos can stop the recovery process before the problem is reintroduced into the database.

The problem is that now Carlos needs to identify which archived redo log file most likely contains the error. Normally, a DBA attempts to estimate the time frame within which the problem occurred and then look at the time and date stamp for the archived redo log files. The alert.log file can be very helpful in determining which archived redo log file might contain the data set that needs to be avoided. Whenever an online redo log file is archived, an entry is made into the alert.log for the database. As shown in Figure 5-4, the entry specifies the time the archive process began, and which archive log file was created.

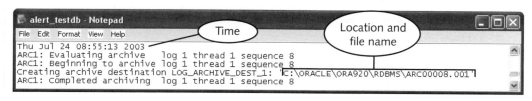

Figure 5-4   Portion of the alert.log file

The accuracy of this approach depends on how often log switches occur. If a log switch occurs only every few hours, then the time estimate does not have to be very precise. However, if log switches occur every few minutes, it is much harder to identify the exact archived redo log file containing the error. In that case, Carlos may have to recover the database several times, using a different archived redo log file as the stopping point each time, to make certain the recovery process did not re-introduce the initial problem back into the database.

After the appropriate archived file has been identified, the database is shut down, and then mounted. The RECOVER DATABASE command with the UNTIL CANCEL keywords is used to recover the database. The UNTIL CANCEL keywords are used to instruct Oracle9i that the user determines the stopping point for the recovery process based on the archived redo log file to be applied to the database files. As shown in Figure 5-5, Oracle9i prompts the DBA to indicate whether the archived redo log file should be applied to the data file by pressing Enter, <RET>, or whether the recovery should be ended by typing CANCEL and pressing Enter. An AUTO option is also available to allow Oracle9i to apply all available archived log files to the data file. This would only be used if stepping through a complete recovery of the database.

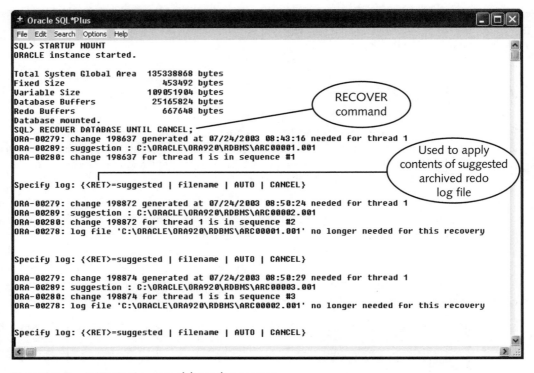

Figure 5-5 Initiating a cancel-based recovery

In the example provided in Figure 5-6, Carlos types CANCEL when prompted as to whether the Arc00007.001 file should be applied. In this case, that file is suspected to be the one containing the problem, so the recovery process is canceled before the transactions in that file are reapplied to the database.

After the recovery process has been canceled, the database can be opened. However, the database should be started with the STARTUP RESTRICT command to prevent users from logging into the database until Carlos has made certain that the database has been recovered properly. After he has verified that the problem was not reintroduced into the database during the recovery process, the database should be shut down immediately and a cold backup performed. The database can then be opened and made available to the users.

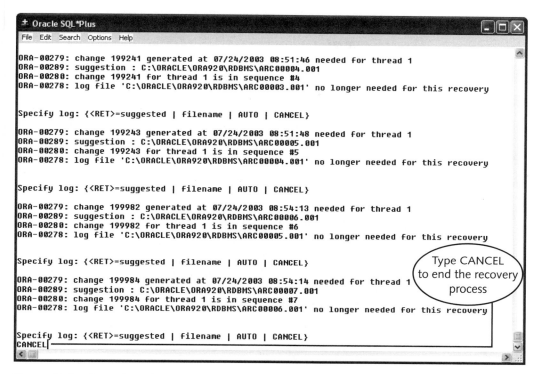

Figure 5-6   Canceling the recovery process

TIME-BASED RECOVERY

As mentioned, a time-based recovery can be performed if Carlos knows when the problematic SQL statement was issued. There is one caveat—the time and date stamp for each transaction is based on the time according to the computer where the Oracle server is located, not the client machine. Therefore, if Carlos is going to perform an incomplete recovery based on the time provided by a user, Carlos should compare the current time as stated by the user to the time currently set on the actual Oracle server.

To perform a time-based recovery, the RECOVER DATABASE UNTIL TIME '*YYYY-MM-DD:HH:MI:SS*'; command must be used. Notice that the TIME option specifies the calendar date, as well as the hour, minute, and second that is to be used to control the recovery process. In most cases, the user is not able to specify the exact second the event occurred, or probably the exact minute, and the *SS* element is set as 00.

Suppose that a user calls Carlos and states that the TRANSACTION table was dropped at 9:18 a.m. Therefore, Carlos is going to attempt to recover the TRANSACTION table by using the time-based recovery approach. Before attempting the steps that Carlos will perform in this section, you need to create the scenario for the simulation.

To create the scenario:

1. Create a cold backup of all database files including the redo log files, control files, and data files. Do not overwrite the cold backup created at the beginning of the chapter. This simulates the regularly scheduled backup performed at Janice Credit Union.

2. Execute the **Ch05script1.sql** file from your Chapter05 folder. This creates the TRANSACTION table in your database and performs a couple of log switches to make certain the transaction is archived from the redo log files.

3. Use the **DESCRIBE** or **SELECT** command to verify that the TRANSACTION table was created.

4. Note the exact time according to your computer by looking in the lower-right corner of your computer screen and recording the time displayed in the taskbar.

5. Execute the **Ch05script2.sql** script file. One of the commands in this file drops the TRANSACTION table. Another command creates a table named DEPOSIT that contains deposit information. (After the simulation is completed, you will see that the TRANSACTION table has been recovered, but that the DEPOSIT table has not.)

After completing these steps, you are now ready to mimic the same recovery process Carlos will be performing.

To recover the database:

1. Type **SHUTDOWN IMMEDIATE** to disconnect any users, roll back any uncommitted transactions, and shut down the database.

2. Perform a cold backup of the database by copying all the database files to a safe location. This backup is required in the event that there is a problem during the recovery process.

3. Restore the affected data file from the last backup of the database created *before* the error occurred. Unless users have been assigned default tablespaces, this is most likely the System01.dbf file.

4. Type **STARTUP MOUNT** to mount the database.

5. Type **RECOVER DATABASE UNTIL TIME '*YYYY-MM-DD:HH:MI:SS*';**, where *YYYY-MM-DD:HH:MI:SS* is the time and date recorded earlier, as shown in Figure 5-7.

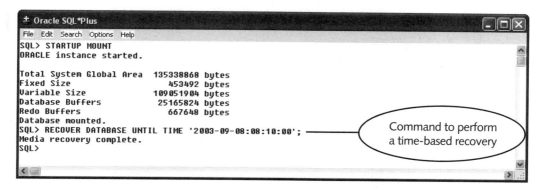

Figure 5-7 Time-based recovery command

6. After the recovery process is completed, type **ALTER DATABASE OPEN RESETLOGS;** to open the database and reset the log files, as shown in Figure 5–8.

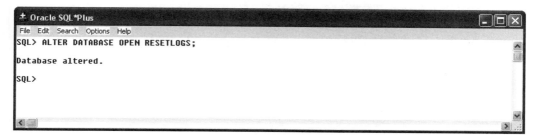

Figure 5-8 Opening the database after time-based recovery

7. After the database is opened, type **SELECT * FROM transaction;** to ensure the table was actually recovered, as shown in Figure 5-9.

8. Type **SELECT * FROM deposit;** to verify that the DEPOSIT table, created after the TRANSACTION table was dropped, was not re-created during the recovery process as shown in Figure 5-10.

5

Figure 5-9   Verification of restored table

```
Oracle SQL*Plus
File  Edit  Search  Options  Help
SQL> SELECT * FROM deposit;
SELECT * FROM deposit
              *
ERROR at line 1:
ORA-00942: table or view does not exist

SQL>
```

Figure 5-10   Verification of data loss

After Carlos has recovered the lost TRANSACTION table, the database must be shut down and then a cold backup performed. After all the database files have been backed up, the database can be opened and made available to the database users.

 You can monitor which log files are being updated to the database and other statistics regarding the current recovery process by selecting the contents of the V$RECOVERY_STATUS view.

CHANGE-BASED RECOVERY

A transaction's SCN is the central concept for a change-based recovery. Whenever a transaction is committed, it is assigned an SCN that is recorded in the redo logs, control file, and data files. This allows Oracle9*i* to keep all transactions in sequential order. In addition, the SCNs are used to ensure that the database is in a consistent state. If the

SCN of the transaction can be identified, then all transactions up to the problematic transaction can be applied to the database. This enables the DBA to identify exactly what data needs to be re-entered into the database after the recovery process is complete.

Suppose, for example, that Carlos receives an error message from the Oracle9*i* database that one of the online redo log files can not be opened as shown in Figure 5-11. In this case, Carlos needs to perform an incomplete recovery to recover the database. However, before attempting the recovery process, Carlos needs to know what transactions were contained in the unavailable redo log file.

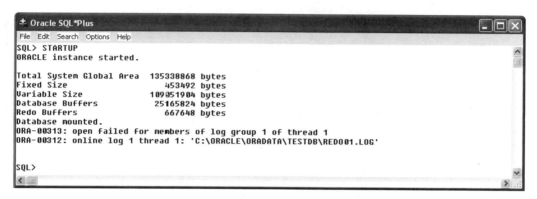

Figure 5-11   Error generated due to an unavailable online redo log file

The lowest and highest SCN for the transactions contained in each redo log file are stored in the control file. Carlos can access these values through the V$LOG_HISTORY view as shown in Figure 5-12. The lowest SCN in the fourth online redo log file shown in the results is 5739731 and the highest SCN is 5909697. The highest SCN is derived by taking the NEXT_CHANGE#, which represents the SCN at which a log switch occurred, and subtracting one. This basically tells Carlos that the redo log file contains all transactions between 5739731 and 5909697. To determine exactly which file is being referenced by the value in the SEQUENCE# column of the V$LOG_HISTORY view, cross reference the value to the sequence number contained in the V$LOG view, as shown in Figure 5-12. However, this information does not help Carlos identify exactly which transaction is assigned a particular SCN. To determine this information, Carlos can use the LogMiner utility.

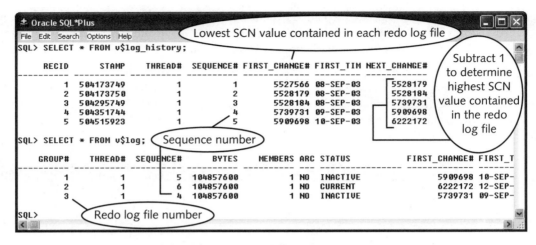

Figure 5-12   Contents of the V$LOG_HISTORY view

LogMiner

LogMiner is a utility included in Oracle9*i* that can be used to analyze the contents of the online and archived redo log files. In actuality, it is an Oracle9*i* package containing several procedures required to analyze and view the contents of the redo log files, as shown in Table 5-1. Any changes made to data or the data dictionary are stored in the redo log files. LogMiner can be used to access these files, and their contents can be examined through the V$LOGMNR_CONTENTS view to determine the SCN, time, the user, and SQL command associated with each transaction.

Table 5-1   Procedures within the LogMiner package

Procedure	Purpose
DBMS_LOGMNR_D.BUILD	Used to extract information from the data dictionary for analysis
DBMS_LOGMNR.ADD_LOGFILE	Used to identify the online and/or archived redo log files to be included in the analysis
DBMS_LOGMNR.START_LOGMNR	Used to perform the analysis of the specified files. Results are returned to the V$LOGMNR_CONTENTS view. The DDL_DICT_TRACKING option must be specified to include DDL operations in the analysis.
DBMS_LOGMNR.END_LOGMNR	Used to end the LogMiner session, and release all resources held by the session

There are several steps required for a LogMiner session. First, Carlos must identify the source for analyzing the data dictionary information using the PL/SQL procedure DBMS_LOGMNR_D.BUILD. There are various options available. The data dictionary can be extracted and stored in the redo log files, in an online catalog, or in a flat file that is separate from the database. The flat file approach is most popular because it does not consume as many system resources as the other approaches.

After the contents of the data dictionary have been extracted, the redo log files to be analyzed are specified using the DBMS_LOGMNR.ADD_LOGFILE procedure. After identifying the desired redo log files, the actual analysis of the files' contents are performed using the DBMS_LOGMNR.START_LOGMNR procedure, which records the results of the analysis into the V$LOGMNR_CONTENTS view. This view can then be queried by the user to determine the time, SCN, SQL command, user name, and so on, for each transaction.

Note, however, by default, LogMiner analyzes only DML operations. If DDL operations need to be included in the analysis, Carlos needs to include the DDL_DICT_TRACKING option of the DBMS_LOGMNR.START_LOGMNR procedure. After Carlos has completed his analysis of the redo log files and is ready to recover the database, the LogMiner session is terminated by executing the DBMS_LOGMNR.END_LOGMNR procedure to close the redo logs and release any database and system resources allocated to LogMiner.

Not all problems require a recovery of the database. The V$LOGMNR_CONTENTS view contains a column named SQL_UNDO that displays the necessary command to be executed to reverse an SQL operation. For example, if a user accidentally deleted one or two rows from a table, LogMiner can be used to determine the SQL commands necessary to add those records back to the database. It would make more sense to simply add the transactions back to the database manually as opposed to performing an incomplete recovery and subsequently losing a large number of transactions that would still have to be re-entered. It is part of the DBA's responsibility to determine when an incomplete recovery, versus some alternative action, is required.

If these various procedures seem confusing, don't be too alarmed. The following sections will demonstrate how to use the LogMiner utility. The basic scenario is that someone has dropped the DEPOSIT table and Carlos needs to recover the database to a state immediately prior to when the table was dropped. Carlos will use LogMiner to identify the SCN of the specific transaction that dropped the table.

Extracting A Data Dictionary

The first step Carlos needs to take to analyze the redo log files is to extract the dictionary to a flat file. To do this, he uses the DBMS_LOGMNR_D.BUILD procedure, which basically identifies the name of the destination file and its location. However, before this procedure can be executed for the first time, Carlos must include the UTL_FILE_DIR parameter in the Init.ora file to identify the location of any destination files. When the DBMS_LOGMNR_D.BUILD procedure is executed, the location specified in the procedure is compared to the location specified by the UTL_FILE_DIR parameter; if there is no such parameter, the server returns an error message.

To extract the data dictionary to a flat file:

1. Type **SHUTDOWN** to shut down the database.

2. Type **CREATE PFILE FROM SPFILE;** as shown in Figure 5-13, to create a text version of the spfile.

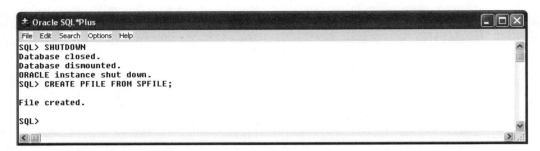

Figure 5-13 Creating a pfile

3. From the operating system, use Notepad to open the pfile created in Step 2. By default, this file is assigned the file name INIT<*databasename*>.ORA and is stored in the DATABASE folder of the *Oraclehome*\ directory.

4. On the first blank line at the end of the file, type ***.utl_file_dir = '<** *location of the flat file*\>', as shown in Figure 5-14, where <*location of the flat file*\> is the destination location for the file that will store the data dictionary information.

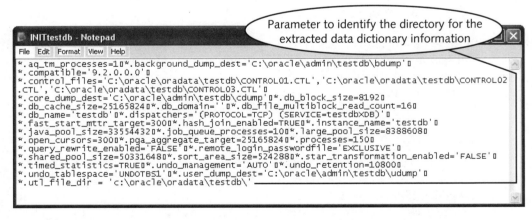

Figure 5-14 Adding the UTL_FILE_DIR parameter to the Init.ora file of the TESTDB database

5. Click **File** from the Notepad menu bar, and then click **Exit** to exit Notepad. Click **Yes** to save the changes.

6. Type **CREATE SPFILE FROM PFILE;** as shown in Figure 5-15, to create a new spfile containing the changes made in Step 4.

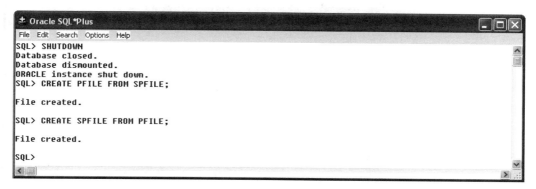

Figure 5-15   Creating a new spfile

7. Type **STARTUP**, as shown in Figure 5-16, to open the database based on the parameters contained in the new spfile.

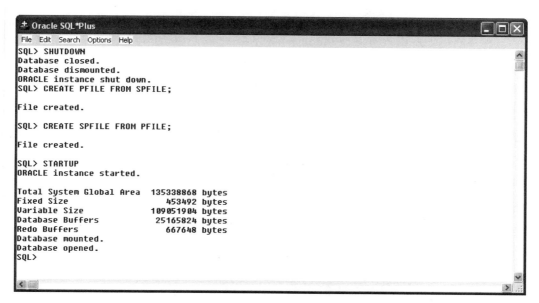

Figure 5-16   Starting the database

8. Type **EXECUTE DBMS_LOGMNR_D.BUILD ('<*filename*>', '<*location of the flat file*\\>');**, as shown in Figure 5-17, where <*filename*> is to be assigned to the resulting file and <*location of the flat file*\\> is the location specified by the UTL_FILE_DIR parameter created in Step 4.

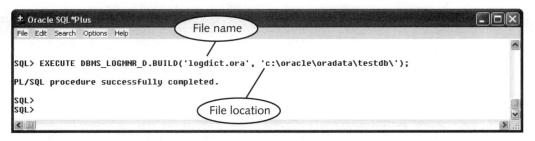

Figure 5-17   Extracting the data dictionary to a flat file

Now that Carlos has extracted the data dictionary information, he can identify the log files that need to be analyzed.

Specifying Log Files For Analysis

In this section, Carlos will identify which redo log files need to be included in the analysis. In the following example, only the online redo log files have been specified. However, any of the database's archived redo log files can also be included in the analysis by simply referencing the name of the files in the DBMS_LOGMNR.ADD_LOGFILE procedure.

When using the DBMS_LOGMNR.ADD_LOGFILE procedure to identify redo log files, there are two basic options that can be used. The first is the DBMS_LOGMNR.NEW option and the second is the DBMS_LOGMNR.ADDFILE option. The DBMS_LOGMNR.NEW option is used to create a new list of files to be analyzed. If a list had previously been identified, the NEW keyword indicates that the old list is to be overwritten. The ADDFILE keyword is used to add files to an existing list.

In the following steps, Carlos uses the NEW keyword to create a list of files to be analyzed and to add the Redo01.log file to that list. The ADDFILE keyword is subsequently used to add the Redo02.log and Redo03.log files to that same list.

To work with the NEW keyword:

1. Type **EXECUTE DBMS_LOGMNR.ADD_LOGFILE(LOGFILENAME=> '<*location*>\\redo01.log', OPTIONS => DBMS_LOGMNR.NEW);**, where <*location*> is the location of the file to be included in the analysis. As shown in Figure 5-18, if the command is longer than the width of the screen, then a hyphen must be placed at the end of the first line to indicate that the command is continued on the next line. If the hyphen is not included, Oracle9*i* returns an error message.

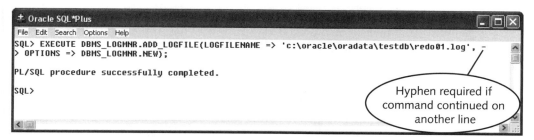

Figure 5-18   Specifying a new file list for analysis by LogMiner

2. Type **EXECUTE DBMS_LOGMNR.ADD_LOGFILE(LOGFILENAME=>
   '<*location*>\redo02.log', OPTIONS => DBMS_LOGMNR.ADDFILE);**,
   as shown in Figure 5-19, where <*location*> is the location of the second redo
   log file to be included in the analysis.

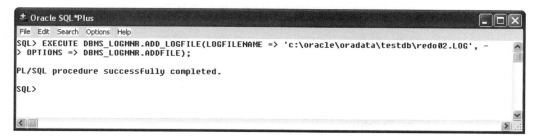

Figure 5-19   Adding the Redo02.log file to the analysis list

3. Type **EXECUTE DBMS_LOGMNR.ADD_LOGFILE(LOGFILENAME=>
   '<*location*>\redo03.log', OPTIONS => DBMS_LOGMNR.ADDFILE);**,
   as shown in Figure 5-20, where <*location*> is the location of the third redo
   log file to be included in the analysis.

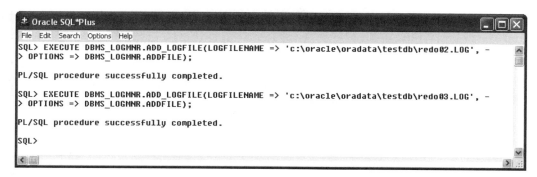

Figure 5-20   Adding the Redo03.log file to the analysis list

At this point, Carlos has added all three of the database's online redo log files to the analysis list. However, if the database had more than three online redo log files, or if he had wanted to include any of the archived redo log files to the analysis list, then Step 3 would be repeated for each file.

Performing the Redo Log Analysis

So far Carlos has identified the source file that will contain the information extracted from the data dictionary and the redo log files that will be included in the analysis. The next step is to actually have LogMiner analyze the contents of the specified files. Various options are available for performing the analysis. For example, Carlos can specify that only the transaction between a specific range of SCNs be analyzed.

As previously shown in Figure 5-12, Carlos was able to determine the range of SCN values contained in each online redo log file. In this section, Carlos first performs an analysis based on a specified range of SCN values. However, as previously mentioned, the analysis only references DML operations contained in the log files. Because the purpose of the analysis is to determine the transaction that dropped the DEPOSIT table, Carlos needs to specify that the analysis must also include DDL operations using the DBMS_LOGMNR.DDL_DICT_TRACKING option of the DBMS_LOGMNR.START_LOGMNR procedure. Therefore, he performs the analysis again specifying this option. However, during this analysis, he omits the range of SCN values which causes LogMiner to analyze the entire contents of all the files previously placed on the analysis list.

To begin the analysis:

1. Type **EXECUTE DBMS_LOGMNR.START_LOGMNR(DICTFILENAME =>
   '<name and location of dictionary file>', STARTSCN =>
   '<starting value>', ENDSCN => '<ending value>');**, as shown
   in Figure 5-21, where *<name and location of dictionary source file>* is the name
   and location of the dictionary flat file created previously using the
   DBMS_LOGMNR_D.BUILD procedure, and *<starting value>* and *<ending
   value>* are the starting and ending SCN values, respectively. If you are uncertain about the valid SCN values you can specify for your database, query the
   V$LOG_HISTORY view to determine the SCN values contained in the
   redo log files and select a range that is currently contained within those files.

2. Type **EXECUTE DBMS_LOGMNR.START_LOGMNR(OPTIONS => DBMS_
   LOGMNR.DDL_DICT_TRACKING);** as shown in Figure 5-22 to instruct
   LogMiner to analyze all the contents of the redo log files previously specified,
   including any DDL operations.

```
Oracle SQL*Plus
File  Edit  Search  Options  Help
SQL> EXECUTE DBMS_LOGMNR.START_LOGMNR(DICTFILENAME => 'c:\oracle\oradata\testdb\logdict.ora', -
> STARTSCN => 5527566, ENDSCN => 5528184);

PL/SQL procedure successfully completed.

SQL>
```

Figure 5-21   Analysis based on a range of SCN values

```
Oracle SQL*Plus
File  Edit  Search  Options  Help
SQL> EXECUTE DBMS_LOGMNR.START_LOGMNR(OPTIONS => DBMS_LOGMNR.DDL_DICT_TRACKING);

PL/SQL procedure successfully completed.

SQL>
```

Figure 5-22   Including DDL operations in LogMiner analysis

When LogMiner analyzes the contents of the redo log files, the results are returned to the V$LOGMNR_CONTENTS view. Although Carlos performed two separate analyses of the redo log files, only the last analysis is viewable through the V$LOGMNR_ CONTENTS view because the contents are overwritten by the most recent analysis.

 Analysis of the redo log files can also be restricted based on a specified time range, a session number, user name, etc. For more information regarding LogMiner, refer to Chapter 9 of the *Oracle9i Database Administrator's Guide Release 2 (9.2)*. This documentation is available through the Online Technology Network (OTN) at *http://otn.oracle.com*.

Displaying Results from the LogMiner Analysis

To display the contents of the V$LOGMNR_CONTENTS view, Carlos can simply issue a SELECT statement. The following SQL statement can be used to identify the SCN and SQL statement of any transaction that references the DEPOSIT table: `SELECT scn, sql_redo FROM v$logmnr_contents WHERE seg_name = 'DEPOSIT';`. After the statement is executed, the SCN value and SQL statement that was previously used to drop the DEPOSIT table is displayed, as shown in Figure 5-23. Your display will be different if more than one SQL statement in your redo log files references the DEPOSIT table.

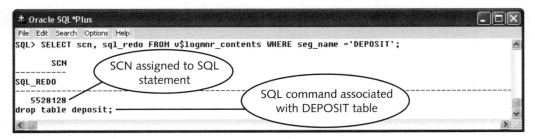

Figure 5-23   Results of the SELECT statement

Ending a LogMiner Session

Now that Carlos has retrieved the SCN value associated with the SQL command that was used to drop the DEPOSIT table, he no longer needs to use the LogMiner utility to analyze any other information in the redo log files. Therefore, before attempting to perform an incomplete recovery of the database, Carlos releases all the database and system resources currently allocated to LogMiner and closes all the redo log files being accessed by LogMiner. Carlos uses the following command to end the current LogMiner session: `EXECUTE DBMS_LOGMNR.END_LOGMNR;`, as shown in Figure 5-24. After the LogMiner session has been terminated, none of the information from the previous analysis is available. If Carlos realizes he needs more information, the entire process for initializing a LogMiner session needs to be repeated, with the exception of adding the UTL_FILE_DICT parameter to the Init.ora file, unless this parameter was deleted at a later time.

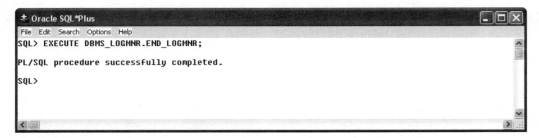

Figure 5-24   Ending a LogMiner session

Performing a Change-Based Recovery

In the previous section, Carlos used the LogMiner utility to determine the SCN assigned to the transaction that dropped the DEPOSIT table. Now he can use this SCN to perform a change-based recovery and minimize the amount of data loss. The exact steps performed during the recovery process are the same as the steps used earlier in the chapter for the time-based recovery, with the exception of the RECOVER DATABASE command. To perform a change-based recovery, the RECOVER DATABASE UNTIL CHANGE *SCN*; command must be used, where *SCN* indicates the actual SCN assigned to the erroneous transaction, as shown in Figure 5-25.

Figure 5-25 Command to perform a change-based recovery

After Carlos executes the command in Figure 5-25, the database is recovered to the point immediately prior to when the DEPOSIT table was dropped. The DROP TABLE command and any subsequent database changes are no longer included in the database.

> After completing the steps in this chapter, restore the cold backup created at the beginning of the chapter, which can reset the database to NOARCHIVELOG mode and remove any reference to the TRANSACTION and DEPOSIT tables.

CHAPTER SUMMARY

- An incomplete database recovery occurs when all the available transactions are not applied to the data file(s) during the recovery process.

- Data loss always occurs during an incomplete recovery.

- An incomplete recovery is performed by using the UNTIL CHANGE, UNTIL TIME, or UNTIL CANCEL keywords with the RECOVER DATABASE command.

- A cold backup should always be performed before attempting an incomplete recovery, as well as after recovering the database.

- A cancel-based recovery is the least flexible recovery procedure and of the three options normally results in the most data loss.

- A time-based recovery requires the DBA to know the time that the problem transaction was completed. The accuracy of the estimated time of the actual event determines the accuracy of the recovery process because the recovery process applies all transactions that occurred before the specified time. A time-based recovery is specified by the RECOVER DATABASE UNTIL TIME 'YYYY-DD-MM:HH:MI:SS'; command, where the date/time model is substituted with the correct date and time that the transaction occurred.

- Of the three options for an incomplete recovery, a change-based recovery normally results in the least amount of data loss for an incomplete recovery. A change-based recovery is specified by the RECOVER DATABASE UNTIL CHANGE SCN;, where SCN represents the transaction containing the problem transaction.

❏ The SCN of a transaction can be determined by using the LogMiner utility.

❏ A cancel-based recovery is specified by the RECOVER DATABASE UNTIL CANCEL; command. The DBA needs to manually specify when the recovery process should end by entering CANCEL and pressing Enter.

Syntax Guide

SQL Commands	Description	Example
ALTER DATABASE OPEN RESETLOGS	Used to discard any redo information that was not referenced during the recovery process; must be executed after performing an incomplete recovery	`ALTER DATABASE OPEN RESETLOGS;`
EXECUTE	Used to execute a PL/SQL procedure	`EXECUTE DBMS_ LOGMNR.ADD_LOGFILE (LOGFILENAME=> 'c:\ foldera\REDO02.LOG', OPTIONS => DBMS_ LOGMNR.ADDFILE);`
RECOVER DATABASE UNTIL CANCEL	Initiates a cancel-based incomplete recovery; requires DBA to manually halt the recovery process	`RECOVER DATABASE UNTIL CANCEL;`
RECOVER DATABASE UNTIL CHANGE SCN	Initiates a change-based incomplete recovery; requires DBA to specify the appropriate system change number	`RECOVER DATABASE UNTIL CHANGE 5689712;`
RECOVER DATABASE UNTIL TIME 'YYYY-DD-MM:HH:MI:SS'	Initiates a time-based incomplete recovery; requires DBA to specify the appropriate time that the problem occurred	`RECOVER DATABASE UNTIL TIME '2003-07-24:09:18:00';`
INIT.ORA Parameter	**Purpose**	**Example**
UTL_FILE_DICT	Used to specify the destination of the extracted data dictionary	`utl_file_dict = 'c:\foldera\'`
LogMiner Procedures	**Purpose**	**Example**
DBMS_LOGMNR.ADD_ LOGFILE	Used to identify the online and/or archived redo log files to be included in the analysis. Use DBMS_ LOGMNR.NEW to create a new list and DBMS_LOGMNR.ADDFILE to add files to an existing list.	`EXECUTE DBMS_ LOGMNR.ADD_LOGFILE (LOGFILENAME=> 'c:\foldera\ REDO02.LOG', OPTIONS => DBMS_ LOGMNR.ADDFILE);`

LogMiner Procedures	Purpose	Example
DBMS_LOGMNR.END_LOGMNR	Used to end the LogMiner session and release all resources held by the session	`EXECUTE DBMS_LOGMNR.END_LOGMNR;`
DBMS_LOGMNR.START_LOGMNR	Used to perform the analysis of the specified files. Results are returned to the V$LOGMNR view. The DDL_DICT_TRACKING option must be specified to include DDL operations in the analysis.	`EXECUTE DBMS_LOGMNR.START_LOGMNR(OPTIONS => DBMS_LOGMNR.DDL_DICT_TRACKING);`
DBMS_LOGMNR_D.BUILD	Used to extract information from the data dictionary for analysis	`EXECUTE DBMS_LOGMNR_D.BUILD('logdict2.ora', '<c:\foldera\>');`

5

REVIEW QUESTIONS

1. Which type of incomplete recovery offers the least flexibility in terms of controlling exactly what data is recovered?

2. Which of the following is an example of a situation requiring an incomplete recovery?

   a. A data file is lost and there is no backup copy of the file that can be used as a replacement.

   b. A data file for a database in NOARCHIVELOG mode is unavailable and must be recovered.

   c. A data file contains an error that must be removed from the database so it can only be recovered to a certain point.

   d. all of the above

3. Because archived redo log files are used to perform an incomplete recovery, the database must be in _____ mode.

4. Which of the following commands can be used to perform an incomplete recovery?

   a. RESTORE DATABASE

   b. RECOVER DATABASE

   c. RECOVER DATA

   d. RESTORE DATA

5. What command is used to specify that the DBA will manually stop the recovery process?

6. The Init.ora file contains the time and date stamp to indicate when each redo log file was archived and the name of the archived file. True or False?

7. Which of the following should be completed before attempting an incomplete recovery?

   a. hot backup

   b. cold backup

   c. deletion of the unnecessary archived redo log files

   d. mounting of the database

8. The LogMiner utility can be used to determine the SCN associated with a specific transaction. True or False?

9. The DBMS_LOGMNR_D.BUILD procedure is used to identify the redo log files to be included in an analysis of the redo log files. True or False?

10. Which of the following types of incomplete recoveries most likely results in the greatest data loss?

   a. time-based

   b. cancel-based

   c. change-based

   d. there is no difference

11. Which of the following can be used to determine the SCN for a particular transaction?

   a. Alert.log file

   b. control file

   c. LogMiner utility

   d. DataDredge utility

12. The database should be in what state when the recovery process is performed?

13. What effect does resetting the log files have on future recovery attempts?

14. Transaction loss always occurs during a(n) _____ recovery.

15. Which keyword is used with the DBMS_LOGMNR.START_LOGMNR procedure to specify that DDL operations should be included in the analysis performed by LogMiner?

16. Which of the following incomplete recovery types require the DBA to specify a system change number?

   a. change-based

   b. time-based

   c. cancel-based

   d. all of the above require the system change number

17. Which of the following incomplete recovery types result in the least amount of data loss?

   a. change-based

   b. time-based

   c. cancel-based

   d. there is no difference

18. What is the purpose of an SCN during the incomplete recovery process?

19. If only a portion of an archived redo log file is to be applied during the recovery process, which type(s) of incomplete recovery is being performed?

20. What information is needed by the DBA to perform a time-based recovery?

HANDS-ON ASSIGNMENTS

 If an error occurs while performing one of the following assignments, simply use the cold backup created at the beginning of the chapter to restore the database files.

Assignment 5-1 Determining the SCN Range of the Online Redo Log Files

The purpose of this assignment is to determine the range of SCN values assigned to specific online redo log files.

1. Log into the database, and enable the SYSDBA privileges.

2. Query the V$LOG_HISTORY view to determine the range of SCN values contained in each sequence number.

3. Query the V$LOG view to determine the online redo log file associated with each sequence number.

4. Using the information obtained in Steps 3 and 4, determine the range of SCN values currently stored in each online redo log file.

Assignment 5-2 Performing a Change-Based Recovery

In this assignment, you will be performing an incomplete recovery based on an SCN. The particular SCN for this recovery is estimated from information derived from the V$LOG_HISTORY and V$LOG views. Because the database is recovered to a point prior to the creation of the DEPOSIT table, that table will no longer exist after the recovery process is completed.

1. Using the information obtained in Hands-On Assignment 5-1, determine the last SCN value currently contained in the online redo log files.

2. Execute the **Ch05script2.sql** file from the Chapter05 folder to create the DEPOSIT table.

3. Issue a **SELECT** statement to view the contents of the DEPOSIT table and to verify that the table exists.

4. Perform a change-based recovery using the value obtained in Step 1 as the stopping point for the recovery process.

5. Verify that the DEPOSIT table no longer exists.

Assignment 5-3 Performing a Cancel-Based Recovery

The purpose of this assignment is to recover a database at a point prior to when a specific archived redo log file is updated to the database. This type of recovery might be required if none of the online redo log files are available for the recovery process or at least one of the archived redo log files is unavailable.

1. Create a cold backup of the database. Do not overwrite the files from the cold backup created at the beginning of this chapter.

2. Query the V$LOG view to determine the current online redo log file.

3. Execute the **Ch05script2.sql** file in the Chapter05 folder to create the DEPOSIT table.

4. Type **ALTER SYSTEM SWITCH LOGFILE;** to cause a log switch to occur, which results in archiving the redo log file identified in Step 2.

5. Query the V$ARCHIVED_LOG view to determine the name of the archived redo log file created in Step 4. (It is the last file name displayed in the list.)

6. Restore the cold backup created at the beginning of this chapter.

7. Perform a cancel-based recovery and stop the process immediately before the contents of the archived redo log file identified in Step 5 are updated to the database by typing **CANCEL** when prompted during the recovery process.

8. Use a SELECT statement or the DESCRIBE command to verify that the table no longer exists.

Assignment 5-4 Performing a Time-Based Recovery

In this assignment, you will perform a time-based recovery.

1. Use a SELECT statement or the DESCRIBE command to verify that the TRANSACTION table does not exist in the database. If the table exists, use the DROP TABLE command to remove the table.

2. Record the exact time according to your computer as shown on the taskbar in the lower-right corner of your screen.

3. Wait one minute after the time recorded in Step 2, and then execute the **Ch05script1.sql** file from the Chapter05 folder to create the TRANSACTION table.

4. Use a SELECT statement or the DESCRIBE command to verify that the TRANSACTION table has been created.

5. Perform a time-based recovery of the database using the time recorded in Step 2 as the stopping point for the recovery process.

6. Verify that the TRANSACTION table no longer exists in the database.

Assignment 5-5 Using LogMiner to Identify a Specific DML Operation

In the following steps, you will be creating and using a LogMiner session to identify the specific command used to "accidentally" remove a row from the TRANSACTION table. This type of situation can occur at Janice Credit Union if a user meant to reverse a transaction for a particular customer, but entered the wrong transaction number.

1. Execute the **Ch05script1.sql** file from the Chapter05 folder.

2. Use a SELECT statement or DESCRIBE command to verify that the TRANS-ACTION table exists.

3. Type **DELETE FROM transaction WHERE trans# = 1987682;** to remove a row from the database table.

4. Use a SELECT statement to verify that the transaction referenced in Step 3 is no longer contained in the TRANSACTION table.

5. If necessary, add the UTL_FILE_DICT parameter to the INIT.ORA file for the database, and specify the location for the flat file that will contain the data dictionary information to be used by the LogMiner utility.

6. Using the DBMS_LOGMNR_D.BUILD procedure, extract the data dictionary information into a flat file in the same location identified by the UTL_FILE_DICT parameter in the INIT.ORA file.

7. Add the online redo log files to the list of files to be analyzed using the NEW and ADDFILE options of the DBMS_LOGMNR.ADD_LOGFILE procedure.

8. Use the DBMS_LOGMNR.START_LOGMNR procedure to analyze the files specified in Step 7.

9. Type **SELECT SCN, sql_redo FROM V$LOGMNR_CONTENTS WHERE seg_name = 'TRANSACTION';** and locate the statement executed in Step 3 in the SQL_REDO column of the results. Terminate the LogMiner session unless you are continuing with Assignment 5-6.

Assignment 5-6 Using Results of a LogMiner Session to Undo a DML Operation

In Assignment 5-5, Carlos identified the SQL statement that was executed to "accidentally" remove a row from the TRANSACTION table. However, LogMiner can display the exact command executed, as well as the command to undo the executed command. Because only one row was deleted, Carlos would rather identify the single transaction and reverse that transaction. Why? By avoiding an incomplete recovery, valid transactions committed after the problematic transaction occurred are not lost.

1. Continuing with the LogMiner session created in Assignment 5-5, type **SELECT SCN, sql_undo FROM V$LOGMNR_CONTENTS WHERE SCN = <integer>;** , where the *<integer>* is the SCN value previously identified in Step 9 of Assignment 5-5.

2. To add the previously deleted row back to the table, enter and execute the INSERT statement listed in the SQL_UNDO column for the transaction identified in Step 1.

3. Use a SELECT statement to query the TRANSACTION table and verify that the previously deleted transaction has been added back to the TRANSACTION table.

4. Terminate the LogMiner session unless you are continuing with Assignment 5-7.

Assignment 5-7 Using LogMiner to Locate DDL Operations

In this assignment, you will be required to use LogMiner to locate the CREATE TABLE command that was previously used to create the TRANSACTION table in Assignment 5-5.

1. Execute the DBMS_LOGMNR.START_LOGMNR procedure with the correct option to analyze any DDL operations that are contained with redo log files that are currently specified for the analysis.

2. Query the V$LOG_CONTENTS view to locate any operations that reference the TRANSACTION table. When the results are displayed, record the value of the SCN for the SQL statement that was previously used to create the TRANSACTION table for later use in Assignment 5-8.

3. Terminate the current LogMiner session using the DBMS_LOGMNR.END_LOGMNR procedure.

Assignment 5-8 Performing a Change-Based Recovery Based on an SCN Obtained through LogMiner

In this assignment, you will perform an incomplete recovery that recovers the database to the point immediately prior to when the command to create the TRANSACTION table was executed.

1. Use a SELECT statement or the DESCRIBE command to verify that the TRANSACTION table currently exists.

2. Perform a change-based recovery by specifying the SCN value in Step 2 of Assignment 5-7.

3. Use a SELECT statement or the DESCRIBE command to verify that the TRANSACTION table no longer exists.

CASE PROJECTS

Case 5-1 Table Does Not Exist

Shortly after lunch on Thursday, Carlos receives a call that a user attempted to access the PAYROLL table from the human resources database; however, a message indicating that the table did not exist was displayed. The PAYROLL table is considered a critical component of the human resources database and must be recovered at all costs. Unfortunately, the only information the user can provide is that the table was last accessed on Thursday of last week when the previous payroll checks were printed. At this time, the user cannot verify whether any other activities occurred in the database since that time.

Assuming that Carlos has logged into the database and verified that the table no longer exists, identify the appropriate steps for Carlos to take to determine when and how the table was removed from the database and how to recover the lost table with minimal data loss.

Case 5-2 Identify Recovery Steps

Continuing with the procedure manual you began in Chapter 2, prepare lists that identify the necessary steps to accomplish the following:

❏ Perform a time-based recovery.

❏ Perform a change-based recovery.

❏ Perform a cancel-based recovery.

❏ Locate a specific SCN for a DDL operation using LogMiner.

Provide a statement identifying the main advantage and disadvantage for each of the different recovery approaches.

6

OVERVIEW OF RECOVERY MANAGER

After completing this chapter, you should be able to do the following:

- ◆ List some of the features provided by using Recovery Manager (RMAN) to perform backup and recovery operations
- ◆ Identify the four basic components of RMAN
- ◆ Identify the required components of RMAN
- ◆ Explain the purpose of the Media Management Library (MML) interface
- ◆ Start the command-line interface of RMAN
- ◆ Identify the target database and catalog mode when starting RMAN
- ◆ List the contents of the RMAN repository
- ◆ Identify the two possible locations for storing the RMAN repository
- ◆ Configure RMAN to perform automatic backup of the control file
- ◆ Explain the purpose of channel allocation

In previous chapters, Carlos performed backup and recovery procedures using the operating system to copy and restore files, and issued the RECOVER command to manually recover the database through SQL*Plus. However, this creates a huge administrative burden for Carlos.

Normally, a DBA keeps several days worth of backups available in case a problem with the database files is not detected immediately. For example, a user may have dropped a table, but the event may not be noticed for a couple of days. For Carlos to recover the table, the most recent backup may not be an option. Therefore, all backup sets of the database must be organized in a manner so that Carlos can quickly locate the necessary files to perform an incomplete recovery.

Many DBAs such as Carlos organize their backup copies of the database files into folders, or even tapes, named for the date and time the backup occurred. In addition, Carlos needs to use various utilities, such as DBVERIFY, to manually test the validity of the backup copies. However, to simplify the burden of managing backup and recovery operations of the Oracle9i database, Carlos has the option of using Recovery Manager (RMAN).

RMAN is a utility that can be used to backup, restore, and recover a database. It is automatically installed during the Oracle9i database installation process. With RMAN, a DBA would be able to perform **incremental backups** which can save time by copying only changed data (i.e., backup only used data blocks), and test the validity of backup files. Because the backup and recovery operations performed by RMAN are conducted through a server session, they are commonly referred to as Server Managed Backup and Recovery operations.

This chapter provides an overview of the RMAN utility, including its components and configuration. Subsequent chapters will focus on using RMAN to perform backup operations, complete and incomplete recoveries, and maintenance requirements for the utility.

THE CURRENT CHALLENGE IN THE JANICE CREDIT UNION DATABASE

As mentioned, with user-managed backup procedures, Carlos and his staff must remember which database files to backup, when to perform the backup, as well as where the valid backups are stored (i.e., file names and location). Remember that without a valid copy of the control file, data files, and any archived redo log file, it may be impossible to recover a database in the event of media, or even user, error.

One option is to use scripts that are executed to specify which files to backup and where to store the copies. However, as an alternative, Carlos has been considering the advantages of using the Recovery Manager (RMAN) utility to control the backup process. RMAN automatically identifies the most recent backup copy of the database files, backs up the control files whenever structural changes occur, restores data files during recovery operations, and so on. It basically becomes "one-stop shopping" for backup and recovery operations of an Oracle9i database.

SET UP YOUR COMPUTER FOR THE CHAPTER

Before attempting any changes demonstrated in this chapter, remember to perform a cold backup of all database files, including the data files, all copies of the control file, and any archived redo log files. In the event an irreversible error occurs when attempting the material in this chapter, the backup can be used to return the database to its original state.

NOTES ABOUT DUAL COVERAGE IN THIS CHAPTER

This chapter focuses on the command-line approach to using RMAN, known as the RMAN Executable. The GUI approach, requires access to RMAN through the Enterprise Manager Console. Using that approach, you are actually interacting with the Backup Wizard for backup operations and the Recovery Wizard for recovery operations. The GUI interface will be presented in later chapters covering those specific topics.

RMAN FEATURES

There are different features of RMAN that Carlos takes into consideration when trying to determine whether Janice Credit Union should begin using RMAN to perform backup and recovery procedures versus user-managed procedures. These features include:

- Providing report features such as identification of files no longer needed for recovery purposes, which files need to be backed up, and which files have not been backed up in a certain timeframe

- Supporting incremental backups which can save time when compared to backing up the entire database during each backup session, even for databases in NOARCHIVELOG mode; however, it may take longer to recover a database because all the incremental backups need to be applied to the last full backup copy of the database

- Testing backup and recovery processes without actually performing the processes on the actual database

- Simplifying the creation of test databases based on the actual production database

- Eliminating the need to place tablespaces in backup mode to perform online backups

- Reducing time required by backing up just data blocks that have been written to; empty data blocks can be ignored, which saves time during both backup and recovery operations

- Providing error checking by computing checksums for every block during the backup process; checks for corrupt blocks during both backup and recovery operations

- Storing scripts of repetitive tasks that can be executed when needed; requires use of a recovery catalog

At a first glance, Carlos wonders why a DBA would elect not to use RMAN to simplify the backup and recovery process. In essence, the burden of tracking backup copies, identifying the current backup set, and so on, has been shifted from the user to the software. Therefore, Carlos thinks it's worth his time to begin to examine exactly what is required to implement RMAN and how it works.

RMAN COMPONENTS

RMAN has four basic components, or parts: the RMAN Executable, the target database, the recovery catalog, and the Media Management Library (MML) interface. Only the RMAN Executable and target database are required. What does that mean to a DBA such as Carlos? It means he has the flexibility of using only a minimum configuration of the utility to support backup and recovery operations, or he can add more complexity and include the Media Management Library interface to support the use of tape libraries or the recovery catalog to store scripts.

The following list describes each of the four basic components that are illustrated in Figure 6-1.

- **RMAN Executable**: A user can interact with RMAN using either a command line or a graphical user interface (GUI). The command-line interface can be accessed directly from the operating system, while the GUI is available through the Enterprise Manager Console when the Oracle Management Server (OMS) has been configured. The RMAN Executable allows the user to issue commands to perform backup and recovery operations, as well as view the utility's configuration, display information about the database, and so on.

- **Target database**: The database that is to be backed up or recovered is referred to as the **target database**. RMAN refers to the information contained in the control file of the target database to determine the location and size of the physical files to be copied.

- **Repository**: A **repository** is used to store the metadata regarding the target database and backup and recovery operations. This repository can be stored in the control file of the target database or in a special schema referred to as the recovery catalog (as shown in Figure 6-1). It is recommended that the recovery catalog be stored in a different database than the target database. In the event the control file of the database becomes inaccessible, RMAN is unable to access the necessary repository information to recover the database. When a recovery catalog is used, there are additional backup and recovery options, such as managing scripts, that are available to Carlos.

- **Media Management Library (MML) interface**: This interface allows RMAN to interact with third-party media managers to read from and write to sequential storage media such as magnetic tapes. Some large organizations use automated tape libraries (ATLs) to automatically change magnetic tapes during backup and recovery operations and to track the contents of the tapes. The MML interface is used to communicate between the Oracle server session and the media manager software.

 The RMAN utility can also connect to an auxiliary database that is created from the backup of the original database or is created temporarily for the purpose of performing tablespace point-in-time recovery (TSPITR).

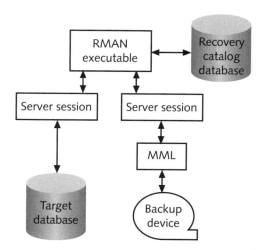

Figure 6-1   Recovery Manager components

RMAN EXECUTABLE

The RMAN Executable provides access to the RMAN utility and allows Carlos to enter the commands necessary to perform backup and recovery operations. The RMAN Executable can be accessed by a command-line interface. A command-line interface, as you recall, requires the user to manually type the command that is to be executed. The command-line interface can be started directly from the operating system using the Run option of the Start menu. Many experienced DBAs, such as Carlos, prefer to use this type of interface because less time is spent navigating through a series of windows and menus, which is required with a GUI. However, it does require Carlos to know the correct syntax of any command that needs to be executed; otherwise, an error message is returned.

When an RMAN session is started, a user process is created for RMAN and two default server processes are started on the Oracle9*i* server. The first server process provides a connection to the target database and allows the DBA (via RMAN) to execute SQL commands. In addition, it enables RMAN to perform the roll forward stage of the recovery process and provides access to the control file of the target database. The second server process is actually a polling connection, which is used by RMAN to monitor for remote procedure calls (RPC). Additional processes can be started depending on the operation(s) to be performed during the session.

If a catalog mode is not specified and a command is executed that requires connection to a repository, the control file of the target database is used by default. After the connection to a repository is made, this option cannot be changed dynamically.

There are various choices you can make when starting up the command-line interface. Each is discussed in turn.

Starting RMAN Without Specifying Parameters

The basic approach is to start the Recovery Manager from the operating system, then issue backup, recovery, or configuration commands at the RMAN> prompt. After RMAN is started, the target database and other settings are then specified.

To start RMAN without specifying any connection requirements:

1. Click the **Start** button on the Windows XP task bar.

2. When the Start menu appears, click the **Run** command on the right side of the menu.

3. When the Run dialog box appears, type **rman** in the dialog box as shown in Figure 6-2, and then click **OK**.

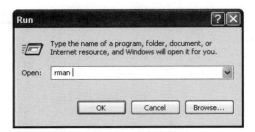

Figure 6-2   Command to start Recovery Manager

4. After a window appears displaying the RMAN> prompt, type **show all;** and press **Enter** to display the utility's current configuration, as demonstrated in Figure 6-3.

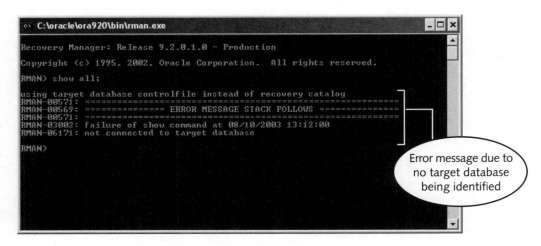

Figure 6-3   Current configuration of RMAN

Because you did not specify whether a Recovery Catalog was to be used as the repository when RMAN was started, the control file of the target database is used to store the metadata. This is the default option in RMAN. After making a connection to the repository, this option cannot be dynamically changed. The RMAN session needs to be terminated and then restarted to change the repository connection.

In addition, notice the error messages displayed in Figure 6-3. These error messages occurred because you did not specify the name of the target database. Basically, the user is not connected to a target database, which is one of the requirements for using the RMAN utility.

To make the connection:

1. Type **connect target** / at the RMAN> prompt, and press **Enter**. When the command is executed, the information regarding the target database is displayed, as shown in Figure 6-4.

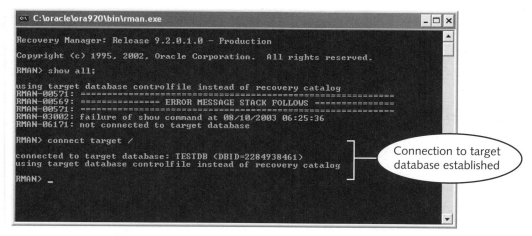

Figure 6-4 Specifying the target database in RMAN

How did RMAN know that the target database was TESTDB? The current value assigned to the ORACLE_SID variable in the Oracle9*i* environment is the TESTDB database. If you want to connect to a different database, the ORACLE_SID has to be reset to the desired database, or the correct connect string needs to be provided.

After reviewing the messages displayed by RMAN, the utility can be closed by typing either EXIT or QUIT at the RMAN> prompt and pressing Enter.

Starting RMAN Parameters

As an alternative, the target database, database user, and so on, can be specified as arguments or parameters when RMAN is started. This approach is preferred because the user can specify whether the utility should be connected to the Recovery Catalog.

In the following step sequence, Carlos is starting RMAN and specifying that the target database is the current default database in Oracle9*i*. The user connection is specified as a slash (/) to indicate that a privileged SYSDBA account is to be used. The NOCATALOG option is explicitly included to instruct RMAN that a Recovery Catalog should not be used during this connection.

To start RMAN:

1. Click the **Start** button.

2. When the Start menu appears, click the **Run** command on the right side of the menu.

3. When the Run dialog box appears, type **rman target / nocatalog** in the dialog box, as shown in Figure 6-5, and click **OK**.

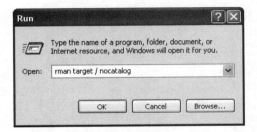

Figure 6-5 Starting RMAN with target database connection and catalog mode indicated

4. After a window appears displaying the RMAN> prompt, type **show all;** and press **Enter** to display the utility's current configuration, as shown in Figure 6-6.

Because RMAN is connected to a target database, several configuration messages are displayed. However, each configuration variable is assigned a default value. As Carlos becomes more familiar with RMAN in subsequent chapters, the RMAN configuration can be changed to meet the needs of any backup policy he implements to specify the number of backup copies to be retained, frequency of backups, and so on.

Notice the first configuration parameter listed in Figure 6-6, CONFIGURE RETENTION POLICY TO REDUNDANCY 1. This is a default configuration indicating that only one backup copy of the control file and each data file should exist. If there is more than one copy, the extra copies are marked as obsolete and can be deleted at any time using the DELETE OBSOLETE command.

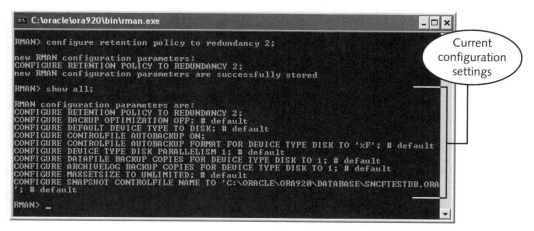

Figure 6-6 Current RMAN configurations

However, as a precaution, suppose Carlos wants to make certain that there are always two copies of the control file and data files available. All he needs do is change the integer for the retention policy to a value of 2 by issuing the command CONFIGURE RETENTION POLICY TO REDUNDANCY 2 at the prompt. If Carlos later decides to reset the retention policy to 1, he can either reissue the command using 1 as the integer value for the policy, or issue the command CONFIGURE RETENTION POLICY CLEAR indicating that the default value of 1 should be assigned to the parameter.

To change, display, and then reset the parameter to its default value:

1. Type **configure retention policy to redundancy 2;** at the RMAN> prompt. After the policy has been changed, the new value of the parameter is displayed, as shown in Figure 6-7.

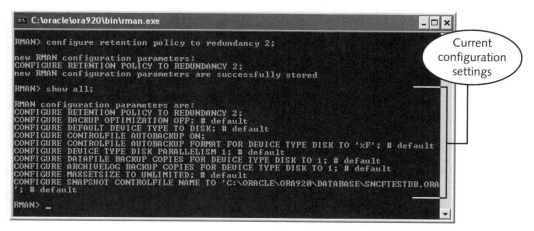

Figure 6-7 Changing RMAN configuration parameter

2. Type **show all;** to view the current values of all configuration parameters. The retention policy is now assigned the value of 2.

3. Type **configure retention policy clear;** to reset the parameter back to its default value. As shown in Figure 6-8, the old value is displayed, followed by a message indicating that the parameter has been reset to its default value. As previously mentioned, for this particular parameter, that value is 1.

Figure 6-8 Resetting configuration parameter to default value

RMAN REPOSITORY

Recall that information about the target database is stored in a repository. That repository contains the following information:

- **Backup sets**: All the output files from one backup operation including the date and time the files were created

- **Backup pieces**: Each file in a backup set

- **Image copies**: Exact replicas of the database files; includes empty data blocks

- **Target database schema**: The structure of the target database

- **Configuration settings**: How long the backup sets should be stored before they are overwritten, automatic channels, and so on

In essence, the repository houses all the information regarding previous backup operations, the target database, and the RMAN settings. Why is all of this necessary? Consider that if RMAN were not able to access the structure of the database, how would the data files be identified and copied? If RMAN could not distinguish among the various backup sets, how would the most recent copy be identified when needed for recovery operations?

In essence, the repository is what provides RMAN with its functionality, separating it from user-managed backup and recovery operations.

As mentioned, this repository can be stored in the control file of the target database or in a recovery catalog. The following sections describe the nuances associated with each of these approaches.

Control File

By default, the control file of the target database is used to store backup and recovery information. There are two issues to keep in mind when a control file is used to store the repository.

The first issue is the size of the control file. Information regarding RMAN operations is stored in the reusable section of the control file. The size of this section is dynamic and grows depending on the frequency of the backup process. To manage the size of the control file, the CONTROL_FILE_RECORD_KEEP_TIME parameter can be used to specify the number of days RMAN data should be kept in the control file before it expires and can be overwritten. By default, the value assigned to the parameter is seven days.

Note that when information needs to be added to the control file, any available free space is first used. Then, any information other than the value specified by the CONTROL_FILE_RECORD_KEEP_TIME parameter is overwritten. If there is still not enough space available, the size of the control file increases to provide the necessary space. However, there is a maximum size to a control file. This maximum size is determined by the operating system of the computer where the control file is stored.

The second issue to keep in mind is that if all copies of the control file are unavailable, then any information regarding RMAN backups is lost. Therefore, Oracle9i documentation recommends that RMAN be configured to perform an automatic backup of the control file. With this configuration, RMAN backs up the control file anytime a BACKUP or COPY command is executed, or if there are any structural changes to the database. By default, the automatic backup of the control file is disabled.

Given this information, Carlos decides to go ahead and configure RMAN to create an automatic backup of the control file whenever changes occur. You can use the following steps to help him make this change.

To configure RMAN:

1. If necessary, open the RMAN utility and connect to the target database with a privileged SYSDBA account.

2. Type **configure controlfile autobackup on;** and press **Enter**. After the configuration has been changed, the message shown in Figure 6-9 is displayed.

Figure 6-9   RMAN configuration change

Now that the configuration change has been made, the control file is automatically backed up when a structural change occurs in the database or if the database is backed up or copied. In the event all multiplexed copies of the control file become unavailable, the backup copy created by RMAN can be used to mount and recover the database.

Recovery Catalog

Use of a recovery catalog allows all features of RMAN to be accessible. In addition to the information that would otherwise be contained in a control file, a recovery catalog can also be used to store scripts of frequently executed RMAN commands. Furthermore, if all copies of the target database's control file are lost, backup files of the database can still be accessed without having to mount the target database. A recovery catalog also provides the ability to backup and recover multiple databases. Therefore, Oracle Corporation recommends that a recovery catalog be used to store RMAN information about multiple databases, rather than in the control file of the respective database.

CHANNEL ALLOCATION

A channel is a link to the target database via a server session. When a channel is allocated, one server process is created. To issue backup and recovery commands in RMAN, a **channel** must be allocated. The channel can be allocated manually using the ALLOCATE CHANNEL command every time a backup or recovery operation is performed. As an alternative, RMAN can be configured to automatically allocate a channel. Channel allocation will be demonstrated in Chapters 7 and 8.

The channel also identifies the storage device that is to be used with the backup or recovery operation. As shown in Figure 6-10, the channel allows communication between RMAN and the target database via a server session. Any necessary read or write operations are also performed by the server session.

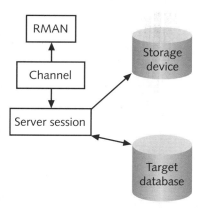

Figure 6-10 Channel usage

Channels can be allocated manually or automatically. Normally a DBA configures RMAN to automatically allocate a channel whenever a BACKUP, RESTORE, or DELETE command is executed. However, a channel specifies the device to be used for the read-write operations. There may be occasions when the destination device should be changed. In that case, most DBAs manually allocate the desired channel. Basically, if a channel is not specified, the automatic channel allocations stored by RMAN are used. However, if the channel is manually allocated, it overrides any automatic settings.

Chapters 7 through 11 of this book will demonstrate how to use RMAN to perform various operations. The purpose of this chapter is simply to provide an overview of the Recovery Manager utility.

CHAPTER SUMMARY

- ☐ Recovery Manager (RMAN) provides more flexibility and options than user-managed backup and recovery procedures.

- ☐ RMAN consists of four components: RMAN executable, target database, a repository, and an MML interface.

- ☐ Only the RMAN executable and target database are required for every RMAN configuration.

- ☐ The target database can be specified when RMAN is started, otherwise the default database for the Oracle9*i* environment is assumed.

- ☐ A user must log into the target database with a SYSDBA privileged account.

- ❑ The repository can be stored in the control file of the target database or in a recovery catalog.
- ❑ Use of a recovery catalog allows the DBA to store frequently executed scripts and manage multiple databases.
- ❑ The amount of information regarding backup information stored in the control file can affect the size of the file.
- ❑ A channel must be allocated before backup and recovery operations are executed through RMAN. The allocation can be configured to occur automatically, or it can be manually allocated.

SYNTAX GUIDE

Command	Description	Example
RMAN	Command issued in the operating system to launch the command-line interface for Recovery Manager	C:\> rman
CONFIGURE	Used to specify a configuration setting for RMAN	RMAN> configure controlfile autobackup on;
SHOW	SQL command that can be executed in RMAN to display current configuration values	RMAN> show all;
RMAN Configuration Setting	Description	Example
CONTROLFILE	Indicates whether the control file of the target database should be automatically backed up whenever the database structure is changed or when a BACKUP or COPY command is executed	configure controlfile autobackup on;
RETENTION POLICY ...REDUNDANCY	Indicates the number of copies of the control file and each data file that RMAN should currently have stored	configure retention policy to redundancy 2;

RMAN Parameter	Description	Example
TARGET /	Indicates that a connection should be made to the target database using a SYSDBA privileged account	`rman target /`
NOCATALOG	Indicates that backup and recovery information should be contained in the control file of the target database	`rman nocatalog`

INIT.ORA Parameter	Description	Example
CONTROL_FILE_ RECORD_KEEP_ TIME	Indicates how many days worth of backup and recovery information must be kept before it is overwritten	`CONTROL_FILE_RECORD_ KEEP_TIME = 1`

6

REVIEW QUESTIONS

1. Which of the following is a required component of RMAN?

    a. target database

    b. channel

    c. recovery catalog

    d. Media Management Library interface

2. If a recovery catalog is not available, the RMAN repository is automatically stored in the _____ of the target database.

3. The _____ command is used to start the RMAN command-line interface.

4. Which of the following parameters is used to determine how long backup information should be retained in the control file of the target database?

    a. RETENTION_TIME

    b. MAX_CONTROL_SIZE

    c. CONTROL_FILE_RECORD_KEEP_TIME

    d. RMAN_MAX_TIME

5. Which parameter indicates that RMAN should connect to the default database?

    a. DEFAULT

    b. DATABASE

    c. TARGET

    d. CONNECTION

6. What is an incremental backup?

7. What is a backup set?

8. Which of the following stores the metadata regarding the target database and backup and recovery operations?

    a. channel

    b. repository

    c. data bank

    d. none of the above

9. Which of the following tasks can be completed with Recovery Manager, but is not possible during user-managed backups?

    a. copy just used data blocks and not empty data blocks

    b. place online tablespaces in backup mode

    c. practice or test the backup process

    d. all of the above

10. Which of the following is required for RMAN to manage stored scripts in the repository?

    a. recovery catalog

    b. target database

    c. channel

    d. all of the above

11. Which of the following parameters specifies that RMAN should not use a recovery catalog?

    a. NORECOVERY

    b. NOCATALOG

    c. NOCAT

    d. none of the above

12. Which of the following commands is used to display configuration settings for RMAN?

    a. DISPLAY

    b. SHOW

    c. LIST

    d. TERMOUT

13. A channel must be allocated before which of the following commands can be executed?

    a. CONFIGURE

    b. SHOW

    c. TARGET

    d. BACKUP

14. What is an image copy?

15. Where are time and date of each backup set stored?

16. Each file in a backup set is referred to as a backup _____.

17. When connecting to the target database, the user account must have _____ privileges.

18. Which of the following commands is entered from the operating system to start the RMAN command-line interface?

    a. RECOMAN

    b. RMAN

    c. START RMAN

    d. none of the above

19. When a channel is allocated, it interacts directly with the _____.

20. Unless a value is assigned to a configuration setting, RMAN uses the _____ value for that parameter.

HANDS-ON ASSIGNMENTS

Assignment 6-1 Opening RMAN Without Initial Target Database Connection

In this assignment, you access RMAN and subsequently establish a connection to a target database.

1. Click the **Start** button on the task bar at the bottom of the screen.

2. When the Start menu appears, click the **Run** command on the right side of the menu.

3. When the Run dialog box appears, type **rman** in the dialog box, and click the **OK** button.

4. Establish a connection with the target database.

5. Exit the Recovery Manager.

Assignment 6-2 Displaying RMAN Configuration Settings

In this assignment, you display the current configuration settings in RMAN.

1. Start the Recovery Manager, and connect to the target database.

2. Display all configuration settings for RMAN using the SHOW command.

3. Determine which values are default and which have been assigned by a user.

4. Exit the Recovery Manager.

Assignment 6-3 Clearing RMAN Configuration Settings

In this assignment, you reset the RMAN configuration settings back to their default values.

1. Start the Recovery Manager, and connect to the target database.
2. Clear any previous control file configuration settings by entering the command `configure controlfile autobackup clear;`. This resets the setting and disables automatic backup of the control file.
3. Exit the Recovery Manager.

Assignment 6-4 Determining RMAN Repository Location

In this assignment, you determine the location of the RMAN repository based on messages displayed by RMAN.

1. Start the Recovery Manager, and connect to the target database without specifying whether a recovery catalog should be used.
2. Once RMAN is started, determine whether the control file or a recovery catalog is being used for the RMAN repository.
3. Exit the Recovery Manager.

Assignment 6-5 Automating Backup of Target Database Control File

In this assignment, you are required to change the appropriate RMAN configuration setting to specify that the control file must be backed up whenever a structural change occurs within the database.

1. Start the Recovery Manager, and connect to the target database.
2. Configure RMAN to automatically backup the control file whenever the structure of the database is changed.
3. Display all configuration settings for RMAN.
4. Exit the Recovery Manager

Assignment 6-6 Specifying the Number of Copies for the Control File and Data Files

In this assignment, you specify that extra copies of the control file and data files must be created by RMAN whenever a backup operation is performed.

1. Start the Recovery Manager, and connect to the target database.
2. Configure RMAN to automatically create three copies of the control file and data files whenever a backup operation occurs.
3. Display the new configuration settings for RMAN.
4. Reset the configuration change made in Step 2 to its default value.
5. Exit the Recovery Manager

CASE PROJECTS

Case 6-1 Recovery Manager, User-Managed Operations

Carlos is trying to determine whether the backup and recovery operations should be user-managed or managed through the Recovery Manager. Write a memo that discusses the advantages and disadvantages of each approach to be circulated to employees of the IT Department soliciting their input for the decision. For additional reference, you may want to consult the *Oracle9i Recovery Manager User's Guide Release 2 (9.2)*, which is available in the documentation section of the Oracle Technology Network at *http://otn.oracle.com*.

Access to the Oracle Technology Network, and subsequently the online Oracle9i database documentation, requires you to become an OTN member. There is no fee to become a member.

Case 6-2 Automating RMAN Access

Create a batch or script file (that can be executed through the operating system) that starts the Recovery Manager and logs the user into the target database with a CARLOS privileged account. Create an icon that resides on the desktop that executes the script when double-clicked by the user.

7

PERFORMING BACKUP OPERATIONS WITH RECOVERY MANAGER

**After completing this chapter,
you should be able to do the following:**

- Identify the tablespaces and data files associated with the target database using the REPORT command
- Manually allocate a channel to perform backup operations
- Perform a full backup of a tablespace using the BACKUP command
- Include a "snapshot" of the control file when backing up a tablespace
- Distinguish between a full backup and a Level 0 backup
- Explain the difference between differential and cumulative incremental backup strategies
- Perform a Level 0 backup of multiple tablespaces using the BACKUP command
- Describe the difference between an image copy and a backup of a data file

Carlos can create a copy, or a backup, of the database files through the operating system. For a database that is in NOARCHIVELOG mode, Carlos simply shuts down the database and then makes a copy of each data file and each mirrored copy of the control file. If a database is in ARCHIVELOG mode, Carlos then has the option of performing a hot backup of the data file by placing a tablespace in backup mode, and then copying the file. However, with user-managed backup procedures, it is up to Carlos to track the location of the backup files, delete invalid backup copies, and so on.

Use of the Recovery Manager (RMAN) utility provides Carlos with more flexible backup options. It also eases the administrative burden of tracking files and deleting obsolete database backup files. With user-managed backup procedures, the backup copy of a database file is created with the operating system's COPY command. In other words, an exact duplicate of the file is created. However, with RMAN, Carlos has the option of backing up only the changed data blocks within a file. Furthermore, RMAN supports incremental backups, which can be used to backup only the data blocks that have been changed after a previous backup was performed.

In this chapter, the various backup strategies supported by RMAN will be examined. This includes the ability to perform various levels of incremental backups. In addition, the REPORT command will be used to identify, based on specified criteria, data files that have not been backed up. Also introduced is the COPY command, which is used to create image copies of database files.

THE CURRENT CHALLENGE IN THE JANICE CREDIT UNION DATABASE

Previously, all backup copies of the database files were created through the operating system's COPY command. Not only is Carlos required to track the location of each set of backup files, he also has to remember to delete any backup copies that become invalid when an incomplete recovery is performed, and so on. In addition, keeping multiple copies of the database files is beginning to consume large amounts of storage space. Therefore, Carlos is trying to determine whether backup and recovery operations would be more efficient by using RMAN.

SET UP YOUR COMPUTER FOR THE CHAPTER

Before performing any of the tasks demonstrated in this chapter, create a cold backup of all database files. In the event an error occurs and the database becomes unrecoverable, this backup can be used to restore and recover the database to its original state.

After completing the backup, place the database in ARCHIVELOG mode before attempting the steps in this chapter. Refer to Chapter 2 if you do not remember how to change the archiving mode for the database.

NOTES ABOUT DUAL COVERAGE WITHIN THIS CHAPTER

The backup procedures discussed in this chapter will be presented using both the command-line and GUI approaches. When using the command-line approach, you will be interacting directly with the RMAN Executable using the Windows XP operating system. The GUI approach uses a wizard, which can be accessed through the Enterprise Manager Console. However, the wizards for backup and recovery operations are available only when connected to the Oracle Management Server (OMS). The OMS is a component of the

Oracle Enterprise Manager that allows a DBA centralized control over databases, users, and so on. Because the OMS is typically used to manage multiple databases in a network environment, sections using the GUI approach will be presented using the Windows 2000 Server operating system.

BASIC RMAN BACKUP FEATURES

RMAN is a utility available with the Oracle9i DBMS that can be used to perform backup and recovery operations. There are different backup options available with RMAN that cannot be performed with the user-managed backup procedures demonstrated in Chapter 3. Each backup option is discussed here in turn.

Backing Up Only Changed Data Blocks

In addition to creating exact copies of the current data files, called image copies, RMAN can also back up only the used portions of the database files. When database files are created in Oracle9i, the initial space for some files is pre-allocated. In other words, a file may only contain 80 KB of data, but it actually consumes 5 MB of disk space.

For example, extra space for the control file is used when data is written to the file. This makes write operations more efficient because the size of the file does not have to be increased every time additions are made to the file. Although the pre-allocated space can make write and read operations more efficient, creating backup copies of empty data blocks can create a lot of wasted storage space. Therefore, rather than create an image of the database file during backup operations, RMAN can just copy a file's used data blocks.

Backup Levels

In addition to RMAN's ability to copy only used data blocks, the utility also offers various levels of backup that can be used to increase the speed of the backup process. For example, RMAN can track which data blocks have previously been backed up. This enables RMAN to offer the option of backing up only changed data blocks; this is known as an **incremental backup**. A DBA can implement a backup strategy that requires all occupied data blocks to be backed up once a week, and changed data to be backed up on a daily basis.

Note that in the case of the financial database for Janice Credit Union, it takes hours to perform a full backup for such a large database. This incremental approach can save a lot of time. However, the downside is that if a media failure occurs, recovery may take longer because first a baseline backup, or the initial backup for the series of incremental backups, must be restored. Then all incremental backups must be applied in the correct sequence. To reduce the burden placed on the administrator, RMAN is able to determine which backup copies are required, and the correct sequence, in order to recover the database. The exact levels of backup available in RMAN are discussed later in this chapter.

REPORT COMMAND

After reviewing the basic information about Recovery Manager, Carlos is ready to perform some standard backup operations. Because Carlos has never worked with RMAN, he uses the TESTDB database rather than one of the production databases.

One of the first things Carlos needs to determine is whether any of the database files actually need to be backed up. Although Carlos has never worked with the utility, some of the previous DBAs have used RMAN for backup and recovery operations. Luckily, RMAN includes a REPORT command that can be used to provide an analysis of the information contained in the RMAN repository. There is a variety of information that can be retrieved with the REPORT command. In this particular case, Carlos is concerned about whether backup is needed for any of the database files. Therefore, he works with the NEED BACKUP keyword of the REPORT command, as detailed in Table 7-1.

Table 7-1   NEED BACKUP keyword

Keyword	Description
NEED BACKUP	Lists all data files that need to be backed up. The keyword assumes that the most recent backup is used in recovery. Unless specified, "1 copy" is the default value, based on the default retention policy, indicating that the information is based on the last backup performed by RMAN.
Parameter	**Description**
INCREMENTAL = n	Indicates that no more than the stated number of incremental backups should be used in recovery. If more than n incremental backups are required, then a full backup needs to be created for the data file, and its file name is included in the output of the REPORT command.
DAYS = n	Indicates the maximum number of days worth of archived redo log files that are needed to recover a data file. A higher value indicates that more archived files are required and the recovery process will take longer.
REDUNDANCY = n	Indicates the number of copies of the file that should exist. If the actual number of copies is less than the indicated value, then the data file needs to be backed up.

Because Carlos has never used the Recovery Manager to back up any of the data files in the TESTDB database, he does not specify any parameters with the REPORT command. The NEED BACKUP keywords are sufficient to determine which data files need to be backed up.

To determine which data files need to be backed up:

1. Start RMAN from the operating system, and specify that Recovery Manager should connect to the target database with a SYSDBA privileged account.

2. At the RMAN> prompt, type **report need backup;** and press **Enter**. Because no backup procedures have been performed by RMAN, all data files are displayed, as shown in Figure 7-1.

```
C:\oracle\ora920\bin\rman.exe                                    _ □ ×

Recovery Manager: Release 9.2.0.1.0 - Production

Copyright (c) 1995, 2002, Oracle Corporation.  All rights reserved.

connected to target database: TESTDB (DBID=2284938461)
using target database controlfile instead of recovery catalog

RMAN> report need backup;

RMAN retention policy will be applied to the command
RMAN retention policy is set to redundancy 1
Report of files with less than 1 redundant backups
File #bkps Name
------ ----- ------------------------------------------------
1      0     C:\ORACLE\ORADATA\TESTDB\SYSTEM01.DBF
2      0     C:\ORACLE\ORADATA\TESTDB\UNDOTBS01.DBF
3      0     C:\ORACLE\ORADATA\TESTDB\CWMLITE01.DBF
4      0     C:\ORACLE\ORADATA\TESTDB\DRSYS01.DBF
5      0     C:\ORACLE\ORADATA\TESTDB\EXAMPLE01.DBF
6      0     C:\ORACLE\ORADATA\TESTDB\INDX01.DBF
7      0     C:\ORACLE\ORADATA\TESTDB\ODM01.DBF
8      0     C:\ORACLE\ORADATA\TESTDB\TOOLS01.DBF
9      0     C:\ORACLE\ORADATA\TESTDB\USERS01.DBF
10     0     C:\ORACLE\ORADATA\TESTDB\XDB01.DBF
11     0     C:\ORACLE\ORADATA\TESTDB\OEM_REPOSITORY.DBF

RMAN>
```

Figure 7-1   Report of all data files needing backup

Luckily, the names of the data files tend to correlate to the name of the actual tablespaces used in the database. Sometimes an inexperienced DBA may create tablespaces and assign the associated data file a name that does not follow the standard naming convention of including the tablespace name in the file name. On occasions when it is difficult to tell which data file contains data from which tablespace, Carlos can have RMAN display the data file(s) belonging to each tablespace in the TESTDB database. To do so, he can again use the REPORT command. This time, he uses the SCHEMA keyword to indicate the correlation between the tablespaces and the data files.

Technically, the SCHEMA keyword is used to present a "snapshot" of the structure of the database based on an SCN, a particular point in time, or a log sequence. Refer to the *Oracle9i Recovery Manager Reference Manual* to view the appropriate AT clause for each of these "snapshot" parameters.

To determine which data file belongs to which tablespace:

1. At the RMAN> prompt, type **report schema;** and press **Enter**. Your output should resemble Figure 7-2.

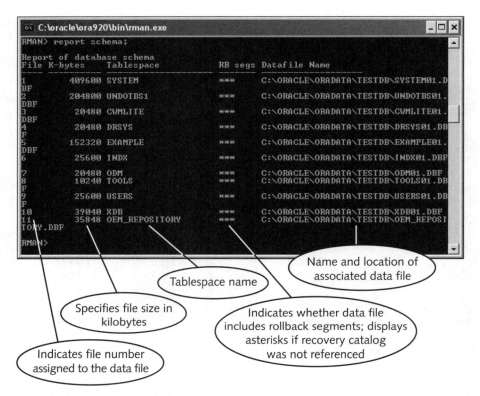

Figure 7-2 Schema of the TESTDB database

2. Study the output carefully. Note that in this case, the fourth column displays a series of asterisks (***). This is because RMAN is not referencing information contained in a recovery catalog. If a recovery catalog is used to generate the report, "YES" is displayed if the data file contains rollback segments and "NO" is displayed if it does not.

Now that Carlos has verified which data files need to be backed up, he can begin the backup process using RMAN.

COMMAND TO BACK UP A TABLESPACE

Carlos decides to first back up the TOOLS tablespace because it is the smallest, and therefore requires the least amount of time to copy during this practice session. The command Carlos executes is shown in Figure 7-3.

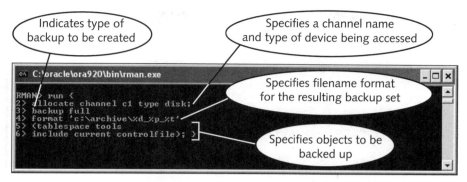

Figure 7-3   Command to back up the TOOLS tablespace using RMAN

In the following sections, you examine the code that Carlos executes.

RMAN Command Types

In Recovery Manager, there are two types of commands: stand-alone and job. A **stand-alone command** is a single command that is executed immediately. Recall that if a user starts RMAN without including the target database parameter, the CONNECT command is used later to connect to the database. CONNECT is an example of a stand-alone command.

A **job command** consists of a group of commands that are executed in sequential order. The group of commands is provided in a block. The block is identified using curly braces. In the example given in Figure 7-3, the block consists of the ALLOCATE command and the BACKUP command. A semicolon is used to terminate each command. The RUN command is used to execute a block of commands.

The stand-alone and job command types are comparable to SQL commands and PL/SQL blocks. SQL commands are executed as soon as the statement is terminated. However, the statements included in a PL/SQL block are executed after the entire block is terminated.

ALLOCATE Command

Recall that a channel is required when performing backup and recovery operations in RMAN. Because an automatic channel has not already been configured, a channel must be manually allocated before the backup of the TOOLS tablespace can begin. The basic syntax of the ALLOCATE command is given in Figure 7-4.

The CHANNEL keyword is used to specify the name of the channel. An industry standard is to assign a name such as c1 or disk1. The actual name itself is not important, as long as when that particular channel is referenced within the same command block that the same name is specified—including matching the case exactly.

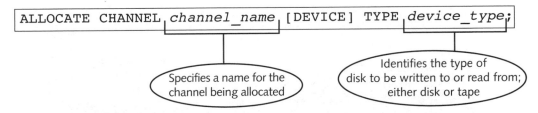

Figure 7-4   ALLOCATE command syntax

The DEVICE TYPE keywords are used to identify the type of device being used for the read/write operation. The square brackets included in the syntax of the command indicate that the DEVICE keyword is optional. If the device uses a direct access method, such as a hard drive, then the DISK keyword is included. However, if the storage device uses the sequential access method, then the SBT_TAPE or SBT keyword is used. SBT is the abbreviation for System Backup to Tape.

BACKUP Command

Regardless of the type of backup you are performing, the BACKUP command is used to specify the parameters for the backup being performed. Various options can be included to indicate the type of backup to be generated. You can specify a full backup or a Level n incremental backup, a name format for the output file, the maximum size of each output file, whether the archived log files should be backed up, and so on. Table 7-2 identifies the most commonly used options available for the BACKUP command.

Table 7-2   Selected options of the BACKUP command

Option	Description
DATABASE	Specifies that all data files and a copy of the control file should be included; does not include any archived redo log files
FORMAT	Used to indicate the format of the name to be assigned to the output file
FULL	Copies all used data blocks; default type is INCREMENTAL is not specified
INCLUDE CURRENT CONTROLFILE	Specifies that a snapshot of the control file should be included in the backup set; used when AUTOBACKUP is on
INCREMENTAL LEVEL = n [CUMULATIVE]	Copies all used data blocks since the last n or lower backup was performed. Indicates a differential backup, by default. Include the keyword CUMULATIVE to specify that a cumulative incremental backup should be performed.
MAXSETSIZE = n	Specifies the maximum size (in bytes) for the backup set. Used when backup set may be larger than the storage medium. Allows one set to reside on one tape, or disk, rather than have the set spread across multiple tapes or disks.

Table 7-2   Selected options of the BACKUP command (continued)

Option	Description
ARCHIVELOG ALL	Makes a backup of the archived redo log files during the backup procedure
TABLESPACE	Specifies the tablespace to be included in the backup procedure; if more than one tablespace is to be included, the tablespace names are separated by a comma

As shown in Table 7-2, there are different options available with the BACKUP command. For example, when a backup is created, the resulting file is called a **backup set**. A backup set can consist of data files, the control file, or an archived redo log file. The problem is that the set normally consists of one binary physical file, called a **backup piece**. If the size of the backup set is larger than the size of the storage medium, then the backup process returns an error message. Therefore, if Carlos knows that he has limited space available on a particular drive or on a tape, but he needs to make certain that space is not exceeded, he can use the MAXSETSIZE to place a limit on the total size of the backup set and prevent the error from occurring. The remaining options provided in the table will be demonstrated throughout this chapter.

The BACKUP command with the FULL option is used in Figure 7-3 to indicate that all used data blocks should be backed up, regardless of whether those blocks have been previously backed up. A **full backup** of the data files can be performed to create a copy of all used data blocks. The term "full" should not be confused with "whole." A **whole backup** is a backup of all the files needed in the event that database recovery becomes necessary. This includes data files, control files, and archive logs, if applicable. A full backup can consist of one data file or several. The term applies to the fact that all used data blocks are copied.

Any empty blocks are not included. The FORMAT option indicates the desired format of the name assigned to the output file. The various elements that can be used to format the file name are provided in Table 7-3. Notice in the example provided in Figure 7-3, the FORMAT option can also be used to specify the storage location for the file. In this case, the output file is stored on the hard drive in the Archive folder.

Table 7-3   Format elements

Format Elements	Description
%d	Represents the database name
%p	Represents the backup piece number within the backup set
%s	Represents the database set number
%t	Represents the number of seconds that have passed since a fixed point established by Oracle9i software
%T	Represents the time in the format of YYYYMMDD

The backup specification portion of the command previously provided in Figure 7-3 is enclosed in parentheses and not only identifies the tablespace to be backed up but also explicitly states that a current snapshot of the control file should be included in the backup set.

In RMAN, the control file is automatically backed up anytime the SYSTEM tablespace is backed up.

Performing a Full Backup of a Tablespace with RMAN—the Command-Line Approach

Carlos now needs to enter and execute the command to create a backup of the TOOLS tablespace.

To back up the TOOLS tablespace:

1. From the operating system, start RMAN and connect to the target database with a DBA privileged account, if necessary.

2. Type **run {** at the RMAN> prompt to begin the block, and then press **Enter**.

3. At the 2> prompt that appears on the next line, type **allocate channel c1 type disk;** to manually allocate the required channel, and then press **Enter**.

4. At the 3> prompt, type **backup full** to indicate that a full backup is to be performed, and then press **Enter**.

5. At the 4> prompt, type **format '<*location*>\%d_%p_%t'**, where <*location*> is the appropriate destination for the file copy, to specify a name format for the output file, and then press **Enter**.

6. At the 5> prompt, type **(tablespace tools** to specify that the TOOLS tablespace should be backed up, and then press **Enter**.

7. At the 6> prompt, type **include current controlfile);}** to ensure that a backup is made of the control file, and then press **Enter**.

Because the last line included the closing curly brace for the block, the BACKUP FULL command is automatically executed and results similar to those shown in Figure 7-5 are displayed.

If an error message is displayed while you are typing a block, the error that generated the message may affect subsequent attempts to enter the block correctly. If this occurs, the quickest way to resolve the problem is to exit and then restart RMAN.

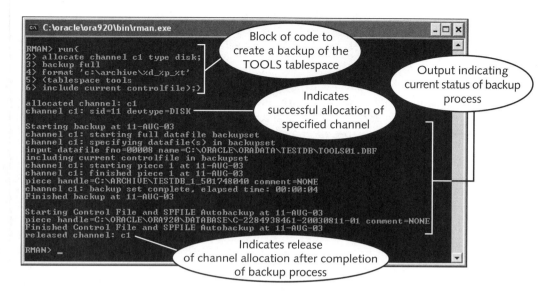

Figure 7-5   Displayed output of the BACKUP FULL command

Now that the TOOLS tablespace has been backed up, a report of all data files that need to be backed up no longer includes the data file for this tablespace, as shown in Figure 7-6. Once a data block belonging to the TOOLS tablespace has been changed, it will reappear in the report.

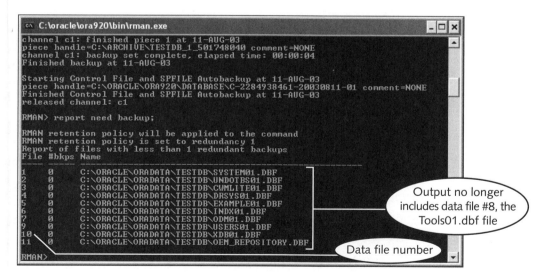

Figure 7-6   Updated report

Performing a Full Backup of a Tablespace with RMAN—the Enterprise Manager Approach

In this section, Carlos will use the GUI approach to create a backup of a single tablespace—the SYSTEM tablespace. When accessing RMAN through a GUI interface, Carlos uses the Backup Wizard available through the Enterprise Manager Console. Based on the responses provided by Carlos, the wizard generates the RMAN script, or block of code, necessary to perform the desired action.

Recall that the Backup Wizard can be accessed only if Carlos is connected to an Oracle Management Server (OMS). It is the OMS that provides Carlos, or any privileged DBA, a centralized location to administer multiple databases, users, and so on. It also provides enhanced features, such as GUI interfaces for backup and recovery operations.

To create a backup of the SYSTEM tablespace:

1. Start the Enterprise Manager Console (EMC), and connect to the Oracle Management Server, as shown in Figure 7-7, using the connection information provided by your instructor.

Figure 7-7 Connecting to the Oracle Management Server through the Enterprise Manager Console

2. From the Navigation tree displayed in the left pane of the window, click the + (plus sign) next to the word "Databases."

3. When the list of available databases is displayed, click the name corresponding to the database you intend to back up.

4. Click **Tools** on the menu bar of the EMC.

5. From the Tools submenu, click **Database Tools**.

6. From the Database Tools submenu, click **Backup Management**.

7. From the Backup Management submenu, click **Backup**. When performing this step, your screen should look similar to Figure 7-8.

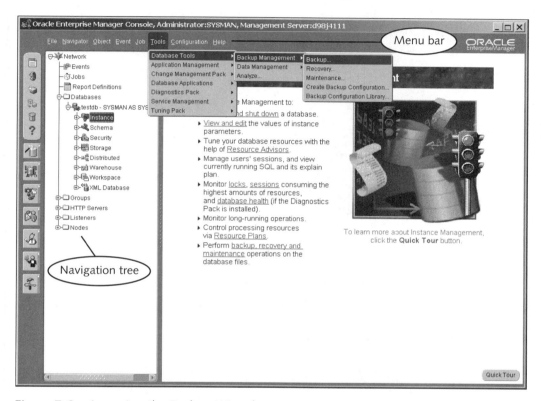

Figure 7-8   Accessing the Backup Wizard

8. If the Introduction window shown in Figure 7-9 appears, click the **Next** button in the lower-right corner of the window.

9. When the window appears requiring you to select a strategy choice, click the **Customize backup strategy** option button, as shown in Figure 7-10.

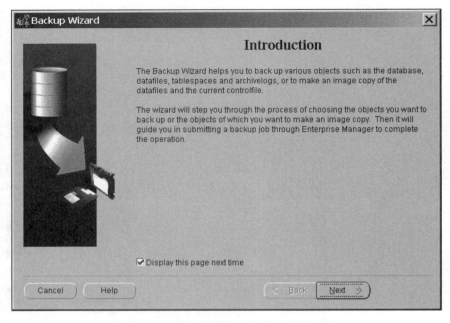

Figure 7-9 Introduction window of the Backup Wizard

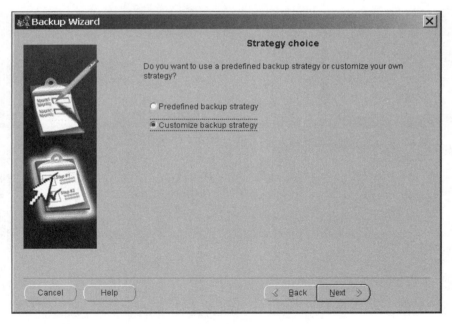

Figure 7-10 Strategy choice of the Backup Wizard

10. Click **Next** to proceed to the next window.

11. When the window appears prompting you to choose the object to back up, click the **Tablespaces** option button, as shown in Figure 7-11.

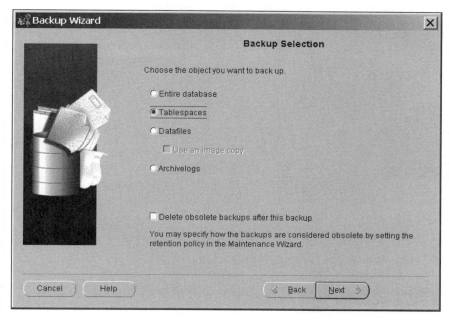

Figure 7-11 Choosing an object for the backup procedure

12. Click the **Next** button to proceed to the next window.

13. When the Tablespaces window appears, click the **SYSTEM** tablespace from the Available tablespaces list provided on the left side of the window.

14. Click the **Single Selection Arrow** button, as shown in Figure 7-12, to add the tablespace name to the Selected Tablespaces list on the right side of the window.

15. Click the **Next** button to proceed to the next window.

At this point you can click the Finish button to submit the job if no changes need to be made to the default backup configuration or policy. Backup configurations and retention policies will be discussed in Chapter 10. However, to display all the available options through the Backup Wizard, the Next button was chosen in Step 15 instead.

16. When the Archived Logs window shown in Figure 7-13 appears, click the **No** option button, if necessary, to indicate that the archived redo log files should not be included in the backup procedure.

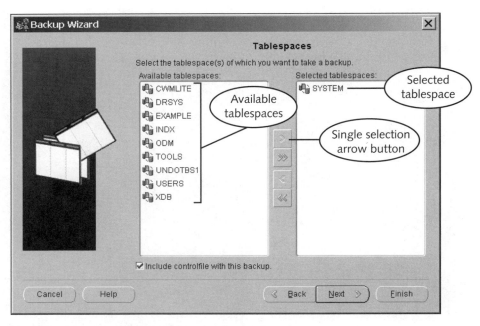

Figure 7-12   Selecting a tablespace for backup

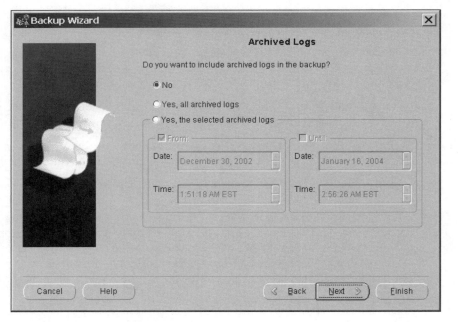

Figure 7-13   Specifying inclusion preference for the archived redo log files

You will be provided with the option to include the archived redo log files in the backup process only if your database is in ARCHIVELOG mode.

17. Click the **Next** button to proceed to the next window.

18. When the Backup Options window shown in Figure 7-14 appears, click the **Full backup** option button, if necessary, in the upper portion of the window. Click **Next**.

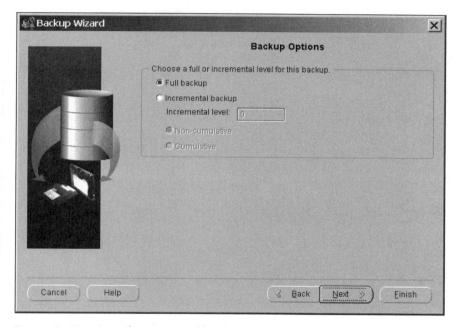

Figure 7-14   Specifying type of backup

19. When the Configuration window appears, as shown in Figure 7-15, make any changes specified by your instructor. Then click the **Next** button to proceed to the next window.

20. When the Override Backup and Retention Policy window shown in Figure 7-16 appears, make any changes specified by your instructor. To advance to the next window, click the **Next** button.

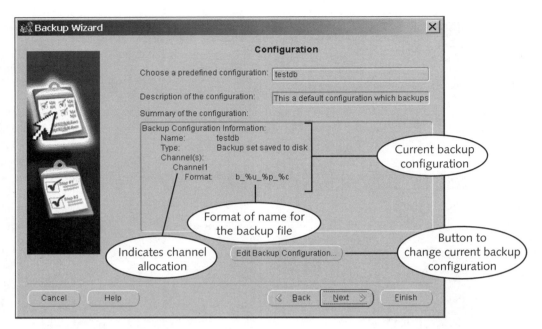

Figure 7-15   Configuration window

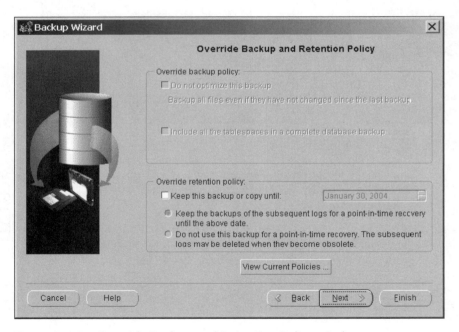

Figure 7-16   Override Backup and Retention Policy window

21. When the Schedule window appears, click the **Immediately** option button, if necessary and as shown in Figure 7-17, to indicate that the job should be executed as soon as it is submitted. Click **Next**.

Figure 7-17   Schedule options

22. When the Job Information window appears, verify that the **Submit the job now** option button is selected. Then click the **Finish** button to submit the job without changing the default job name or description.

23. When the Summary window shown in Figure 7-18 appears, verify that the information is correct, and click the **OK** button.

24. When the dialog box shown in Figure 7-19 appears, the job has been submitted and you can click the **OK** button to close the window.

To verify that the database has been backed up, you need to view the Active pane of the Jobs objects listed in the Navigation tree to determine whether the backup procedure was actually executed. If the job is not listed, or the status of the job is listed as Failed on the subsequent pane, notify your instructor before continuing with this chapter.

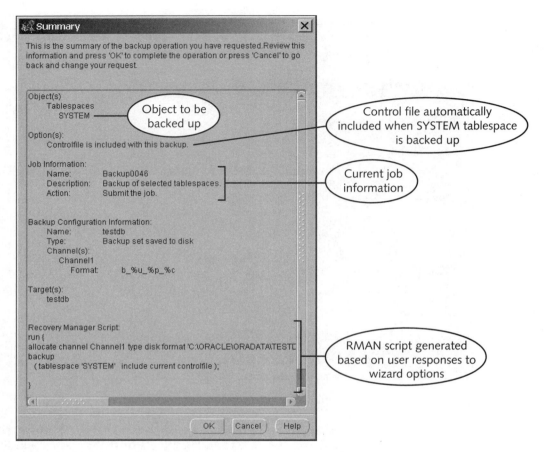

Figure 7-18   Summary information

Figure 7-19   Successful submission of the backup job

INCREMENTAL BACKUP

A data block is the smallest structure that can be referenced by Oracle9*i*. A changed data block is a data block whose contents have been altered since the last backup occurred. The concept of "changed data blocks" serves as the basis for incremental backups.

There are five different levels of incremental backups, Level 0 through Level 4. The baseline, or starting point, for incremental backups is a Level 0 backup. A Level 0 backup backs up the same data blocks as a full backup. However, a full backup *cannot* be used in an incremental backup strategy. All subsequent backup levels, Levels 1 through 4, are based on data blocks that changed after a previous incremental backup was performed.

There are two types of incremental backups. A differential, or non-cumulative, incremental backup copies only data blocks that have changed since the last Level *n*, or lower level, backup was performed. For example, a Level 3 differential backup only copies data blocks changed after the last Level 3 or lower backup was performed. By default, incremental backups are differential; the CUMULATIVE keyword is required to create cumulative incremental backups.

Figure 7-20 provides an example of the contents for a series of differential incremental backups. Suppose on Sunday evening, Carlos creates a Level 0 backup of the data files, and then creates a Level 1 backup on Monday evening. The Level 1 backup consists only of data blocks that have been changed since the Level 0 backup was created on Sunday. If, on Tuesday evening, Carlos performs another Level 1 backup, it contains only the changed data blocks since the last Level *n*, or lower, backup occurred. In this case, only the data blocks changed after the Monday night backup are included. The same is true for the Level 2 backups created on Wednesday and Thursday nights.

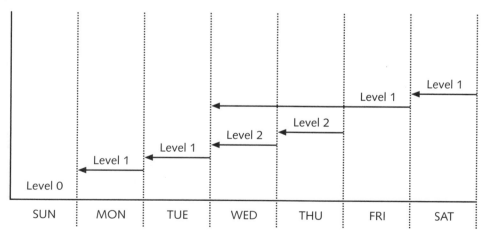

Figure 7-20 Contents of various levels of differential incremental backups

Notice, however, the Level 1 backup created on Friday night. Because the last Level *n*, or lower, backup would be the Level 1 backup created on Tuesday night, the Friday night backup contains all changes that have occurred since Tuesday night. The subsequent Level 1 backup created on Saturday contains all changes made after the Level 1 backup was performed on Friday night.

The second type of incremental backup is a cumulative incremental backup. A cumulative incremental backup copies only data blocks that have changed since the last Level *n*-1, or lower, backup was performed. A Level 3 cumulative backup copies the data blocks that have changed after the last Level 2, 1, or 0 backup was performed. As shown in Figure 7-21, when the Level 1 backup is created on Tuesday night, it contains all the changed data blocks since the last Level *n*-1, or lower, backup was created. In this case, that is all the changed data blocks since Sunday night when the Level 0 backup was created. The Level 2 backup created on Thursday night contains all the data blocks that have been altered since the last Level 1 backup was created on Tuesday. Why? Because 2 - 1 = 1, so when performing a Level 2 cumulative incremental backup, RMAN backs up all changed data blocks since the most recent Level 1 backup, or Level 0 if a prior Level 1 backup does not exist. Notice that on Saturday night, when a Level 1 backup is created, it contains all data blocks that have changed since the Level 0 backup was created on Sunday night.

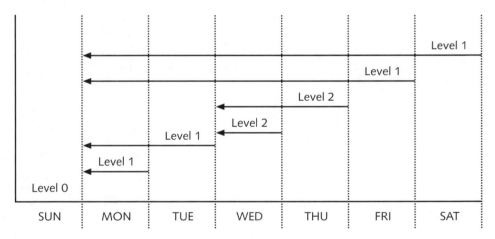

Figure 7-21 Contents of various levels of cumulative incremental backups

 If a backup level higher than 0 is specified and no valid Level 0 backup exists, a Level 0 backup is automatically performed.

Recovery Implications of Incremental Backup Procedures for Janice Credit Union

The basic trade-off between differential and cumulative incremental backup strategies relates to the time it takes to back up versus to recover the database. Relative to the last time the Level 0 backup was created, differential backups are usually quicker because there are few changed data blocks included in the backup. However, the recovery process may take longer because more backup sets need to be applied to the Level 0 backup. For example, in Figure 7-20, if a database failure occurred after the Friday night backup was created, then the backups taken on Sunday, Monday, Tuesday, and Friday nights are required to recover the database. However, if a Level 0 backup, or a full backup, had been created every night, then only the Friday night backup would be required to recover the database.

With the cumulative backup strategy, each backup set above a Level 0 backup may be larger, and take longer to create, but the recovery process is quicker because fewer backup sets are required to recover the database. In essence, the choice of backup strategy is a case of "pay now or pay later." Using the backup scheme presented in Figure 7-21, if a database failure occurs after the backup is created on Friday night, only the Level 0 backup from Sunday and the Level 1 backup from Friday night are required to recover the database. A DBA such as Carlos needs to determine which element is more important, the amount of time spent backing up the database or the amount of time spent recovering the database. When a database is considered to be a critical component of daily operations, such as the financial database, the focus should be placed on speed of recovery, rather than speed of backup operations.

 The contents of relevant archived redo log files still need to be applied to the database to perform a complete recovery, regardless of the type of incremental backup strategy.

The Use of the SCN in Incremental Backups

For an open database, incremental backups can be performed only if the database is in ARCHIVELOG mode. To create incremental backups of a database that is in NOARCHIVELOG mode, the database must be shut down in a consistent state. Why? Incremental backups use the SCN to identify the data blocks that have been changed. As you recall, the SCN is a value assigned to each transaction. A backup of a database that is in NOARCHIVELOG mode can be performed only if the SCN in all data files and the control file are consistent. This basically means that the database needs to be shut down first, using any shutdown option except ABORT. Then the database is mounted and the incremental backup is performed. This is also true when creating a full backup for a database that is in NOARCHIVELOG mode.

In Figure 7-22, the error message returned when a user attempts to perform a full backup of an opened database that is in NOARCHIVELOG mode is displayed.

Figure 7-22   Failed attempt to back up an opened database in NOARCHIVELOG mode

The BACKUP Command in Action—the Command-Line Approach

This section looks at a few examples of the BACKUP command. You can try out these examples on your computer by simply connecting to the target database through RMAN and entering the code shown. However, remember to change the location of the copied files to the appropriate location for your computer (i.e., correct drive and folder reference). In Figure 7-23, a Level 0 differential backup is created of the SYSTEM, TOOLS, and USERS tablespaces. Because the resulting backup set includes the SYSTEM tablespace, a copy of the control file is automatically created.

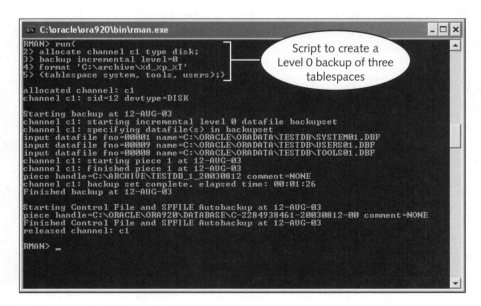

Figure 7-23   Level 0 backup

In Figure 7-24, a backup of the archived redo log files is performed. An incremental backup cannot be performed with an archived redo log file; it must be a full backup. Because a full backup is the default, the FULL keyword has been omitted in this example.

Figure 7-24   RMAN backup of archived redo log files

The BACKUP Command in Action—the GUI Approach

The sequence of steps required to create an incremental backup using the Backup Wizard are basically the same as when Carlos created the full backup of the SYSTEM tablespace. The main difference is when the wizard prompts you for the desired backup option, as shown in Figure 7-25, the Incremental Backup option button needs to be clicked, rather than the Full Backup option button. Carlos can then select the desired level of backup after the drop-down list becomes available. In addition, this same window can be used to specify whether the type of incremental backup to be performed is non-cumulative (i.e., differential), or cumulative.

During the creation of a backup of the archived redo log files through the Backup Wizard, when you are prompted for the object to be backed up, simply click the Archivelogs option button at the end of the object list, as shown in Figure 7-26.

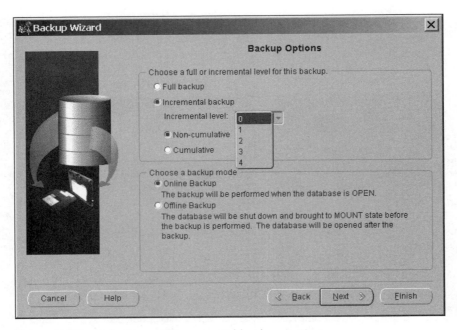

Figure 7-25   Selecting an incremental backup option

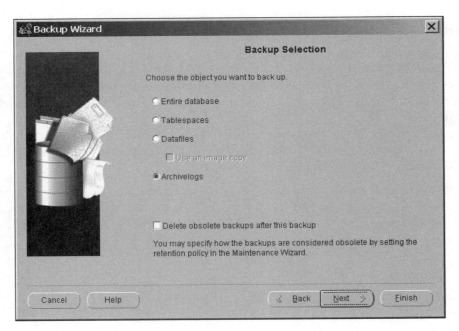

Figure 7-26   Selecting a backup of the archived redo log files in the Backup Wizard

As shown in Figure 7-27, the Backup Wizard then prompts Carlos to determine which archived redo log files should be included in the backup. In this case, all archived redo log files have been selected. However, if some of the redo log files are already contained in a previously created backup, then a date and time range can be specified to indicate which files should be included.

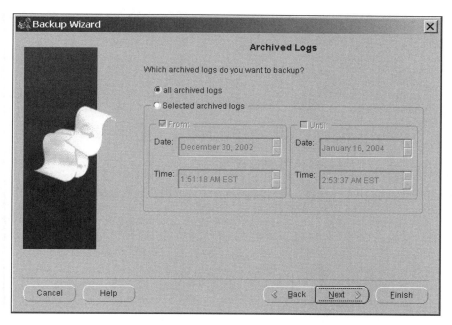

Figure 7-27   Identify the archived redo log files to be included in the backup

CREATING IMAGE COPIES USING THE COPY COMMAND

RMAN can create exact duplicates of database files, called image copies, using the COPY command. The COPY command can be used with data files, archived redo log files, and control files. When RMAN creates an image copy of a file, it copies all data blocks, including empty data blocks. It is the same process as using the operating system's command to copy files, except the user is not required to manually place the tablespace in backup mode. In addition, RMAN automatically checks the image copy for corruption when it is created.

An image can be used as part of a full or incremental backup strategy. However, in an incremental strategy, it can only be created at Level 0 because all data blocks are copied. A DBA might consider creating image copies of database files to speed up the recovery process because no additional processing is required to use an image backup other than applying the contents of any relevant archived redo log files. In addition, image copies can be stored only to a disk, not to a sequential access medium such as magnetic tape. This also provides quicker access to the file because it is stored on a direct access device.

The basic syntax for the COPY command is shown in Figure 7-28:

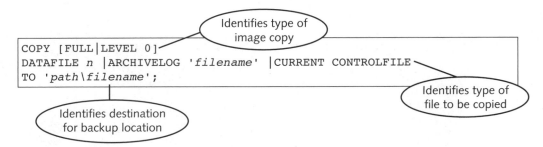

Figure 7-28 Basic COPY command syntax

As mentioned, the image can be a full backup or a Level 0 backup for an incremental backup strategy. If no backup type is specified, FULL is the default value. Next, the type of file being copied is identified. If a data file is the source, then the file number is specified. The file number for a data file can be obtained through RMAN using the **report schema;** statement. If an archived redo log file is to be copied, the location and name of the file must be included. If an image of the control file needs to be created, the keywords CURRENT CONTROLFILE are included in the command. After the appropriate source information is provided, the DBA must then specify the location for the resulting output file after the TO keywords.

Carlos uses the command shown in Figure 7-29 to create a Level 0 backup of the data file for the TOOLS tablespace. The number of the data file was previously obtained when he had RMAN display the schema of the TESTDB database in Figure 7-2. The name of the output file created by the COPY command is Toolsimg.dbf; this helps Carlos to identify the contents of the file at a later time, if necessary.

```
C:\oracle\ora920\bin\rman.exe                                        _ □ ×
RMAN> run {
2> allocate channel c1 type disk;
3> copy level 0
4> datafile 8 to 'c:\archive\toolsimg.dbf';}

allocated channel: c1
channel c1: sid=9 devtype=DISK

Starting copy at 12-AUG-03
channel c1: copied datafile 8
output filename=C:\ARCHIVE\TOOLSIMG.DBF recid=1 stamp=501848624
Finished copy at 12-AUG-03

Starting Control File and SPFILE Autobackup at 12-AUG-03
piece handle=C:\ORACLE\ORA920\DATABASE\C-2284938461-20030812-02 comment=NONE
Finished Control File and SPFILE Autobackup at 12-AUG-03
released channel: c1

RMAN>
```

Figure 7-29 Creating a Level 0 backup of a datafile

To create an image copy using the Backup Wizard, Carlos selects a data file as the object to be backed up, and then specifies that an image copy be created, rather than a full or incremental backup.

In Chapters 8 and 9, additional backup procedures will be demonstrated prior to recovering a database or tablespaces. Different types of backups will be performed, such as full backup of a NOARCHIVELOG database, to provide you with demonstrations of the full range of backup procedures available through RMAN.

 If you are not completing the hands-on assignments at this time, the database should be placed back in NOARCHIVELOG mode after completing the steps demonstrated in this chapter.

7

CHAPTER SUMMARY

- ❑ RMAN can be used to back up changed data blocks within a file, or to create image copies of a data file.

- ❑ The REPORT command in RMAN can be used to identify the data files that need to be backed up. Parameters can be included to provide the criteria for including files in the output.

- ❑ A stand-alone command is an RMAN command that is executed immediately while a job command is submitted as a block of statements and is executed using the RUN keyword.

- ❑ The ALLOCATE command is used to manually allocate a channel during backup and recovery operations.

- ❑ A full backup is used to create a backup copy of all used data blocks within a data file.

- ❑ A differential incremental backup can be used to back up only data blocks that have changed since a previous incremental Level n backup was performed. A cumulative incremental backup backs up data blocks that have changed since the previous Level $n-1$ backup was performed.

- ❑ A Level 0 backup serves as the baseline for the incremental backup strategy. It backs up the same data blocks as a full backup.

- ❑ A full backup cannot be used in an incremental backup strategy.

- ❑ On average, cumulative incremental backups take longer than differential backups; however, recovery is faster.

- ❑ An image copy is an exact copy of a file, including all empty data blocks, and is created with the COPY command.

SYNTAX GUIDE

Command	Description	Example
REPORT [option]	Generates a report based on the specified option	REPORT NEED BACKUP;
BACKUP Command Options	**Description**	**Example**
NEED BACKUP	The keyword assumes that the most recent backup is used in recovery. Unless specified, 1 is the default value, indicating that the information is based on the last backup performed by RMAN.	REPORT NEED BACKUP;
SCHEMA	Identifies the tablespaces and associated data files of the database	REPORT SCHEMA;
Parameters for NEED BACKUP Option	**Description**	**Example**
INCREMENTAL = n	Indicates that no more than the stated number of incremental backups should be used in recovery. If more than n incremental backups are required, then the data file needs a full backup to be performed, and the name of the data file should be displayed in the requested report.	REPORT NEED BACKUP INCREMENTAL = 3;
DAYS = n	Indicates the maximum number of days worth of archived redo log files that are needed to recover a data file. A higher value indicates that more archived files are required and that the recovery process takes longer.	REPORT NEED BACKUP DAYS = 2;
REDUNDANCY = n	Indicates the number of copies of the file that should exist. If the actual number of copies is less than the indicated value, then the data file needs to be backed up.	REPORT NEED BACKUP REDUNDANCY = 3;
Options for BACKUP Command	**Description**	**Example**
DATABASE	Specifies that all data files and a copy of the control file should be included; does not include any archived redo log files	BACKUP DATABASE;
FORMAT	Used to indicate the format of the name to be assigned to the output file	BACKUP FORMAT 'c:\archive\%d_%p_%t' (TABLESPACE tools);
FULL	Copies all used data blocks; default type if INCREMENTAL or FULL is not specified	BACKUP FULL DATABASE;

Options for BACKUP Command	Description	Example

INCLUDE CURRENT CONTROLFILE	Specifies that a snapshot of the control file should be included in the backup set; used when AUTOBACKUP is on	`BACKUP FULL`
		`(TABLESPACE tools`
		`INCLUDE CURRENT`
		`CONTROLFILE);`

INCREMENTAL LEVEL = *n* [CUMULATIVE]	Copies all changed data blocks since the last *n* or lower backup was performed. A differential backup is the default. Include the keyword CUMULATIVE to specify that a cumulative incremental backup be performed.	`BACKUP INCREMENTAL`
		`LEVEL = 2 CUMULATIVE`
		`FORMAT 'c:\archive\`
		`%d_%p_%t'`
		`(TABLESPACE tools);`

MAXSETSIZE = *n*	Specifies the maximum size (in bytes) for the backup set. Used when the backup set may be larger than the storage medium. Allows one set to reside on a tape rather than have the set spread across multiple tapes.	`BACKUP FULL DATABASE` `MAXSETSIZE = 200 M;`
ARCHIVELOG ALL	Creates a backup of the archived redo log files during the backup procedure	`BACKUP` `ARCHIVELOG ALL;`

TABLESPACE	Specifies the tablespace to be included in the backup procedure. If more than one tablespace is to be included, the tablespace names are separated by a comma.	`BACKUP (TABLESPACE`
		`system, tools,`
		`users);`

Options for COPY Command	**Description**	**Example**

FULL|LEVEL 0	Indicates whether the image copy is to be included as part of an incremental backup strategy	`COPY LEVEL 0 DATAFILE`
		`8 TO 'c:\imagefolder\`
		`copyof8';`

DATAFILE *n*	Indicates that an image copy is to be created of a data file. The file number assigned to the data file must be specified.	`COPY FULL DATAFILE`
		`8 TO 'c:\imagefolder\`
		`copyof8';`

ARCHIVELOG *'filename'*	Indicates that an image copy is to be created of an archived redo log file. The file name of the archived redo log file must be specified.	`COPY ARCHIVELOG`
		`'c:\oracle\oradata\`
		`ARC0001' TO`
		`'c:\imagefolder\`
		`copyARC1';`

CURRENT CONTROLFILE	Indicates that an image copy is to be created of the target database's control file	`COPY CURRENT`
		`CONTROLFILE TO`
		`'c:\imagefolder\`
		`copycrl';`

7

REVIEW QUESTIONS

1. Which of the following can be used in RMAN to display the data file associated with each tablespace in the target database?

   a. `REPORT ALL;`

   b. `REPORT DATAFILES;`

   c. `REPORT SCHEMA;`

   d. `REPORT STRUCTURE;`

2. What is the difference between a full backup and a Level 0 backup?

3. Which of the following commands is used in RMAN to create an image of a data file?

   a. IMAGE

   b. DUPLICATE

   c. COPY

   d. CP

4. Which of the following keywords can be included in the BACKUP command to back up the control file?

   a. ADD CONTROLFILE

   b. INCLUDE CURRENT CONTROLFILE

   c. INCLUDE CONTROL

   d. ADD CONTROL

5. Which of the following is a valid statement?

   a. When a backup copy of an archived redo log file is created through RMAN, the copy does not include the empty data blocks contained in the source file.

   b. An archived redo log file does not need to be included in an incremental backup strategy because the contents of the file do not change.

   c. If you attempt to perform a Level 2 backup that includes a control file, an error message is returned.

   d. all of the above

6. If a Level 0 backup is performed on Wednesday afternoon, and subsequent differential incremental Level 1 backups are performed daily, which backup sets are required to recover a database following a media failure on Saturday morning?

7. Which of the following commands is used to manually allocate a channel for a backup operation?

   a. ALLOCATE CHANNEL

   b. CREATE CHANNEL

   c. CHANNEL ALLOCATE

   d. CREATE LINK

8. Which of the following commands identify all data files that do not have a valid backup copy available?

   a. IDENTIFY

   b. VALIDATE

   c. BACKUP

   d. REPORT

9. Which of the following commands is used to execute a block of job commands in RMAN?

   a. EXECUTE

   b. START

   c. RUN

   d. none of the above

10. A Level 1 cumulative backup copies all changed data blocks since the last _____ backup was performed.

    a. Level 0

    b. Level 1

    c. Level 2

    d. full

11. The DISK type is specified in the ALLOCATE CHANNEL command if the destination device is a _____.

    a. sequential access device

    b. tape drive

    c. direct access device

    d. none of the above

12. What is the benefit to using an incremental backup strategy rather than performing frequent full backups through RMAN?

13. What automatically occurs if the SYSTEM tablespace is backed up with the BACKUP command?

14. How does RMAN determine whether data blocks have been changed?

    a. It references the SCN in the control file and data file headers.

    b. It compares the contents of each data block to the data block stored in the last Level 1 backup.

    c. It queries the operating system.

    d. It tracks all DML operations performed on the target database.

15. What option can be included in a BACKUP command to ensure that the backup set is not larger than the destination storage media?

    a. MAXDESTSIZE

    b. MAXIMUMSIZE

    c. MAXSETSIZE

    d. none of the above

16. Can a full backup be performed for a database that is in NOARCHIVELOG mode?

17. Level 0 is the lowest level available for cumulative backups in RMAN, while Level _____ is the highest level.

    a. 3

    b. 4

    c. 5

    d. 6

18. A Level _____ backup is the only option available when creating an image of a data file as part of an incremental backup strategy.

19. Which of the following parameters can be used with the REPORT command to identify the datafiles that would need more than three incremental backup sets to be used in the event recovery becomes necessary?

    a. MAXSETS

    b. INCREMENTAL

    c. DAYS

    d. REDUNDANCY

20. Which of the following is an example of a stand-alone command?

    a. FETCH

    b. BACKUP

    c. CONNECT

    d. ALLOCATE

HANDS-ON ASSIGNMENTS

Before performing the following assignments, create a folder, or directory, to be used to store the output files generated by each task. Unless specified otherwise, you can delete the backup copies after completing each assignment to conserve the storage space on your hard drive. After completing all the assignments, delete the folder you previously created. In addition, if you are performing several assignments in sequence, then you do not need to exit and then reconnect to RMAN for each individual assignment.

Assignment 7-1 Identifying Data Files and Tablespaces

In this assignment, you are required to identify the file number associated with a data file.

1. Start RMAN and connect to the target database with a privileged SYSDBA account.
2. Execute the correct REPORT command to identify which tablespace contains the datafiles associated with the target database.
3. Identify the file number associated with the largest data file in the target database.
4. Exit RMAN.

Assignment 7-2 Identifying Datafiles Meeting Specific Backup Criteria

In this assignment, you are required to use the REPORT command to identify the data files that meet a stated criteria.

1. Start RMAN and connect to the target database with a privileged SYSDBA account.
2. Execute a REPORT command that identifies which data files have not been backed up in the last five days.
3. Exit RMAN.

Assignment 7-3 Creating a Level 0 Backup of a Data File

In this assignment, you create a Level 0 backup of a data file.

1. Start RMAN and connect to the target database with a privileged SYSDBA account.
2. Execute a BACKUP command that creates a Level 0 backup of the SYSTEM tablespace. Recall the following:
   a. The BACKUP command must be included in a block.
   b. A channel must be allocated before the BACKUP command is executed.
   c. A block must be enclosed in curly braces.
3. Exit RMAN.

Assignment 7-4 Creating a Level 0 Backup of a Database

In this assignment, you create a backup of a database. Be aware of any time constraints because this operation can take up to 30 minutes, depending on the size of your database files and the processing speed of your computer.

1. Start RMAN and connect to the target database with a privileged SYSDBA account.
2. Execute a BACKUP command that creates a Level 0 backup of all the data files and the control file. (*Hint:* Specify the DATABASE option in the BACKUP command.)
3. Exit RMAN.

Assignment 7-5 Creating a Level 2 Differential Backup of a Database

In this assignment, you create a Level 2 differential backup of a database. You should use the same target database previously backed up in Assignment 7-4.

1. Start RMAN and connect to the target database with a privileged SYSDBA account.
2. Execute a BACKUP command that creates a Level 2 differential incremental backup of the SYSTEM and USERS tablespaces.
3. Exit RMAN.

Assignment 7-6 Creating a Level 2 Cumulative Backup

In this assignment, you create a Level 2 cumulative backup of specific data files.

1. Start RMAN and connect to the target database with a privileged SYSDBA account.
2. Execute a BACKUP command that creates a Level 2 cumulative backup of the SYSTEM and TOOLS tablespaces.
3. Exit RMAN.

Assignment 7-7 Creating a Full Backup of a Database

In this assignment, you create a full backup of all the database files using RMAN.

1. Start RMAN and connect to the target database with a privileged SYSDBA account.
2. Execute a BACKUP command that creates a full backup of the database files.
3. Exit RMAN.

Assignment 7-8 Creating a Full Backup of the SYSTEM Tablespace

In this assignment, you create a full backup of a single tablespace.

1. Start RMAN and connect to the target database with a privileged SYSDBA account.
2. Execute a BACKUP command that creates a full backup of the data file associated with the SYSTEM tablespace.
3. Exit RMAN.

Assignment 7-9 Creating a Full Backup of all Archived Redo Log Files

In this assignment, you create a backup copy of all the archived redo log files.

1. Start RMAN and connect to the target database with a privileged SYSDBA account.
2. Execute a BACKUP command that creates a copy of each valid archived redo log file associated with the target database.
3. Exit RMAN.

Assignment 7-10 Creating an Image Copy of a Control File

In this assignment, you create an image copy of the control file for your database.

1. Start RMAN and connect to the target database with a privileged SYSDBA account.
2. Execute a COPY command that creates an image copy of the control file of the target database.
3. Exit RMAN.

CASE PROJECTS

Case 7-1 Developing a Backup Strategy

Carlos is developing an incremental backup strategy for the financial database at Janice Credit Union. The credit union is open from 8:00 a.m. until 5:30 p.m., Monday through Friday, with Friday being the most active day of the week for customer deposits and withdrawals. In terms of daily activities, the volume of transactions is usually highest from 11:30 a.m. until 1:00 p.m. and from 4:30 p.m. until closing. Within a monthly timeframe, transactions are usually the heaviest the first two days of the month and again on the 15th of the month because of direct deposits. Based on this information, what would you recommend as an effective backup strategy for the credit union? Make certain you explain the rationale behind the policy.

Case 7-2 Procedure Manual

Continuing with the procedure manual begun in Chapter 2, provide a list of the steps necessary to perform the following types of backups for an entire database:

❐ Full

❐ Differential incremental

❐ Cumulative incremental

8

COMPLETE RECOVERY WITH RECOVERY MANAGER

**After completing this chapter,
you should be able to do the following:**

♦ Identify the benefits of using RMAN

♦ List the basic RMAN recovery procedures

♦ Recover a database that is in NOARCHIVELOG mode

♦ Recover a database that is in ARCHIVELOG mode

♦ Use RMAN to move a data file to a new location

With user-managed recovery, when media failure occurs, the user must locate the most appropriate backup copies of the files, physically copy the files to the correct location, and then instruct Oracle9*i* to recover the database. For a small database that has been backed up recently, this is a fairly simple task. However, for a large database where backup copies of the data files may be spread across several disks or tapes, the procedure can become extremely complex.

Recovery Manager (RMAN) is an Oracle9*i* utility that manages backup and recovery operations. When it is used to perform backup and recovery procedures, and media failure occurs, the utility can quickly determine the most recent set of backup files needed, restore the files to the location specified by the control file, and then recover the database.

In this chapter, Carlos will perform complete recovery through RMAN under various scenarios. First, complete recovery of a database in NOARCHIVELOG mode will be performed. Then, Carlos will recover a database in ARCHIVELOG mode. Finally, Carlos will use RMAN to recover a data file to a new location.

THE CURRENT CHALLENGE IN THE JANICE CREDIT UNION DATABASE

In this chapter, Carlos is examining the procedures necessary to perform a complete database recovery through RMAN. Previously, all backup and recovery procedures were performed through SQL*Plus. This required Carlos and his employees to create backup copies of the database files, and to restore those files, through the operating system. With RMAN, these tasks can be accomplished through the utility program. This shifts the burden of tracking the most recent copies of the database files from the Janice Credit Union employees to RMAN. In order to continue his evaluation of RMAN, Carlos will perform various types of complete database recoveries that he may be required to perform in the event of a database failure.

SET UP YOUR COMPUTER FOR THE CHAPTER

Before attempting any of the procedures demonstrated in this chapter, make certain you manually create a cold backup of all database files, including each data file, each copy of the control file, and any initialization file. In the event an unrecoverable error occurs when completing the steps within this chapter, the cold backup can be used to recover the database. In addition, make certain the database is in NOARCHIVELOG mode at the start of the chapter and is placed back into NOARCHIVELOG mode upon completion of the chapter material.

NOTES ABOUT DUAL COVERAGE WITHIN THIS CHAPTER

In this chapter, procedures for performing a complete database recovery are presented using both the command-line approach and the GUI approach. When using the GUI approach, you will be accessing the Recovery Wizard. However, the Recovery Wizard is available only through the Enterprise Manager Console when you are connected to an Oracle Management Server (OMS). If an OMS is not configured for your system, you will only be able to complete the sections demonstrating use of the command-line approach.

BASIC RECOVERY STEPS WITH RMAN

The exact steps to be performed during database recovery vary based on individual circumstances. For example, if a data file becomes unavailable due to media failure, the file will need to be restored to a new location before the database can be recovered. Regardless of the circumstances, the overall procedure is the same. First, the target database must be mounted—even databases in NOARCHIVELOG mode. (You recall that "mounted" means that the SGA has been allocated and the instance has been started; however, the data files have not been opened.) Of course, as with user-managed recovery, if the database is in ARCHIVELOG mode and only one data file needs to be recovered, then it is possible to open the database, as long as the unavailable file has been taken offline.

The next step is to issue the RESTORE and RECOVER commands to instruct RMAN to copy the necessary files back to their correct location and perform any needed media recovery procedures. Beginning with Oracle9*i*, when the RESTORE command is executed, RMAN automatically determines the best available backup set or image copy to use. Rather than restore all the data files, only the data files necessary to recover the database are replaced. In addition, during the recovery of a database in ARCHIVELOG mode, RMAN identifies the necessary archived redo log files to bring the database to a consistent state without the loss of any committed transactions.

Third, after the database has been recovered, the database is opened. If the database is open during the recovery of a specific tablespace, the tablespace or data file is placed back online.

WHY USING RMAN IS THE PREFERRED BACKUP METHOD

RMAN-managed recovery uses the same basic procedures as user-managed recovery with one major exception: with user-managed recovery, *the user* must restore, or copy, all needed data files to the correct location manually through operating system commands. As previously mentioned, RMAN restores the needed data files when the RESTORE command is executed.

Although user-managed recoveries should be a simple process, realize that in a properly configured database, all files are not stored on the same hard drive. To prevent a single point of failure, or because of the size of some files, the database files may be spread across several drives. The DBA would need to restore each file to the location listed in the control file. In the event of a hard drive failure, the control file would need to be updated with the new location for any affected files. These same procedures can be performed through RMAN—without user intervention.

RECOVERY OF A DATABASE IN NOARCHIVELOG MODE

Previously, Carlos had decided to create a new database for Janice Credit Union that would store information regarding physical assets, such as building improvements, furniture, and computer equipment. One issue he was trying to decide was whether the database should be operated in NOARCHIVELOG or ARCHIVELOG mode.

Of course, when a media failure occurs and a database is in NOARCHIVELOG mode, any changes made to the database after the last closed backup was performed are lost. But what about the actual recovery process through RMAN? When a database is in NOARCHIVELOG mode, RMAN restores the database files to their proper location, as identified in the control file, through execution of the RESTORE command. Then the database can be opened. Note, however, unlike with ARCHIVELOG mode, the RECOVER command does not need to be executed, Why? Because when a database that is in NOARCHIVELOG mode is recovered, *all* database files are restored, even if only one data file was unavailable. When a NOARCHIVELOG database is backed up, the database must be in a consistent state. Therefore, when the backup copies of the database files are restored, the database is in a previous state.

 If the actual control file needs to be recovered, the database must be placed in a nomount state to recover the control file.

In the following sections, the steps necessary to back up and restore a database operating in NOARCHIVELOG log mode will be presented.

Backing up a NOARCHIVELOG Mode Database—Command-Line Approach

Carlos will first verify that the database is actually in NOARCHIVELOG mode. Then a backup will be created with RMAN to make certain a valid copy is available when he attempts to recover the database. After a backup has been created, Carlos will restore the database using RMAN.

1. Log into SQL*Plus with a SYSDBA privileged account.

2. Type `SELECT log_mode FROM v$database;`. If the database is in NOARCHIVELOG mode, you receive the output shown in Figure 8-1. If the database is in ARCHIVELOG mode, the mode must be changed using the procedure outlined in Chapter 2 in the section entitled "Changing Archive Mode" before proceeding to the next step.

3. Type `exit` to exit SQL*Plus.

4. Start RMAN and connect to the target database with a SYSDBA privileged account.

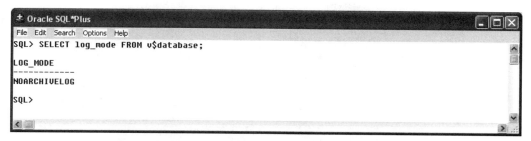

Figure 8-1   Current archiving mode

5. Type **shutdown** at the RMAN> prompt to shut down the database. As the process is being completed, the results of each stage are displayed.

As an alternative, the database can be shut down and mounted in SQL*Plus. In the following examples, the steps are being performed in RMAN because the utility will be used to perform the backup and recovery operations.

6. Type **startup mount** to mount the database.

7. Type the following set of commands at the RMAN> prompt to back up the entire database:

```
run {
allocate channel c1 type disk;
backup database;
}
```

During execution of the command, the files being included in the backup set are displayed, similar to the output shown in Figure 8-2.

8. After the backup process is complete and the RMAN> prompt reappears, type **alter database open;** to open the database.

Now that Carlos is certain he has a valid backup of the database, he can practice using RMAN to perform recovery from a media failure by moving one of the data files. In the following steps, the database is shut down and the Tools01.dbf file is moved to a new location. Realize, however, that in a real-world environment, media failure normally occurs while the database is open. With the media failure being simulated through deletion of a data file, the database must be shut down because the operating system cannot delete a file that is currently being accessed.

Figure 8-2   RMAN backup process

To practice using RMAN:

1. Type **shutdown** at the RMAN> prompt to shut down the Oracle9*i* database.

2. Type **exit** to exit the Recovery Manager utility program.

3. Using your operating system, move the Tools01.dbf data file to a new location.

4. Restart RMAN and connect to the target database with a SYSDBA privileged account. Because the database was previously shut down, the message "(not started)" is displayed, as shown in Figure 8-3.

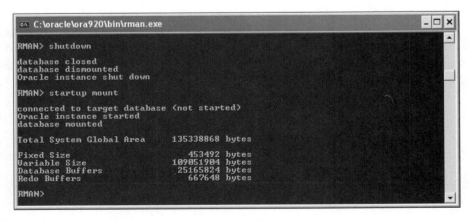

Figure 8-3   Connecting to the target database

5. Type `startup` to start the database. Because Oracle9*i* cannot locate the Tools01.dbf data file, an error message is displayed.

6. Type `shutdown` to close any opened files and to release any memory allocations made when the instance was started.

Restoring a NOARCHIVELOG Mode Database—Command-Line Approach

When RMAN is using the control file as its repository, the database must be mounted to provide access to the control file. After the database is mounted, the RESTORE command can be issued to have RMAN determine the most recent, and valid, backup set, and then copy the files back to the correct location. After the correct files have been copied back to the location specified in the control file, the database can then be opened. In this section, Carlos will mount the database and restore the database files.

To begin restoring the database:

1. Type `startup mount` to mount the database. The status of the mounting process is displayed.

2. Type the following block of commands to restore the database files:

```
run {
allocate channel c1 type disk;
allocate channel c2 type disk;
restore database;
}
```

3. If you receive a message that the SYSTEM tablespace must be recovered, type `recover database;`. This problem can occur because RMAN was referencing the current control file when the database was being restored and the SYSTEM tablespace needs to be updated. This *does not* mean that contents of archived redo log files are being applied to the database because the database is in NOARCHIVELOG mode.

In this example, Carlos allocates two channels to speed up the restoration process. However, as shown in Figure 8-4, only one channel was actually used when the files were copied to their destination. Why? When the backup was created earlier, only one channel was allocated. Therefore, all the data files were written into one backup set. If two channels had been used, two backup sets would have been created and each backup set would have been assigned a separate channel.

To create a new backup of the recovered database:

1. Now that the data files have been restored, the database can be opened by typing `ALTER DATABASE OPEN;` at the RMAN> prompt, as shown in Figure 8-5.

8

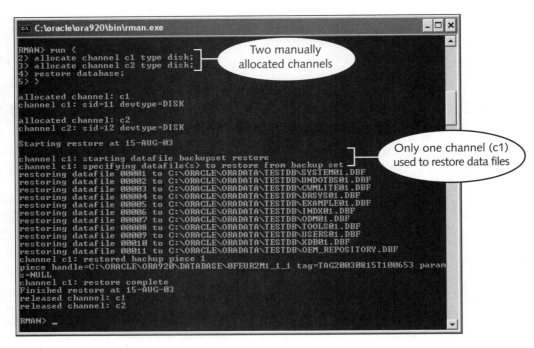

Figure 8-4 Restoring data files

Figure 8-5 Opening the restored database

 As a safety precaution, you would normally want to shut down and reopen the database to make certain there are no problems before making the database available to the users.

2. Type **shutdown** to shut down the database.

3. Type **startup mount** to start the database.

4. Type

```
run {
allocate channel c1 type disk;
backup database;
}
```

to create a new backup of the database. Note that after restoring or recovering a database, you should immediately create a new backup in the event the database needs recovery before the next scheduled backup is performed.

Backing Up a Database in NOARCHIVELOG Mode—GUI Approach

Next Carlos will practice with the GUI approach to using RMAN. Before he can restore a database that is in NOARCHIVELOG mode, he first needs to create a backup of the database. Recall from Chapter 7 that a backup can be created using the GUI approach through the Backup Wizard. However, the wizard can only be accessed if the Enterprise Manager Console is connected to an Oracle Management Server (OMS). If one is not available, you cannot perform the steps in this section.

To back up the database using the Backup Wizard:

1. Connect to the Oracle Management Server through the Enterprise Manager Console with a privileged SYSDBA account.

2. From the object list in the Navigation Pane, click the **+** (plus sign) next to the Databases folder. Then click the **+** (plus sign) next to the database name in the Database list.

3. From the list of options that appears, click the **+** (plus sign) next to the Instance icon.

4. From the list of Instance options that appears, click the word **Configuration**.

5. When the General Configuration for the database appears, verify that the database is in NOARCHIVELOG mode, as shown in Figure 8-6.

6. Click **Tools** on the menu bar at the top of the window, click **Database Tools**, click **Backup Management**, and then click **Backup**.

7. If the Introduction window appears, click **Next**.

8. When the Strategy choice window appears, click the **Customize backup strategy** option button, and then click **Next**.

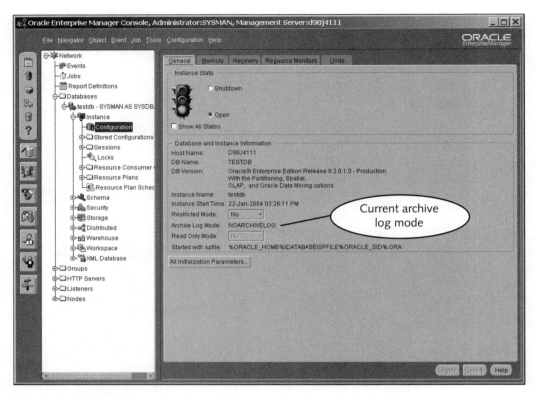

Figure 8-6 Current archive log mode

9. When the Backup Selection window shown in Figure 8-7 appears, click **Next**.

10. When the Backup Options window shown in Figure 8-8 appears, click the **Full backup** option button, if necessary, and click **Next**.

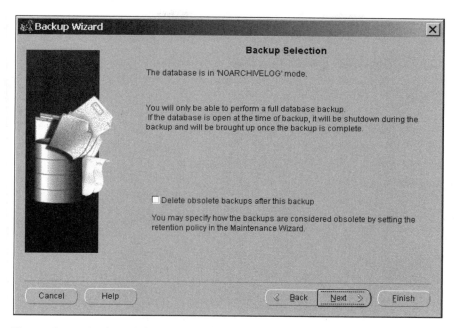

Figure 8-7   Backup Selection window

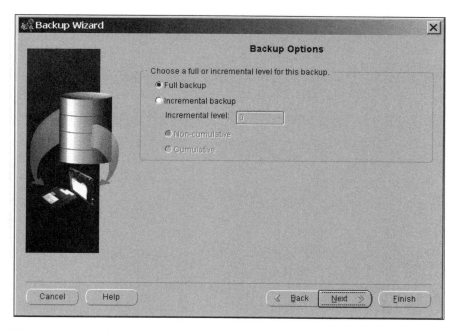

Figure 8-8   Backup Options window

11. When the Configuration window shown in Figure 8-9 appears, click **Next**.

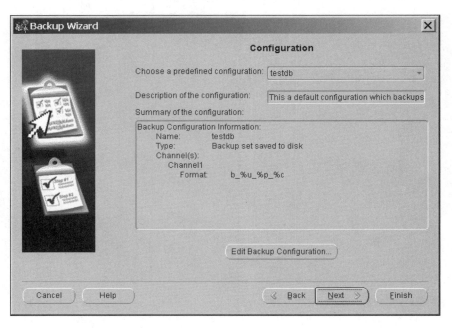

Figure 8-9 Configuration window

12. When the Override Backup and Retention Policy window appears, click **Next**.

13. When the Schedule window appears, click the **Immediately** option button, if necessary, and click **Next**.

14. When the Job Information window shown in Figure 8-10 appears, click **Finish**.

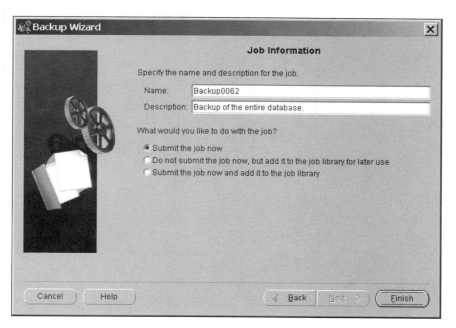

Figure 8-10 Job Information window

15. When the dialog box shown in Figure 8-11 summarizing the job that is to be submitted appears, click **OK**.

16. When a dialog box appears, informing you that the job has been successfully submitted, click **OK**. Check the Jobs pane to verify that the back up was completed successfully.

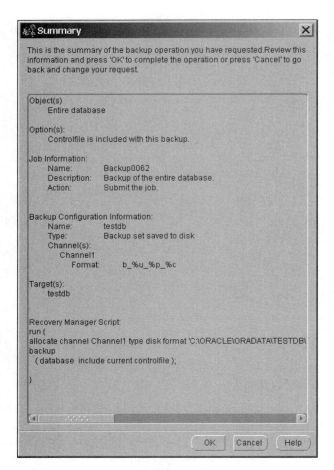

Figure 8-11 Job Summary window

Restoring a Database in NOARCHIVELOG Mode—GUI Approach

Next Carlos will recover the database using the **Recovery Wizard** available through the Enterprise Manager Console. The Recovery Wizard is the GUI interface for Recovery Manager's recovery process. As with the Backup Wizard, the Recovery Wizard can be accessed only if the Enterprise Manager Console is connected to an Oracle Management Server (OMS). If one is not available, you cannot perform the steps in this section.

To create a media failure, Carlos will first move the data file for the SYSTEM tablespace to a new location. To recover the database, he first needs to shut down and then mount the database. Remember that RMAN can only recover a database after it has been mounted, even a database in NOARCHIVELOG mode. After the database has been mounted, Carlos can then use the Recovery Wizard to restore the database files.

To shut down the database, move the System01.dbf data file, and then mount the database:

1. From the General Configuration window, click the **Show All States** check box.

2. After the database states appear as shown in Figure 8-12, click **Mounted**.

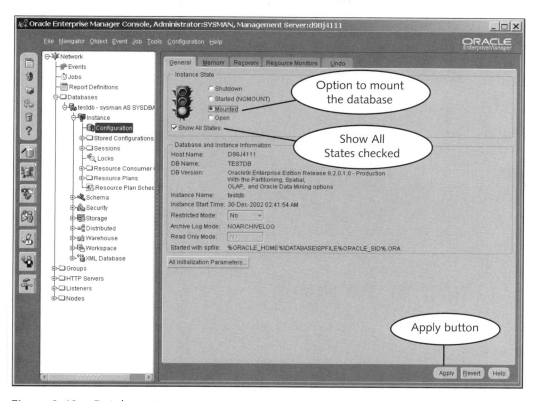

Figure 8-12   Database states

3. Click **Apply**.

4. When prompted with the shutdown options shown in Figure 8-13, click the **Immediate** option button, if necessary. Then click **OK**.

5. After the dialog box shown in Figure 8-14 appears, indicating that the database has been shut down and then mounted, click **Close**.

Figure 8-13   Shutdown Options dialog box

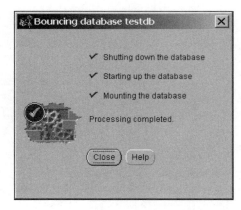

Figure 8-14   Bouncing (shutting down and restarting) the database

6. Using the Windows operating system, move the System01.dbf data file to another location.

Now that Carlos has mounted the database and made the SYSTEM tablespace unavailable by moving its associated data file, he can use the Recovery Wizard to restore the database files.

To restore the database:

1. Click **Tools** on the menu bar, and then click **Database Tools**.

2. Click **Backup Management** on the Database Tools submenu.

3. Click **Recovery** on the Backup Management submenu, as shown in Figure 8-15.

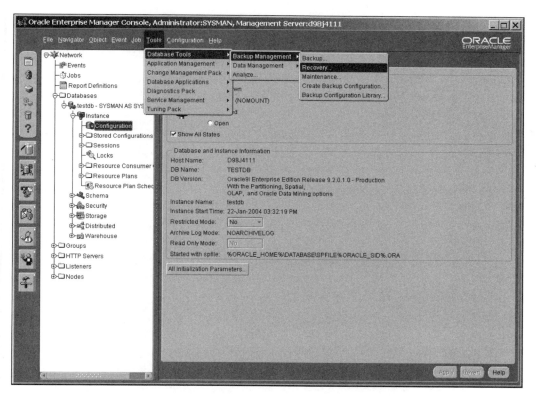

Figure 8-15   Accessing the Recovery Wizard

4. If the Introduction window appears, click **Next**.

5. When the Operation Selection window shown in Figure 8-16 appears, click **Next**.

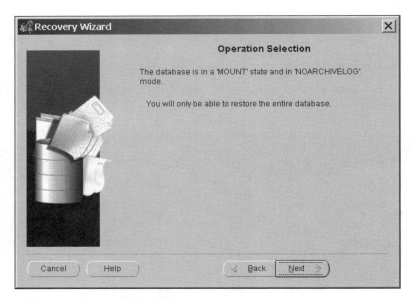

Figure 8-16   Operation Selection window

6. When the Configuration window shown in Figure 8-17 appears, click **Finish**.

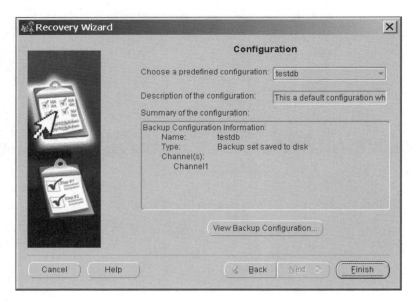

Figure 8-17   Configuration window

7. When the Summary window shown in Figure 8-18 appears, click **OK**.

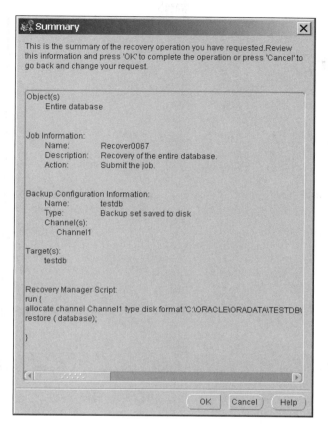

Figure 8-18   Summary window

8. When the dialog box appears indicating that the job has been submitted successfully, click **OK**.

After the job is submitted and executed, the database is recovered and Carlos can then make another backup of the database using the Recovery Wizard. After the backup process is completed, the database can then be opened and made available to the users.

RECOVERING A DATABASE IN ARCHIVELOG MODE

The basic difference between complete recovery of a database that is in ARCHIVELOG mode versus a database in NOARCHIVELOG mode is that the contents of the current archived redo log files are applied to the data files. By "applied," it is meant that any transactions contained in the archived redo log files that are not also contained in the restored data files are written to the data files. Such an application is important to you as a DBA because this places the database in a consistent state.

During the recovery of an ARCHIVELOG database, some of the administrative headaches are eliminated because RMAN automatically determines which archived redo log files are required to recover the database. In addition, recall that a database operating in ARCHIVELOG mode can be open during the recovery process, as long as the unavailable data files are taken offline before the database is opened. The files can be placed online after the recovery process is complete.

Backing up an ARCHIVELOG Mode Database

In the following steps, Carlos will create a media failure of a database by moving the data files for the TOOLS and USERS tablespaces. In this example, the database will first be placed in ARCHIVELOG mode and then a backup of the database files will be created. Because the database is in ARCHIVELOG mode, the archived redo log files should also be backed up.

To create the media failure:

1. Place the database in ARCHIVELOG mode, following the steps presented in Chapter 2 in the section entitled "Changing Archive Mode."

2. Use the SQL command shown in Figure 8-19 to verify that the database is in ARCHIVELOG mode.

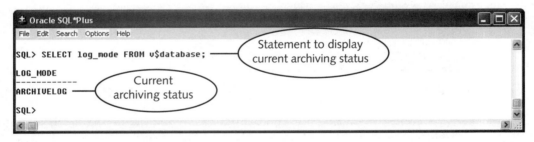

Figure 8-19 Current database archiving mode

3. Start RMAN and connect to the target database with a SYSDBA privileged account.

4. Type the following set of commands at the RMAN> prompt, as shown in Figure 8-20.

```
run {
allocate channel c1 type disk;
allocate channel c2 type disk;
backup database;
backup archivelog all;
}
```

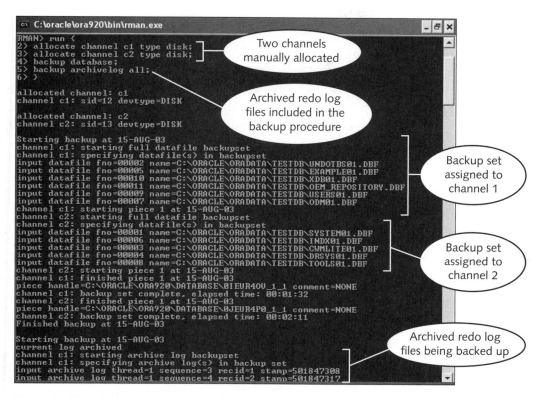

Figure 8-20   Backup process using two channels

Notice in Figure 8-20 that two backup sets were created. Why not just one? When the set of backup commands were issued, there were two channels allocated, c1 and c2. This allowed RMAN to divide the data files into two groups. Each group became a backup set that was assigned to a channel during the backup process. RMAN records information about each backup set, but still recognizes that, together, the sets create a backup of all the data files. In addition, the archived redo log files are grouped as a separate backup set because a separate BACKUP command was issued for those files and, as stated in Chapter 7, each backup set can consist of only one type of file.

Recovering a Database in ARCHIVELOG Mode—Command-Line Approach

In the following steps, the Tools01.dbf and Users01.dbf files are deleted so that Carlos can practice using RMAN to recover a database that is in ARCHIVELOG mode. After the appropriate data files are deleted, the commands to restore and recover the database are executed. Once recovered, the database can then be opened.

To practice using RMAN:

1. Type **shutdown** at the RMAN> prompt.

2. Type **exit** at the RMAN> prompt.

3. Using the operating system, move the Tools01.dbf and Users01.dbf data files to another location to create a media failure.

4. Start RMAN and connect to the target database with a SYSDBA privileged account.

5. Type **startup mount** to mount the database.

6. Type the following set of commands at the RMAN> prompt to recover the database, and then press **Enter**.

```
run {
allocate channel c1 type disk;
restore database;
recover database;
}
```

The set of commands used to recover the database only allocated one channel. Recall that when the database was previously backed up, two channels were allocated and two backup sets were created. Notice in Figure 8-21 that because only one channel is allocated for the recovery process, each set of files is restored using the same channel. The sets are restored in sequence.

After the database has been recovered, it can be opened and made accessible to the users. However, a backup should be made of all the database files in case recovery is necessary before the next scheduled backup is performed.

Restoring an entire database can take a long time, and it seems much longer when users are impatiently waiting. As an alternative, the USERS and TOOLS tablespaces could have been taken offline, and then the database opened, provided that users did not need to access the data contained in those tablespaces. Rather than the entire database being restored, just the individual tablespaces could have been restored using the commands RESTORE TABLESPACE USERS, TOOLS and RECOVER TABLESPACE USERS, TOOLS.

An alternate approach to restore tablespaces is to use the commands RESTORE DATAFILE *'filename'* and RECOVER DATAFILE *'filename'*, where *filename* identifies the file to be restored or recovered. The option to restore only a specific data file is valid only for a database in ARCHIVELOG mode. After completion of the recovery process, the tablespaces can be placed online and made available to the users.

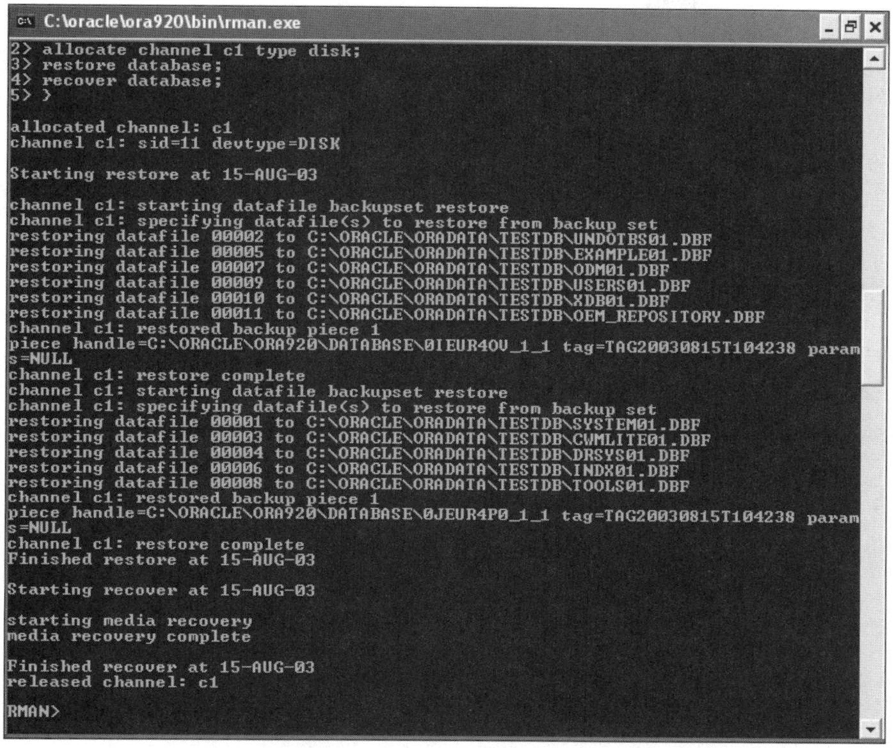

Figure 8-21   Recovery of a database in ARCHIVELOG mode

RECOVERING A DATABASE IN ARCHIVELOG MODE—GUI APPROACH

Carlos can also use the Recovery Wizard to recover a database that is in ARCHIVELOG mode. When recovering a database that is in ARCHIVELOG mode, Carlos has the option of performing the process while the database is open. In the following steps, Carlos will first verify that the database is in ARCHIVELOG mode, then move one of the data files, and finally recover the tablespace associated with that data file using the Recovery Wizard.

To recover the database:

1. From the Enterprise Manager Console, verify that the current Archive Log Mode displayed in the General Configuration for the database is ARCHIVELOG.

2. Use the Backup Wizard to create a backup of the database.

3. Mount the database.

4. Use the operating system to move the System01.dbf data file to another location.

5. Click **Tools** on the menu bar, click **Database Tools**, click **Backup Management**, and then click **Recovery**.

6. If the Introduction window appears, click **Next**.

7. When the Operation Selection window shown in Figure 8-22 appears, click the **Restore and recover** option button, if necessary, and click **Next**.

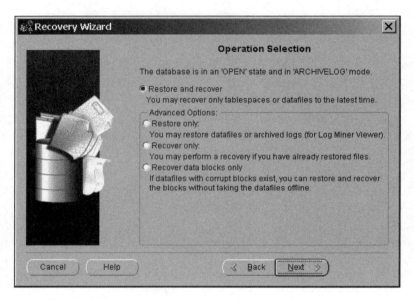

Figure 8-22   Operation Selection window

 Notice that you also have the option of only restoring the data files, only recovering the data files, or only recovering corrupt data blocks.

8. When the Object Selection window shown in Figure 8-23 appears, click the **Tablespaces** option button, and click **Next**.

9. When the Tablespaces window appears, click **SYSTEM** from the Available tablespaces list.

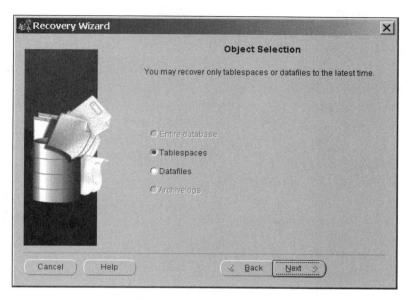

Figure 8-23   Object Selection window

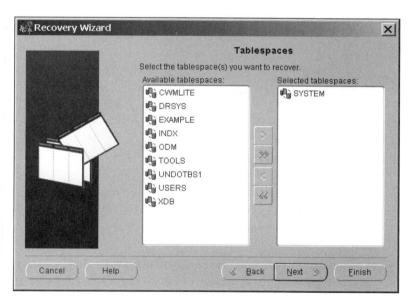

Figure 8-24   Tablespaces window

10. Click the single arrow pointing to the right to move the tablespace name to the Selected tablespaces list, as shown in Figure 8-24, and click **Next**.

11. When the Rename window appears, click **Next**.

12. When the Configuration window appears, click **Finish**.

13. When the Summary window shown in Figure 8-25 appears, click **OK**.

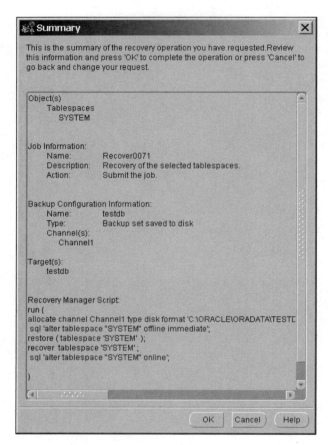

Figure 8-25   Summary window

14. When the dialog box confirming successful submission of the job appears, click **OK**.

After the job has been submitted and executed, the database will be recovered and mounted. At that time, Carlos needs to create a new backup of the database in the event another recovery becomes necessary before the next scheduled backup occurs.

MOVING DATA FILES TO A NEW LOCATION

RMAN can also be used to move data files to new locations during the recovery process and then update the locations to the database's control file. Carlos might need to move data files because of a hard drive failure, or to distribute the data files across several hard drives to avoid a single point of failure. This task requires a DBA to take the corresponding tablespace offline, and then issue the SET NEWNAME FOR DATAFILE command to specify the new location of the data file.

The RESTORE command is used to physically move the file to the new location. After the file has been relocated, the SWITCH command is used to update the location of the file to the control file. The RECOVER command is required to update the contents of the moved file to make certain any DML or DDL operations performed while the tablespace was offline are updated to the tablespace. Finally, the tablespace can be placed back online and made available to users.

Moving Data Files—Command-Line Approach

The following steps demonstrate how Carlos moves the Tools01.dbf data file to a new location through RMAN.

To move the data file:

1. Start RMAN and connect to the target database using a DBA privileged account.

2. Type the following set of commands at the RMAN> prompt:

```
run {
sql 'alter tablespace tools offline immediate';
set newname for datafile 'source_file' TO
'destination_file';
restore tablespace tools;
switch datafile all;
recover tablespace tools;
}
```

The location and file name of the data file's original location should be substituted for *source_file*, and the destination location should be substituted for *destination_file*. While the block of commands are being executed, the status of each command is displayed, as shown in Figure 8-26.

8

Figure 8-26 Moving a data file to a new location

3. Type **sql 'alter tablespace tools online';** to place the table-space back online.

As shown in Figure 8-27, the ALTER TABLESPACE command can be executed through RMAN. When an SQL statement is issued through RMAN, the statement must be enclosed in single quotation marks and be preceded by the command SQL. The SQL command can be used to execute any SQL statement through RMAN except a SELECT statement. SELECT statements cannot be executed through RMAN.

Figure 8-27 Placing a tablespace online

In the preceding step sequence, notice that no channel was manually allocated. Therefore, Oracle9*i* automatically allocated a channel named ORA_DISK_1 to perform the restore operation. The SET NEWNAME FOR DATAFILE specifies the new location for the data file to be moved. The RESTORE command is responsible for the physical movement of the data file from the C:\ drive to the V:\ in the example provided. The SWITCH command specifies that the location of all data files updated through the SET NEWNAME FOR DATAFILE command should be updated to the control file. The TOOLS tablespace is then recovered to update its contents before the tablespace is placed back online.

When RMAN is being used to perform a complete recovery of a tablespace or even the entire database, it automatically determines if the contents need to be updated by the archived redo log files or the online redo log files. If an update is required to complete the recovery process, the necessary files are identified and the contents are applied to the appropriate tablespace(s). This saves Carlos time during the recovery process because he does not need to manage any backup copies of the archived redo log files if they become necessary for recovery of the database.

Moving Data Files—GUI Approach

The procedure for moving a data file through the Recovery Wizard is basically the same as recovering a data file, except that a destination is specified when prompted by the Rename window. In the following example, Carlos will move the System01.dbf data file to a new folder. The procedure would be the same if Carlos decided to move the data file to a new drive.

To move the System01.dbf data file to a new location:

1. Use the Backup Wizard to create a backup of the database.

2. Mount the database.

3. Use the operating system to move the System01.dbf data file to another location.

4. Click **Tools** on the menu bar, click **Database Tools**, click **Backup Management**, and then click **Recovery**.

5. If the Introduction window appears, click **Next**.

6. When the Operation Selection window appears, click the **Restore and recover** option button. Then click **Next**.

7. When the Object Selection window appears, click the **Tablespaces** option button, and click **Next**.

8. When the Tablespaces window appears, click **SYSTEM** from the Available tablespace list.

9. Click the single arrow pointing to the right to move the tablespace name to the Selected tablespaces list, and click **Next**.

10. When the Rename window shown in Figure 8-28 appears, enter the destination where the data file is to reside in the New Name column. This should not be the same location to which the original data file was moved in Step 3.

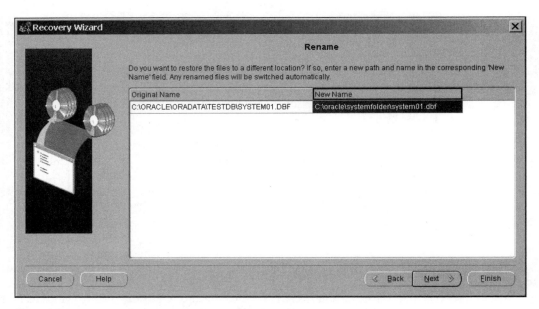

Figure 8-28   Specifying new location for data file

11. Click **Next**.

12. When the Configuration window appears, click **Finish**.

13. When the Summary window shown in Figure 8-29 appears, click **OK**.

14. When the dialog box confirming successful submission of the job appears, click **OK**.

After the job has been submitted and executed, the control file is updated with the new location, and name if applicable, of the data file. At that time, Carlos needs to create a new backup of the database in the event another recovery becomes necessary before the next scheduled backup occurs.

If you are not completing the hands-on assignments at this time, restore the cold backup created at the beginning of this chapter to reset the location of the Tools01.dbf and System01.dbf files and to place the database back in NOARCHIVELOG mode.

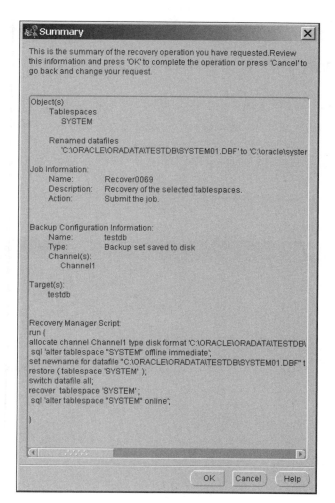

Figure 8-29   Summary window

Chapter Summary

- To recover a database using Recovery Manager, a database in NOARCHIVELOG mode must be mounted, but not opened.

- Beginning with Oracle9*i*, the RESTORE command automatically determines the most appropriate backup set or image copy for restoring the database files.

- With RMAN, the user is not required to manually restore the database files.

- Multiple channels can only be used during the restore process if more than one backup set is being restored.

- If two channels are used to back up a database, RMAN divides the data files into two backup sets.
- When only one channel is allocated to restore multiple backup sets, the backup sets are restored one at a time.
- The SET command is used to specify a new location for a data file.
- The SWITCH command is used to update the control file with the new location for a data file.
- The SQL command can be used to execute SQL statements through RMAN.

SYNTAX GUIDE

Command	Description	Example
SET NEWNAME FOR DATAFILE	Specifies a new location for a data file	`set newname for datafile 'source_file' TO 'destination_file';`
SQL	Specifies that an SQL statement is to be executed	`sql 'alter tablespace tools online';`
SWITCH	Updates the control file with the new location of a data file	`switch datafile all;`

REVIEW QUESTIONS

1. Which of the following commands is used to copy a backup version of a file to the location of the original file?
   a. RECOVER
   b. SWITCH
   c. RESTORE
   d. BACKUP
2. The INCLUDE ARCHIVELOGS option must be specified when recovering a database that is in ARCHIVELOG mode to make certain data loss does not occur. True or False?
3. If the control file is being used as the repository for the Recovery Manager, in what state must a NOARCHIVELOG mode database be during database recovery?
4. Which of the following commands is used to update the control file with the new location of a data file?
   a. RECOVER
   b. SWITCH
   c. RESTORE
   d. BACKUP

5. When recovery of a database is required because of media failure, which command, RECOVER or RESTORE, should be executed first?

6. How can the DBA make valid portions of the database available to users while performing recovery of a data file?

7. If three backup sets are required to restore and recover a database, what is the minimum number of channels that must be either manually or automatically allocated?

   a. one

   b. two

   c. three

   d. four

8. Before recovering a database that is unavailable, the database should always be backed up immediately before beginning the recovery process with Recovery Manager. True or False?

9. Which of the following is a valid statement?

   a. When a database is in ARCHIVELOG mode, RMAN must restore all database files for the recovery process to be successful.

   b. When a database is in NOARCHIVELOG mode, RMAN must restore all database files for the recovery process to be successful.

   c. When a database is in ARCHIVELOG mode, the INCLUDE ARCHIVELOGS clause specifies that a complete database recovery is being performed.

   d. When a database is in NOARCHIVELOG mode, the INCLUDE ARCHIVELOGS clause specifies that a complete database recovery is being performed.

10. The use of multiple channels can reduce the amount of time required to backup a database. True or False?

11. Which of the following commands is used to specify the new location for a data file?

    a. SET NEWNAME FOR DATAFILE

    b. SWITCH

    c. RESTORE

    d. BACKUP

12. When moving a data file to a new location, which of the following commands is responsible for physically placing a copy of the file in the new location?

    a. RECOVER

    b. SWITCH

    c. RESTORE

    d. BACKUP

13. What type of SQL statements cannot be executed through RMAN?

14. If the DBA allocates three channels to restore one backup set, can the additional channels reduce the amount of time taken to restore the backup set?

15. RMAN is able to access the control file of the target database to perform recovery operations when the database is in a(n) _____ state.

    a. non-mounted

    b. mounted

    c. ARCHIVELOG

    d. NOARCHIVELOG

16. RMAN automatically takes a tablespace offline to recover an ARCHIVELOG database. True or False?

17. When RMAN is being used to recover a database that is in NOARCHIVELOG mode, the database cannot be opened until the restore and recovery processes have been completed. True or False?

18. An ALTER TABLESPACE statement used to take a tablespace offline can be executed in RMAN only if it is used with the _____ command.

19. If a data file becomes unavailable due to a hard drive failure, what command is required to instruct RMAN to restore the file to a new location?

20. After a tablespace for a database that is in ARCHIVELOG mode has been recovered, RMAN automatically places the tablespace online. True or False?

HANDS-ON ASSIGNMENTS

Assignment 8-1 Backing Up an ARCHIVELOG Database Using Multiple Channels

In this assignment, you backup a database that is in ARCHIVELOG mode using two channels.

1. Open the database.

2. Verify that the database is in ARCHIVELOG mode.

3. Using RMAN, shut down the database.

4. Mount the database.

5. Back up the database and allocate only two channels for the process.

Assignment 8-2 Recovering an ARCHIVELOG Database Using Two Channels

In this assignment, you recover a database that is in ARCHIVELOG mode using two channels.

1. If necessary, shut down the database using RMAN.
2. Using your operating system, delete the System01.dbf data file.
3. Using RMAN, mount the database.
4. Recover the database, but make certain two channels are allocated for the process.
5. Use RMAN to open the database to verify that the database was recovered successfully.

Assignment 8-3 Recovering a Tablespace for an ARCHIVELOG Database

In this assignment, you recover a tablespace for a database that is in ARCHIVELOG mode.

1. Create a backup of the database and include any archived redo log files.
2. Using RMAN, shut down the database.
3. Using your operating system, delete the Indx01.dbf data file.
4. Mount the database and take the tablespace corresponding to Indx01.dbf data file offline.
5. Open the database.
6. Recover the Indx01.dbf data file.
7. Place the tablespace online.

Assignment 8-4 Recovering a Database After Media Failure

In this assignment, you create a tablespace and store its associated data file on a floppy disk. You will remove the floppy disk to create a media failure. Recovery of the database will require that the data file be restored to another drive.

1. Insert a blank, formatted floppy disk into drive A:.
2. Log into SQL*Plus or access the SQL*Plus Worksheet with a SYSDBA privileged account.
3. Type the following command to create a small tablespace and store the associated data file on the floppy disk:

```
CREATE TABLESPACE myts
DATAFILE 'a:\myts01.dbf' SIZE 100K;
```

4. Use RMAN to create a backup of the database.
5. Eject the floppy disk from the disk drive.
6. Shut down and restart the database.
7. When an error message is returned indicating that the Myts01.dbf data file cannot be located, use RMAN to recover the database. (Hint: You will need to store the unavailable tablespace in a new location.)

Assignment 8-5 Recovering a NOARCHIVELOG Database

In the following assignment, you recover a database that is in NOARCHIVELOG mode using RMAN.

1. If necessary, open the database.
2. Place the database in NOARCHIVELOG mode.
3. Shut down the database.
4. Using RMAN, back up all the database files for the database.
5. Move the System01.dbf, Tools01.dbf, and Indx01.dbf data files.
6. Attempt to open the database. After receiving an error message indicating that the data files cannot be identified, recover the database.

Assignment 8-6 Moving a Data File Through RMAN

In this assignment, you return the Tools01.dbf data file to its original location before it was moved in the chapter. If you did not complete the chapter examples, you need to first complete the steps presented in the section named "Moving Data Files—Command-line Approach."

1. Create a backup of the database using RMAN.
2. Using RMAN, perform the steps necessary to move the Tools01.dbf data file back to its original location.

CASE PROJECTS

Case 8-1 Recovery Strategy

Carlos receives a phone call at 4:00 a.m. from one of his employees and is informed that one of the disk controller cards for the computer that stores the financial database has failed, and that the two hard drives controlled by that card cannot be accessed.

The computer has two disk controller cards. The first card provides access to drives C: and D:. These drives store the operating system, the Oracle DBMS, the initialization file, the archived redo log files, mirrored copies of the control file, and the online redo log files for the financial database. The second card provides access to drives E: and F:, which stores mirrored copies of the control file, the online redo log files, the data files, and archived redo log files. It was the second card that failed.

A backup copy of the whole database was created at midnight. The computer is one of the newest models on the market and its components are not compatible with the other machines in the department. Therefore, there is no other disk controller card immediately available as a substitute for this machine. In addition, there is no other computer that can be used because of storage space limitations on their hard drives. There is enough space left on drives C: and D: to store the data files for the database—but only for a few days.

The credit union opens at 8:00 a.m., but the database needs to be available by 7:30 a.m. for the tellers to begin verifying ATM transactions that occurred during the previous night. Specify the exact steps that need to be taken by the employee to have the financial database available by 7:30 a.m., without any data loss.

Case 8-2 Procedure Manual

Continuing with the procedure manual created in Chapter 1, create lists that identify the steps necessary to perform the following tasks through RMAN:

❐ Make a complete recovery of a NOARCHIVELOG mode database

❐ Make a complete recovery of an ARCHIVELOG mode database

❐ Move a data file

8

9

INCOMPLETE RECOVERY WITH RECOVERY MANAGER

**After completing this chapter,
you should be able to do the following:**

♦ List the types of incomplete recoveries that can be performed through Recovery Manager

♦ Perform a time-based incomplete recovery using the command-line and GUI approaches

♦ Identify the commands necessary to perform a sequence-based incomplete recovery

♦ Perform a change-based incomplete recovery

♦ Identify the sequence and thread number associated with the online redo log files

♦ Use LogMiner Viewer to determine the SCN assigned to a specific transaction

In addition to backup operations and complete database recoveries, Recovery Manager can also be used to perform incomplete database recoveries. As with user-managed incomplete recoveries, the recovery process through RMAN can be based on time (time-based) or a specific SCN (change-based), or it can be canceled prior to the update from a specific redo log file (sequence-based). Each of these approaches will be presented in this chapter using the command-line approach as well as the Recovery Wizard through the Enterprise Manager Console. In addition, the LogMiner's GUI interface—called LogMiner Viewer and used for retrieving the SCN assigned to a particular transaction—will be presented.

The command-line approach to using LogMiner was previously demonstrated in Chapter 5 and is not repeated in this chapter.

THE CURRENT CHALLENGE IN THE JANICE CREDIT UNION DATABASE

There are various reasons that an incomplete recovery may need to be performed. For example, at the end of each workday, the daily transactions are updated to tables used to track weekly and monthly transactions. Then the daily transaction table is cleared for the next workday. However, if the update is not performed correctly, the database transaction could be updated to the wrong tables, providing erroneous results in the credit union's financial statements and for customer accounts. To correct this problem, Carlos would need to perform an incomplete recovery.

In this chapter, Carlos uses Recovery Manager to perform incomplete recoveries caused by two types of problems. One problem is the need to recover the financial database to a previous state because of a user error. Both time-based and change-based incomplete recoveries are performed to correct the error. In addition, a problem caused by a corrupt online redo log file will be resolved by performing a sequence-based recovery. To restore the database to a previous state, Carlos will also access LogMiner through the Enterprise Manager Console to retrieve the necessary information required to perform an incomplete recovery based on the SCN assigned to a specific transaction.

SET UP YOUR COMPUTER FOR THE CHAPTER

Before performing any of the tasks demonstrated in this chapter, make a cold backup of the whole database, including the control files, data files, and any valid archived redo log files. Because this backup can be used to recover the database in the event an error occurs while performing the various recovery scenarios, make certain that backups created in the chapter do not overwrite the original cold backup.

To perform an incomplete recovery, the database must be in ARCHIVELOG mode, preferably with automatic archiving enabled. Therefore, after creating the cold backup, place the database in ARCHIVELOG mode with automatic archiving enabled, using the steps presented in Chapter 2.

NOTES ABOUT DUAL COVERAGE WITHIN THIS CHAPTER

This chapter presents both the command-line and graphical user interface (GUI) approaches for using the Recovery Manager utility. When using the command-line approach, the steps will be demonstrated by accessing RMAN through the Windows XP operating system. In sections covering the GUI approaches, the examples are demonstrated on a system running Windows 2000 Professional or Server. The command-line approach can also be performed in the Windows 2000 environment. However, with the GUI approach, you will only be able to perform the steps if you have Oracle Management Server (OMS) configured. This also includes access to LogMiner Viewer, which is the GUI interface for the LogMiner utility available through the Enterprise Management Console.

INCOMPLETE RECOVERY

Recall from Chapter 5 that a DBA, such as Carlos, performs an incomplete recovery in situations where the database needs to be recovered to a state prior to the introduction of an error. When a database in ARCHIVELOG mode is recovered, any committed transactions that have not been updated to the data and control files are written to the database from the archived and online redo log files. During an incomplete recovery, the DBA decides at what point this update process should be terminated.

An incomplete recovery can be time-based, where the DBA specifies the time of the problematic transaction; change-based, where the SCN of the problematic transaction is used as the termination point; or **sequence-based**, where the DBA specifies the log file sequence number that is used to terminate the recovery process. A sequence-based incomplete recovery is the same as the cancel-based incomplete recovery discussed in Chapter 5. When using RMAN to perform an incomplete recovery, the terminating parameter (the sequence number) is specified using the SET UNTIL command. In the following sections, the SET UNTIL command will be demonstrated for each type of incomplete recovery.

As with user-managed incomplete recovery, when an incomplete recovery is performed through RMAN, the RESETLOGS option must be specified when the database is opened. Basically, the database is recovered to a state prior to the contents of the current online redo log files. Therefore, Carlos wants to make certain that the contents of those files are updated to the database. He also wants to make certain that any subsequent recovery attempts prevent the contents of any archived redo log files that are not valid with the new state of the database from being updated to the database.

When the RESETLOGS option is specified, the contents of any online redo log files that have not been archived are overwritten. In addition, the sequence number assigned to each redo log file is restarted and updated to the data files and control files. For example, suppose there are three online redo log files for the financial database. The sequence number assigned to the first redo log file is 1, while the second redo log file is assigned a sequence number of 2, and the third is assigned 3.

When the next log switch occurs and the first redo log file is overwritten, it is assigned the sequence number of 4. These same sequence numbers are also associated with the archived version of the file. However, use of the RESETLOGS options resets these sequence numbers. After the log files have been reset, the sequence number of the first online redo log file becomes 1, and so on. Then, if another recovery needs to be performed, only the next set of sequence numbers can be included in the recovery. In other words, it prevents any unarchived transactions from being introduced into the database after the recovery process has been completed, and it prevents outdated archived redo log files from being included in a later incomplete recovery procedure.

The basic steps for performing an incomplete recovery through RMAN are the same as when a user-managed incomplete recovery is performed. The database must first be shut down, and then a cold backup is created. If appropriate, the data file, or files, containing the error is deleted. The database is then mounted. An RMAN block of code that specifies the type of recovery, the recovery parameter, and the restore and recover commands is executed. After the database is recovered, it is then opened with the RESETLOGS option. As a safety precaution, after the DBA has verified that the database was correctly recovered, the database should be shut down and a cold backup of the recovered database created. In the following sections, Carlos will implement these steps to recover a database under different scenarios.

TIME-BASED INCOMPLETE RECOVERY

A time-based incomplete recovery is appropriate when the exact time that a problematic transaction occurred is known. For example, suppose a user realizes that a massive amount of data was loaded into the wrong table at 9:13 a.m. If there is no reasonable means of simply deleting the data from the table, the DBA could perform a time-based recovery and recover the database to the state it was in at 9:12 a.m. Of course, this is all based on the assumption that the user knows the exact time the transaction was committed to the database. However, depending on the method used to load the data into the table, there could have been multiple transactions involved and the user may only be aware of the time when the last transaction was committed. In addition, the time of the transaction is recorded based on the time set on the machine that houses the Oracle server, not the time according to the individual user's machine.

With so much uncertainty, why use this approach? If the DBA has no other information, i.e., is unable to determine the SCN of the transaction(s) or the appropriate redo log file(s) containing the transaction(s), this may be the only option. However, if the time is incorrect (for example, the event occurred at 9:10 a.m. rather than 9:13 a.m.), Carlos may have to repeat the process until the database is recovered prior to the point when the data was incorrectly loaded. This is one of the reasons it is crucial to make a backup of the database before attempting an incomplete recovery.

Before Carlos can open the database to make certain it was recovered correctly, he has to issue the RESETLOGS option. Therefore, any prior redo log files become invalid. If the recovery process was not successful (the database was not recovered to the correct point), he cannot attempt the recovery process again because the redo log files are invalid. He would be required to first restore the control file, data files, and redo logs from the backup taken immediately before attempting the incomplete recovery. Then the recovery process could be attempted again.

The RMAN command used to indicate that a time-based recovery is to be performed is SET UNTIL TIME '*date time*'. Recall that with a user-managed, time-based recovery, the time parameter is specified as part of the RECOVER command. However, when the RMAN utility is used to perform the recovery, the SET command is used to specify the recovery parameter before the data files are restored and prior to the RECOVER command. The following is an example of the commands necessary to perform a time-based recovery in RMAN:

```
run {
allocate channel c1 type disk;
set until time 'DD MON YYYY HH:MI:SS';
restore database;
recover database;
sql 'alter database open resetlogs';}
```

The first line indicates that a block of statements is being submitted to the RMAN utility. The second line indicates that one channel is to be allocated for the recovery process. The SET UNTIL TIME command is included on the third line to indicate the time parameter that should be used to terminate the recovery process. However, the '*DD MONYYYY HH:MI:SS*' is replaced with the appropriate date and time. The RESTORE command on the fourth line instructs RMAN to restore the database files from the last valid backup of the database completed by RMAN.

As you examine the preceding code further, note that the RECOVER command given on the fifth line is used to actually perform the recovery process, using the parameter previously specified by the SET command. The ALTER DATABASE command is included to open the database after it has been recovered and to reset the log files. Carlos will use this set of statements in the following section to perform a time-based incomplete recovery.

Performing a Time-Based Incomplete Recovery—the Command-Line Approach

At 1:00 p.m. at the Janice Credit Union, each teller is required to balance his or her cash drawer with the amount of the cash transactions that has been posted to the database. After the cashier has balanced the cash amount, the current transactions for that cashier are updated to the PRIORTRANSACTIONS table. This table contains all transactions that have occurred for the week.

Unfortunately, at 10:23 a.m., one of the cashiers accidentally clicked a button that caused the update to occur. These transactions must be removed or they will be updated to the table again when the cash drawer is balanced at 1:00 p.m. In addition, it causes any queries about the previous day's transactions to return faulty results. Because Carlos knows the time of the update, he can perform a time-based incomplete recovery of the database to remove the invalid entries that were made to the PRIORTRANSACTIONS table.

Before creating the update error, you need to create the PRIORTRANSACTIONS table, and then create a cold backup of the database, using the following steps:

To create the PRIORTRANSACTIONS table:

1. Log into SQL*Plus, and enable the SYSDBA privileges.

2. Type **START** *<location>***\ch09script1.sql**, where *<location>* is the path to the script file contained in the Chapter09 folder of your data disk. This script file creates the PRIORTRANSACTIONS table.

3. Type **SELECT * FROM priortransactions;** to view the current contents of the PRIORTRANSACTIONS table, as shown in Figure 9-1. Note that the table currently contains 14 rows.

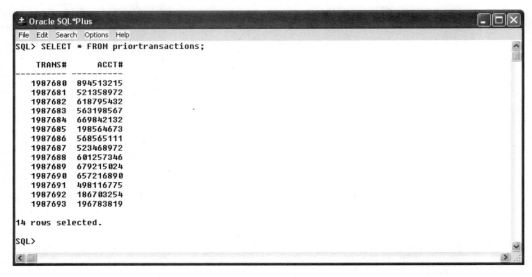

Figure 9-1   Contents of the PRIORTRANSACTIONS table

4. Click the **Start** button on the Windows XP task bar to access the Start menu, and then click **Run**.

5. Type **rman target / nocatalog** in the text box of the Run window to start the RMAN utility.

If you are using the command-line approach in Windows 2000, you will need to type **cmd** in the Run text box to open the Command Prompt window. Then enter the command in Step 5 at the command line.

6. Type **shutdown immediate** at the RMAN> prompt to shut down the database.

7. Type **startup mount** at the RMAN> prompt, as shown in Figure 9-2, to mount the database.

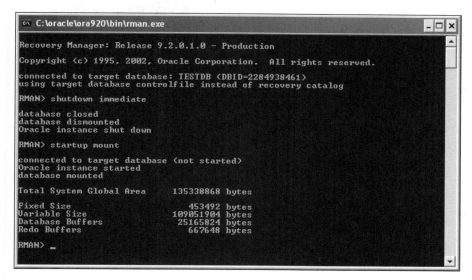

Figure 9-2   Mounting the database in RMAN

8. Type the following block of code at the RMAN> prompt to create a backup of the database, as shown in Figure 9-3:

```
run {
allocate channel c1 type disk;
backup database;
backup archivelog all;
}
```

9. Type **alter database open;** at the RMAN> prompt to open the database.

Figure 9-3 Backing up the database through RMAN (partial output shown)

Now that the table has been created and the database has been backed up and opened, you can simulate an error by loading the invalid entries into the PRIORTRANSACTIONS table using the following steps:

To simulate an error:

1. At the SQL> prompt in SQL*Plus, type **connect <user/password>**, substituting your user name and password in the statement, to reconnect to the database. If your database is accessed through a network or if more that one database is configured on your machine, you also need to include the appropriate host string immediately after the password using this format: *user/password@hoststring.*

2. Enable your SYSDBA privileges.

3. Note the exact time on your computer by looking in the lower-right corner of your task bar. If the time is not displayed, or if the database is not physically stored on your machine, type **SELECT TO_CHAR(SYSDATE, 'hh:mi:ss')** **FROM DUAL;** at the SQL> prompt to retrieve the current time according to the Oracle server.

4. Type **START <*location*>\ch09script2.sql**, where <*location*> is the
path to the script file contained in the Chapter09 folder of your data disk.
This script file adds transactions to the PRIORTRANSACTIONS table.

5. Type **SELECT * FROM priortransactions;** to view the updated contents
of the PRIORTRANSACTIONS table as shown in Figure 9-4. Note that it
now contains 24 rows.

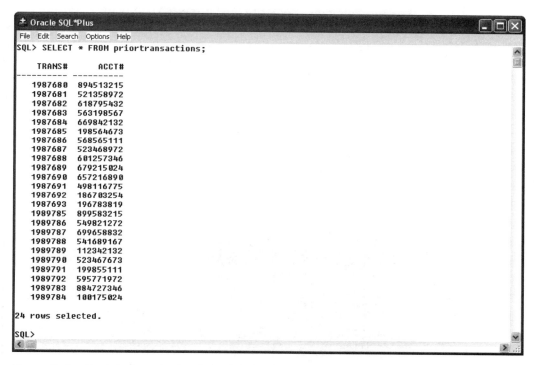

```
± Oracle SQL*Plus
File  Edit  Search  Options  Help
SQL> SELECT * FROM priortransactions;

    TRANS#      ACCT#
---------- ----------
   1987680  894513215
   1987681  521358972
   1987682  618795432
   1987683  563198567
   1987684  669842132
   1987685  198564673
   1987686  568565111
   1987687  523468972
   1987688  601257346
   1987689  679215024
   1987690  657216890
   1987691  498116775
   1987692  186703254
   1987693  196783819
   1989785  899583215
   1989786  549821272
   1989787  699658832
   1989788  541689167
   1989789  112342132
   1989790  523467673
   1989791  199855111
   1989792  595771972
   1989783  884727346
   1989784  100175024

24 rows selected.

SQL>
```

Figure 9-4   Updated contents of the PRIORTRANSACTIONS table

Now that an error has occurred, Carlos needs to perform the following steps to recover
the database to a state prior to when the error occurred.

To recover the database:

1. Type **shutdown immediate** at the RMAN> prompt to shut down the
database.

2. Using Windows Explorer, create a cold backup of the database, including all
copies of the control file, the data files, and any archived redo log files. Make
certain you do not overwrite the cold backup created at the beginning of
the chapter.

3. Using Windows Explorer, delete (or move) the data file that is used to store
user transactions from the database folder. If your database was created using
the default configuration, this would be the data file named System01.dbf.

4. Type **startup mount** at the RMAN> prompt to mount the database.

As with a complete recovery, the command to mount the database can also be provided inside the block of code that is executed to actually perform the recovery process. It should be listed either immediately before or after the SET command.

5. Type the following block of code at the RMAN> prompt:

```
run {
allocate channel c1 type disk;
set until time 'DD MON YYYY HH:MI:SS';
restore database;
recover database;
sql 'alter database open resetlogs';
}
```

Where indicated, substitute the day of the month for *DD*, the three–letter abbreviation of the month for *MON*, the four–digit year for *YYYY*, the hour in which the error occurred for *HH*, the minutes for *MI*, and seconds for *SS*, as shown in Figure 9-5. The time entered should be the time previously noted when the Ch09script2.sql file was executed.

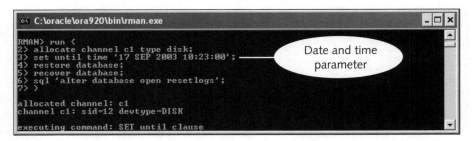

Figure 9-5 Command to perform a time-based recovery in RMAN (partial output shown)

In Figure 9-5, the format for the date and time is based on the format assigned to the NLS_DATE_FORMAT parameter. If you receive an error message indicating a problem with the date format, type **sql 'alter session set nls_date_format = "DD MON YYYY HH:MI:SS" ';** at the RMAN prompt to temporarily change the date format.

6. Type **exit** at the RMAN > prompt to exit the RMAN utility.

7. In SQL*Plus, use the CONNECT command to reconnect with the database.

8. At the SQL> prompt in SQL*Plus, type **SELECT * FROM priortransactions;**, as shown in Figure 9-6, to verify that the incorrect entries are no longer contained in the database.

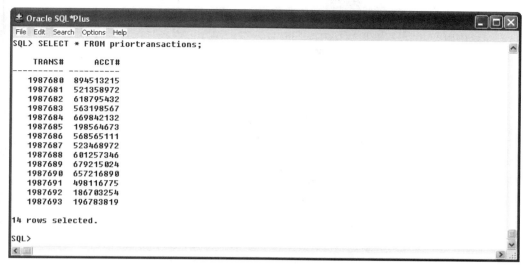

Figure 9-6   PRIORTRANSACTIONS table after recovery

Now that the database has been recovered successfully, Carlos needs to shut down and back up the database in the event a subsequent recovery is necessary before the next scheduled backup occurs.

Performing a Time-Based Incomplete Recovery—the GUI Approach

Previously, the GUI approach to performing a complete recovery with Recovery Manager utilized the Recovery Wizard through the Enterprise Manager Console. Carlos can use this same wizard to perform an incomplete recovery of a database. The Recovery Wizard supports time-based, change-based, and sequence-based incomplete recoveries. The only difference among these types of recoveries in terms of using the wizard is the type of data that needs to be supplied. For example, if a time-based recovery is being performed, then the user does not need to supply an SCN or identify a redo log file.

In this section, Carlos will repeat the procedure necessary to recover the database prior to the erroneous insertion of the current transactions into the PRIORTRANSACTIONS table. As with the previous example using the command-line approach, the PRIORTRANSACTIONS table first needs to be created.

To create the table:

1. Type **start <*location*>\ch09script1.sql**, where <*location*> is the path to the script file, and execute the command to create the PRIORTRANSACTIONS table. Because the table already exists in your database, the script first drops, and then re-creates the table.

2. Type **SELECT * FROM priortransactions;** to verify that there are only fourteen rows in the table.

Now that the table has been created and verified, a backup of the database must be created using the Backup Wizard. Remember that the Backup Wizard is available only if the Enterprise Manager Console is connected to the Oracle Management Server.

To create a backup of the database:

1. Start the Enterprise Manager Console (EMC), connect to the Oracle Management Server using the connection information provided by your instructor, and then verify that the database is in ARCHIVELOG mode.

2. In the navigation list, click the name of the target database. Click **Tools** on the menu bar of the EMC, click **Database Tools**, click **Backup Management**, and then click **Backup**.

3. If the Introduction window appears, click the **Next** button in the lower-right corner of the window.

4. When the window appears requiring you to select a strategy choice, click the **Customize backup strategy** option button, and then click **Next** to proceed to the next window.

6. When the window appears prompting you to choose the object to back up, click the **Entire database** option button, if necessary.

7. Click the **Next** button to proceed to the next window.

8. When the Archived Logs window appears, click the **Yes, all archived logs** option button, if necessary, to include all archived redo log files in the backup procedure.

9. Click the **Next** button to proceed to the next window.

10. When the Archived Log Deletion window appears, click the last option button, if necessary, to indicate that all the archived redo log files are to be deleted after they have been backed up.

11. Click the **Next** button to accept the retention selection and proceed to the next window.

12. When the Backup Options window appears, click the **Full backup** option button in the upper portion of the window.

13. Click the **Online Backup** option button, if necessary, in the lower portion of the screen to specify that the backup will be created while the database is open.

14. Click the **Next** button to proceed to the next window.

15. When the Configuration window appears, make any changes specified by your instructor, and then click **Next**.

16. When the Override Backup and Retention Policy window appears, make any changes specified by your instructor, and click **Next**.

17. When the Schedule window appears, click the **Immediately** option button, if necessary, to indicate that the job should be executed as soon as it is submitted. Then click **Next**.

18. When the Job Information window appears, verify that the **Submit the job now** option button is selected. Then click the **Finish** button to submit the job without changing the default job name or description.

19. When the Summary window shown in Figure 9-7 appears, verify that the information is correct, and click the **OK** button.

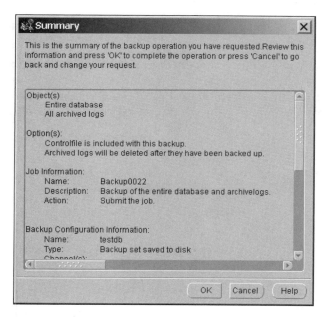

Figure 9-7 Summary information

20. When the dialog box regarding job submission appears, the job has been submitted, and you can click the **OK** button to close the window.

You can verify that the database has been backed up (that the job has actually been executed). You need only open the active pane of the Jobs objects and view the status of the submitted job. If the job is not listed, or the status of the job is listed as Failed on the History pane, notify your instructor before continuing with this section.

Now that the PRIORTRANSACTIONS table has been created and the database has been backed up using RMAN (via the Backup Wizard), the extra rows need to be added to the PRIORTRANSACTIONS table. This creates the scenario that requires an incomplete recovery to be performed.

To add the rows:

1. Reconnect to the database in SQL*Plus. Note the time of your computer by recording the time displayed in the taskbar of the operating system or by typing **SELECT TO_CHAR(SYSDATE, 'hh:mi:ss') FROM DUAL;** in SQL*Plus and executing the command to display the actual time according to the Oracle Server.

2. Type **start <*location*>\ch09script2.sql**, where <*location*> is the path to the script file, and execute the command. This script adds an extra 10 rows to the PRIORTRANSACTIONS table.

3. Type **SELECT * FROM priortransactions;** to verify that there are now 24 rows in the table.

Recall that when a database is being recovered, the database should be mounted, but not opened. Before Carlos can perform a time-based recovery, he needs to change the database to a mounted state. In addition, he needs to move (or delete or rename) the current data file containing the user data to ensure that the file is restored before the recovery process is initiated.

To change the state:

1. From the Navigation list in the left pane of the EMC, click the **+** (plus sign) next to Instance in the object list for the database.

2. From the Instance list, click **Configuration** to display the general configuration for the database.

3. On the General Configuration sheet in the right pane of the EMC, click the **Show All States** check box option.

4. When the available states are displayed, click the **Mounted** option button.

5. Click the **Apply** button at the bottom of the screen to mount the database.

6. When the Shutdown Options dialog box appears, click the **Immediate** option button, if that option is not already selected.

7. Click the **OK** button to close the Shutdown Options dialog box and to begin the shutdown process.

8. The Bouncing database dialog box appears providing you with the status of the shutdown and mounting processes. After the database has been mounted, click the **Close** button at the bottom of the dialog box.

9. After the dialog box has been closed, the Configuration sheet shown in Figure 9-8 is displayed. Verify that the current state of the database is mounted.

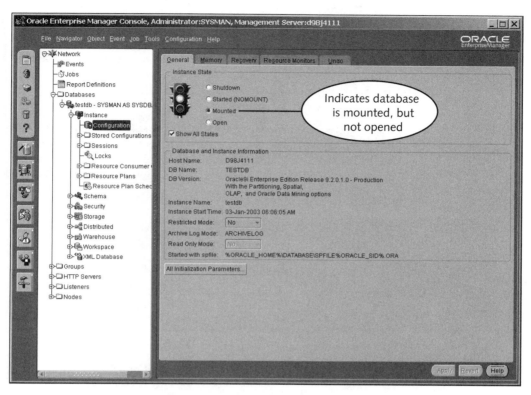

Figure 9-8   Current state of the database

10. Using Windows Explorer, move the current data file containing the erroneous data to a new location. Recall that in a default installation, user data is automatically written to the System01.dbf data file. However, if you have been assigned another tablespace, you should instead move the data file associated with your default tablespace.

Remember that an alternative approach is to first shut down the database, move the data file, and then mount the database.

Now that the database is in a mounted state, Carlos can use the Recovery Wizard to recover the database by performing the following steps:

To recover the database:

1. Click **Tools** from the menu bar, point to **Database Tools**, point to **Backup Management**, and then click **Recovery** to access the Recovery Wizard.

2. If the Introduction window appears, click the **Next** button.

3. When the Operation Selection window appears, click the **Restore and recover** option button, if necessary. Then click **Next**.

4. When the Object Select window appears, select **Entire database**, if necessary. Then click **Next**. When the Range Selection window appears, click the **Recover to a point-in-time in the past** option button to indicate that an incomplete recovery will be performed.

5. Because a time-based recovery is being performed, the time at which the Ch09script2.sql was executed should be entered in the Date text box as shown in Figure 9-9. When entering the time, click the portion of the time parameter that needs to be changed, then type in the correct hour, minute, and so on.

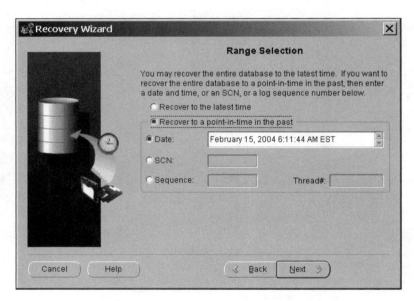

Figure 9-9 Setting the parameter for a time-based recovery

6. After the desired time has been entered, click the **Next** button to proceed to the next window.

7. When the Rename window appears, simply click the **Next** button to indicate that all files should be restored to their original location.

8. If the Configuration window shown in Figure 9-10 containing the configuration information for the backup set that will be used to recover the database appears, click the **Finish** button.

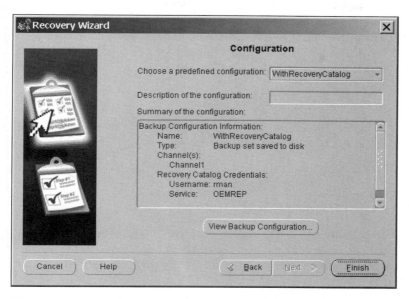

Figure 9-10   Configuration of previous backup set

9. When the Summary window shown in Figure 9–11 is displayed, verify the recovery information, and click the **OK** button to submit the job.

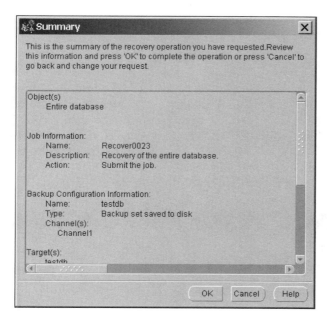

Figure 9-11   Summary of the recovery operation

10. Reconnect to the database and then type **SELECT * FROM priortransactions;** after the database has been recovered to verify that the PRIORTRANSACTIONS table no longer contains the ten extra rows that were erroneously added.

11. When the dialog box appears informing you that the job has been submitted successfully, click **OK**.

As with any recovery procedure, the database should be immediately shut down and backed up in the event another database recovery is required before the next scheduled backup is performed.

SEQUENCE-BASED INCOMPLETE RECOVERY

A sequence-based incomplete recovery is performed when the recovery process needs to be terminated before the transactions contained in a specific redo log file are updated to the database. This approach normally results in the greatest amount of data loss because the entire contents of the specified redo log file are discarded, along with the contents of any subsequent redo log files. In other words, valid portions of the redo log file cannot be applied when making a sequence-based incomplete recovery; it is all or nothing.

So why use this approach? In the event an incomplete recovery is necessary and Carlos cannot determine the exact SCN or time, but he can estimate when the problem occurred, Carlos may be able to narrow down the time frame enough to identify which redo log file contained the error. In addition, if the online redo log files become corrupt, Carlos would use the sequence-based approach to recover the database to a state immediately prior to transactions contained in the redo log files.

For example, suppose one or more of the online redo log files became corrupt, causing the database to hang, or freeze, when the Log Write (LGWR) attempted to record transactions to a file. To determine which log file contains the problem, Carlos first displays the contents of the V$LOGFILE view to check the status of the online redo log files. In Figure 9-12, the last two files listed in the view are marked as STALE in the Status column of the display. This indicates that the files are incomplete, and their contents cannot be updated. Another type of status that should be of concern to a DBA is if the files are marked as INVALID, indicating that the files are unavailable or inaccessible.

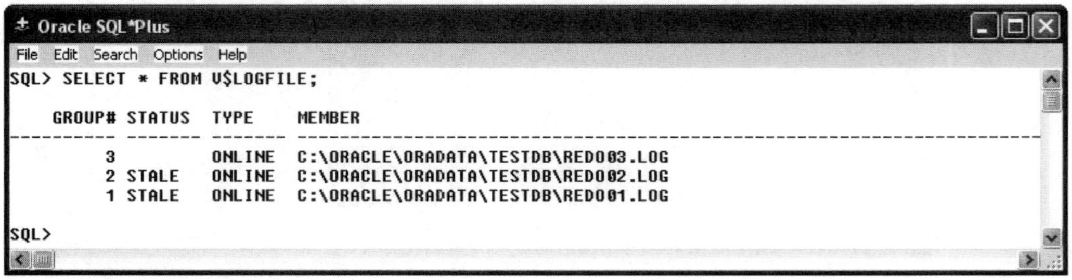

Figure 9-12   Current status of online redo log files

In this case, Carlos needs to recover the database before the contents of the stale redo log files are updated to the database. To perform a sequence-based recovery through RMAN, Carlos uses the SET UNTIL SEQUENCE command. The exact full syntax of the command is SET UNTIL SEQUENCE LOGSEQ = l THREAD = t, where l indicates the sequence number of the redo log file, which is to be used to terminate the recovery process, and t is the process thread number.

To determine the exact sequence number of the stale redo log files, Carlos queries the V$LOG view. As shown in Figure 9-13, the third column in the output identifies the sequence number assigned to each online redo log file. To determine which files were previously identified as stale, simply cross-reference the contents of the Group# column from each view. As displayed in the V$LOGFILE view, group #1 and group #2 are stale. In the V$LOG view, the sequence number for group #1 is 11 and group #2 is 12. The second column of the V$LOG view indicates that both redo log files belong to thread #1. The higher thread value only occurs if the Oracle9i server is configured to run concurrent processes (Oracle Shared Server). Now that Carlos has obtained the necessary data, he can recover the database using the sequence and thread numbers of the first corrupt redo log file as the terminating parameter for the recovery process.

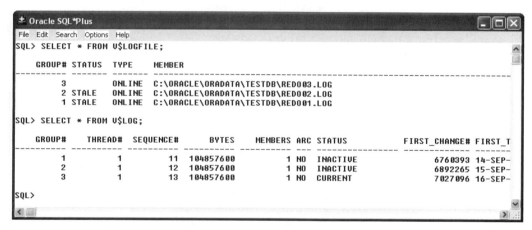

Figure 9-13   Determining sequence values for the online redo log files

Performing a Sequence-Based Incomplete Recovery—the Command-Line Approach

Before creating the problem with the online redo log files, and subsequently recovering the database, a backup of the database needs to be created using RMAN. If RMAN was not used to create a backup at the end of the last section, follow these steps to create the backup:

To create the backup:

1. Start RMAN and connect to the target database. Type **shutdown immediate**, and press **Enter**.

2. Type **startup mount**, and press **Enter**.

3. Type **run {**
   ```
   allocate channel c1 type disk;
   backup database;
   backup archivelog all;
   }
   ```
   and press **Enter**.

4. After the backup process is completed, type **alter database open;** and press **Enter**.

After opening the database, you need to choose one of the online redo log files to "corrupt." In reality, this section simply deletes the redo log file, and then uses an incomplete recovery to recover the database.

To delete and then recover the redo log file:

1. Reconnect to the database, if necessary, and at the SQL> prompt, type **SELECT * FROM V$LOG;** to display the current sequence numbers.

2. Choose one of the redo log files to serve as the basis for the recovery process and record the group, thread, and sequence numbers for that file. Do not select the current online redo log file.

3. Type **SELECT * FROM V$LOGFILE;** to display the name and location of each online redo log file.

4. Use the group number recorded in Step 2 to determine the name and location of the file that should be deleted.

5. Using Windows Explorer, move the online redo log file identified in Step 4 to another location to make the file inaccessible to the Oracle9i database.

Before attempting to recover the database, Carlos creates a cold backup of the database in the event a problem occurs during the recovery process. As in the previous section, this requires Carlos to shut down the database and use the operating system to create duplicates of the control file, the data files, and any valid archived redo log files. Because a cold backup was created at the beginning of the chapter, this step can be omitted. However, the following steps do need to be completed to recover the database:

To recover the database:

1. If necessary, start RMAN and connect to the target database. Then type **shutdown immediate** at the RMAN> prompt to shut down the database.

2. Type **startup mount** at the RMAN> prompt to mount the database.

3. Type this block of code to recover the database:

```
run {
allocate channel ch1 type disk;
set until logseq=l thread=t;
restore database;
recover database;
sql 'alter database open resetlogs';
}
```

where *l* is the sequence number of the deleted redo log file, and *t* is the thread number of the file, as shown in Figure 9-14.

Figure 9-14   Command to perform a sequence-based recovery in RMAN (partial output shown)

After the recovery process has been completed, the V$LOGFILE view can be queried to verify that all online redo log files are available. If the STATUS column is blank, then the files are now available. Because the database was opened with the RESETLOGS option, Carlos also verifies that the sequence number for the redo log files have been restarted by displaying the contents of the V$LOG view.

As shown in Figure 9-15, after the resetting the log files, no rows are displayed in the V$LOG_HISTORY view. This indicates that there currently is no information regarding valid archived redo log files recorded in the control file of the database. At this point, Carlos needs to shut down and back up the database in the event a problem occurs before the next scheduled backup is performed.

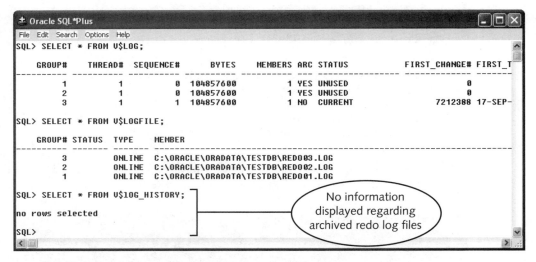

Figure 9-15   Content of views after database recovery

Performing a Sequence-Based Incomplete Recovery—a GUI Approach

The procedure for performing a sequence-based recovery using the Recovery Wizard involves the same steps as a time-based recovery. After creating a backup of the database, Carlos mounts the database and then accesses the Recovery Wizard through the Enterprise Manager Console. However, when prompted to make a range selection, as shown in Figure 9-16, he clicks the Sequence option button. After the Sequence option is selected, Carlos would then enter the appropriate log sequence number into the first text box, and then enter the thread number into the second text box.

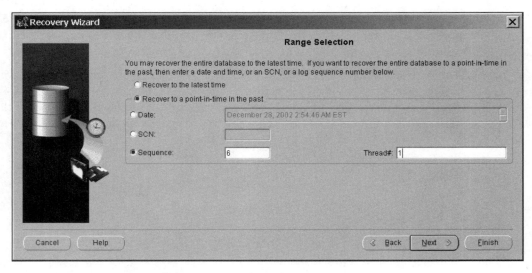

Figure 9-16   Parameters for a sequence-based recovery

The remaining steps of the recovery process are the same as with a time-based recovery. After the job has been submitted and executed, the database is recovered. As with any recovery operation, a backup is performed after the recovery is verified in the event the database needs to be recovered again before the next scheduled backup.

CHANGE-BASED INCOMPLETE RECOVERY

A change-based incomplete recovery uses the SCN of a specific transaction to terminate the recovery process. After the database is recovered, it contains all transactions with a lower SCN than the one specified. However, the transaction assigned to the stated SCN, and any subsequent transactions contained in the redo log files are not reapplied.

To use the change-based approach in RMAN, the command SET UNTIL SCN n is used to identify the terminating SCN value (n) for the recovery process. The other steps involved in the process are the same as when performing a time-based or sequence-based recovery. Therefore, the only other issue a DBA needs to be concerned about is determining the appropriate SCN for the recovery process.

In Chapter 5, Carlos used the LogMiner utility to determine the appropriate SCN used to terminate the change-based recovery. When connected to an Oracle Management Server, a GUI interface for the LogMiner utility is available. It is known as LogMiner Viewer. In the following sections, Carlos will use LogMiner to determine the SCN assigned to a problematic transaction. Because the command-line approach to using LogMiner was demonstrated in Chapter 5, only summary steps will be presented for that approach in this chapter. In the section demonstrating a change-based recovery with RMAN using the GUI approach, all steps for using the LogMiner Viewer will be presented.

 If you do not remember the exact procedure to execute the summarized steps in the following section, refer to Chapter 5.

Performing a Change-Based Incomplete Recovery—a Command-Line Approach

The scenario in the following sections is that a user has accidentally dropped a table named DEPOSITS. A script will be used to create this table, populate it with sample data, and then drop the table. The LogMiner utility will be used to determine the exact SCN assigned to the DROP TABLE command, which is a data definition language (DDL) command that was used to drop the DEPOSITS table.

In this scenario, Carlos knows that the LogMiner utility analyzes only data manipulation language (DML) commands by default. Thus, Carlos must remember to specify that DDL commands are to be included in the redo log analysis. After the SCN has been determined, Carlos uses RMAN to perform a change-based incomplete recovery. Before introducing the problematic transaction into the database, you must make certain a valid backup of the database exists by creating a copy of the database using RMAN.

To create and drop the DEPOSITS table:

1. At the SQL> prompt, type **start <*path*>\ch09script3.sql**, where <*path*> is the location of the script file in the Chapter09 folder. This command creates the DEPOSITS table.

2. Type **SELECT * FROM deposits;** to verify that the table has been created.

3. At the SQL> prompt, type **DROP TABLE deposits;** to drop the DEPOSITS table.

4. Type **SELECT * FROM deposits;** to verify that the table no longer exists.

Now that the table has been dropped, the next task is to determine the SCN that was assigned to the DROP TABLE command.

To determine the SCN:

1. Shut down the database and create a pfile from the database's current spfile.

2. Specify the destination for the extracted data dictionary information using the UTL_FILE_DIR parameter in the newly created pfile.

3. Re-create the database's spfile using the pfile modified in Step 2.

4. Start up the database.

5. Use the DBMS_LOGMNR_D.BUILD procedure to extract the data dictionary information. Remember the location must be the same location specified in the UTL_FILE_DIR parameter.

6. Use the DBMS_LOGMNR.ADD_LOGFILE procedure to identify the log files to be included in the analysis. This step must be repeated for each redo log file if you are not certain which file contains the transaction that dropped the DEPOSITS table.

7. Use the DBMS_LOGMNR.START_LOGMNR procedure to actually perform the analysis. Remember to include the DDL_DICT_TRACKING option because the DROP TABLE command is a DDL operation.

8. To retrieve the SCN assigned to the DROP TABLE command, query the V$LOGMNR_CONTENTS view. To narrow your search, you may want to restrict the rows returned to only those transactions that reference the segment named DEPOSITS (i.e., WHERE seg_name = 'DEPOSITS').

9. After the correct SCN has been determined, use the DBMS_LOGMNR.END_LOGMNR procedure to end the LogMiner session and release all resources currently allocated to the LogMiner utility.

Now that the SCN is known, Carlos can perform a change-based recovery through RMAN.

To create a cold backup, and then recover the database:

1. Shut down the database and create a backup of the mirrored copies of the control file, the data files, and any archived redo log files.

2. Using Windows Explorer, move the data file containing the DEPOSIT table to a new location. Unless a default tablespace has been assigned, the table is stored in the SYSTEM tablespace, with System01.dbf as the associated tablespace.

3. Use RMAN to connect to the target database.

4. Through RMAN, shut down and mount the database.

5. To recover the database, type

```
run {
allocate channel c1 type disk;
set until SCN n;
restore database;
recover database;
sql 'alter database open resetlogs';
}
```

where *n* is the SCN previously obtained through the LogMiner utility, as shown in Figure 9-17.

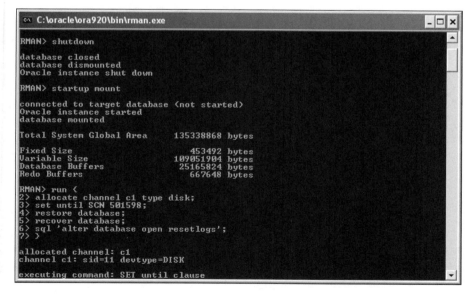

Figure 9-17   Performing a change-based recovery in RMAN (partial output shown)

After the database has been recovered, the contents of the DEPOSITS table can be viewed through SQL*Plus. After determining the successful recovery of the database, a new backup of the database should be created in the event any subsequent problems arise.

Performing a Change-Based Incomplete Recovery—the GUI Approach

In the following section, the LogMiner Viewer is used to determine the SCN assigned to the transaction that is used to drop the DEPOSITS table. After the appropriate SCN is obtained, the Recovery Wizard can then be used to recover the database to the state immediately prior to when the table was dropped.

To determine the SCN:

 If you did not complete the previous section and the DEPOSITS table does not exist, execute the Ch09script3.sql script file from the Chapter09 folder.

1. Using SQL*Plus or the SQL*Plus Worksheet, type **DROP TABLE deposits;** to drop the DEPOSITS table.

2. If necessary, start the Enterprise Manager Console (EMC), and connect to the Oracle Management Server using the connection information provided by your instructor.

3. Click the name of the database to highlight it. Click **Tools** on the menu bar, point to **Database Applications**, and then click **LogMiner Viewer**.

4. When the dialog box appears indicating that there are no saved queries, click **OK**.

5. Click **Object** on the LogMiner Viewer menu bar, and then click **Create Query**.

6. Click the **Redo Log Files** tab located at the top of the Create Query window.

7. In the middle-right portion of the window, click the **Set Dictionary** button, as shown in Figure 9-18.

8. In the Set Dictionary window, click the **Use an existing dictionary file** option button as shown in Figure 9-19. In the corresponding text box, enter the name destination file for the data dictionary. Remember that the location of the file must match the location specified by the UTL_FILE_DIR parameter of the database.

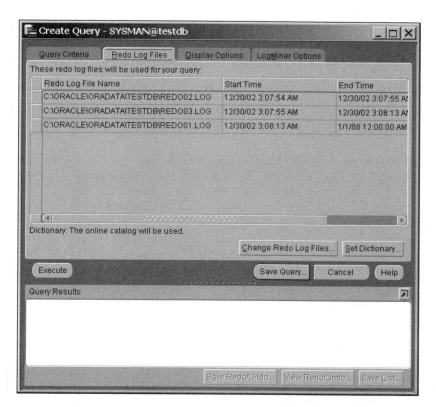

Figure 9-18   Redo Log Files sheet in the Create Query window

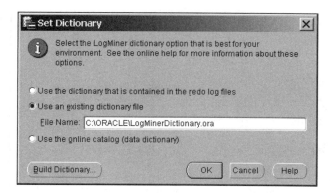

Figure 9-19   Set Dictionary window

9. Click the **Build Dictionary** button in the lower-left corner of the Set
Dictionary window. When the Build Dictionary window opens, click the
Create a separate dictionary file option button, if necessary, and then
click **OK**.

10. When the dialog box confirming whether you want to build the LogMiner dictionary appears, click **Yes** to extract the data dictionary.

11. After the activity dialog box closes, click **OK** from the Set Dictionary window.

12. From the Create Query window, click the **LogMiner Options** tab.

13. When the window shown in Figure 9-20 appears, click the last check box to include DDL changes in the log analysis.

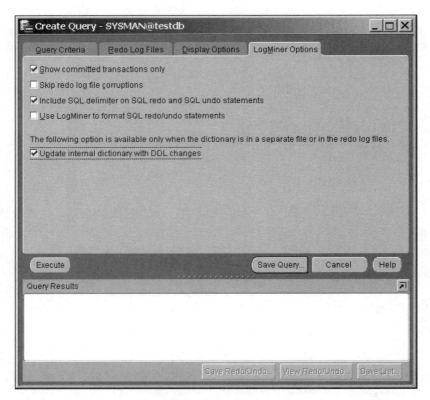

Figure 9-20   LogMiner options

14. Click the **Query Criteria** tab in the upper-left corner of the Create Query window.

15. If necessary, click the **Graphical Filter** option button beneath the tab.

16. In the upper pane of the Query Criteria sheet, click the drop-down arrow in the first portion of the criteria bar, and click **Operation**. In the text box located at the end of the criteria bar, type **DROP**. Your screen should be similar to the one shown in Figure 9-21.

Figure 9-21 Criteria for LogMiner analysis

 In situations where the exact command used to cause an error is unknown, the criteria can be based on other factors provided in the drop-down list, such as user name, table name, and so on.

17. In the middle portion of the Query Criteria, click the **Time Range** option button, if necessary. In the appropriate text boxes, enter the desired start and end time for the analysis. This should be a time frame that contains the command previously used to drop the DEPOSITS table.

18. Click the **Execute** button to execute the query.

19. Locate the assigned SCN for the DROP TABLE command in the query results that are displayed in the bottom pane of the window, and record the value for use during the recovery process.

20. Click the **Close** button in the upper-right corner of the window to close the Create Query window.

Now that Carlos has determined the appropriate SCN, he can recover the database to a state immediately prior to when the DEPOSITS table was dropped. The steps are the same as when performing a time-based or cancel-based recovery, except that an SCN is entered for the terminating parameter.

To recover the database:

1. Use the Backup Wizard from the Enterprise Manager Console to create a copy of the entire database.

2. Mount the database.

3. Using Windows Explorer, move the data file associated with the tablespace that stores user data to a new location. In a default configuration, the user data is stored in the SYSTEM tablespace and its associated data file is named System01.dbf.

4. Access the Recovery Wizard on the Tools menu bar on the Enterprise Manager Console.

5. If the Introduction window appears, click **Next**.

6. When the Operation Selection window appears, click the **Restore and recover** option button, if necessary, and then click **Next**.

7. When the Object Selection window appears, click the **Entire database** option button, if necessary. Then click **Next**.

8. When the Range Selection window appears, click the **Recover to a point-in-time in the past** option button.

9. When the recovery options become available, click the **SCN** option button.

10. Enter the value of the SCN previously obtained through the LogMiner Viewer when the text box becomes available, as shown in Figure 9-22.

11. Click **Next** to proceed to the next window.

12. When the Rename window appears, click **Next**.

13. When the Configuration window appears, click **Finish**.

14. After the Summary window appears, click **OK** to submit the job for execution.

After the job has been executed, the database is recovered and the DEPOSITS table is contained in the database. A SELECT statement can be issued to verify that the DEPOSITS table exists. Afterwards, the database should be shut down, and a new backup created. The database can then be opened and made available to the users.

After completing the material in this chapter, remember to restore the original database from the cold backup created at the beginning of this chapter.

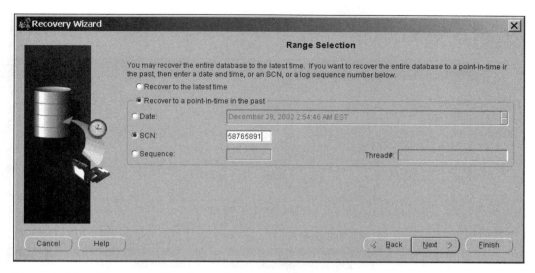

Figure 9-22   Range selection for a change-based recovery

CHAPTER SUMMARY

❏ Recovery Manager (RMAN) supports three types of incomplete recoveries: time-based, sequence-based, and change-based.

❏ A time-based recovery requires the DBA to specify a time value as the terminating parameter for the recovery process.

❏ The Recovery Wizard is used to perform an incomplete recovery through the Enterprise Manager Console.

❏ When a database is recovered as the result of an incomplete recovery, it must be opened with the RESETLOGS option.

❏ A sequence-based recovery uses a log sequence number as the terminating parameter. The recovery concept is similar to a user-managed, cancel-based recovery.

❏ The V$LOG view can be used to determine the sequence number assigned to an online redo log file.

❏ A change-based recovery requires knowledge of the SCN to be used as the terminating parameter for the recovery process.

❏ The SCN assigned to a particular transaction can be determined using LogMiner.

❏ The LogMiner Viewer is the graphical user interface (GUI) for the LogMiner utility.

SYNTAX GUIDE

RMAN Command	Description	Example
SET UNTIL LOGSEQ=*l* THREAD=*t*	Used to specify the log sequence and thread number for a sequence-based incomplete recovery	`set until logseq=10 thread=1;`
SET UNTIL TIME '*DD MON YYYY HH:MI:SS*'	Used to specify the time parameter for a time-based incomplete recovery; the date and time format depend on the value assigned to the NLS_DATE_FORMAT parameter	`set until time '07 SEP 03 10:23:00';`
SET UNTIL SCN *n*	Used to specify the terminating SCN for a change-based incomplete recovery	`set until SCN 59812531;`

Parameter	Description	Example
NLS_DATE_FORMAT	Specifies the format of date and time values recognized internally by Oracle9*i*	`nls_date_format = "DD MON YYYY HH:MI:SS"`

View	Description	Example
V$LOG	Displays information contained in the control file regarding the redo log files, including the sequence number and archiving status	`SELECT * FROM V$LOG;`
V$LOGFILE	Displays the current status and location of the online redo log files	`SELECT * FROM V$LOGFILE;`

REVIEW QUESTIONS

1. Of the three types of incomplete recoveries that can be performed through RMAN, a sequence-based recovery normally results in the fewest number of transactions being reapplied. True or False?

2. An incomplete recovery can be performed through the Enterprise Manager Console using the _____ Wizard.

3. When performing an incomplete recovery, which of the following steps is completed last?

   a. Shut down the database.

   b. Open the database, and reset the redo log files.

   c. Recover the database.

   d. Restore the database files.

4. Which of the following types of incomplete recovery uses an SCN as the terminating parameter?

   a. sequence-based

   b. change-based

   c. cancel-based

   d. time-based

5. The _____ parameter is used to specify the destination location for extracted data dictionary information when using the LogMiner Viewer.

6. Provide two reasons why a DBA would perform an incomplete rather than a complete recovery of a database.

7. To perform a sequence-based incomplete recovery in RMAN, the SET UNTIL SCN command is used to specify the terminating parameter for the recovery process. True or False?

8. Which of the following tasks is required after an incomplete recovery, in order to open the database?

   a. The database must be backed up.

   b. The RMAN utility must be exited.

   c. All archived redo log files must be renamed.

   d. The redo log files must be reset.

9. Due to the granularity of the terminating parameter, which of the following types of incomplete recoveries normally result in the least number of lost transactions?

   a. sequence-based

   b. change-based

   c. cancel-based

   d. time-based

10. When a log sequence number is used to specify the stopping point of an incomplete recovery, only the valid transactions within that log file are updated to the database. True or False?

11. The _____ view can be used to determine the sequence number assigned to a specific online redo log file.

12. When analyzing the contents of the redo log files, which type of SQL statements are included in the analysis by default?

13. The _____ procedure is used to specify the redo log files to be included in the LogMiner analysis.

9

14. Which of the following parameters determines the format of the date and time specified for a time-based recovery?

    a. TO_CHAR

    b. NLS_DATE_FORMAT

    c. NLS_DATE_MODEL

    d. DATE_FORMAT

15. Only the current online redo log file can be analyzed through the LogMiner utility. True or False?

16. An incomplete recovery is only performed if a data file becomes unavailable. True or False?

17. To perform an incomplete recovery, the database should be in a(n) _____ state.

    a. closed

    b. dismounted

    c. opened

    d. mounted

18. Using the _____ option when opening the database after a recovery results in restarting the log sequence numbers of the redo files.

19. When analyzing the contents of the redo log files, which of the following steps is completed first?

    a. Identify the archived redo log files to be analyzed.

    b. Identify the online redo log files to be analyzed.

    c. Identify the types of SQL statements to be analyzed.

    d. Identify the location for the data dictionary information.

20. An incomplete recovery can only be performed on a database that is in _____ mode.

HANDS-ON ASSIGNMENTS

As you are completing these assignments, remember that an incomplete recovery can be performed only for a database that is in ARCHIVELOG mode.

Assignment 9-1 Performing a Time-Based Recovery Through RMAN

In this assignment, you perform a time-based recovery and observe the effect of resetting the log files after completing the recovery.

1. Use RMAN to create a backup of the database, and then open the database.

2. Type the following command to create a new database table:

```
CREATE TABLE tablea (ColA NUMBER(5), ColB DATE);
```

3. Type the command **SELECT SYSDATE FROM DUAL;** and note the current time according to the computer.

4. Type the command **ALTER SYSTEM SWITCH LOGFILE;** to cause a log switch to occur.

5. Type the command **SELECT * FROM V$LOG;** and note the current sequence number assigned to each log file.

6. Type **DROP TABLE tablea;** to drop the table previously created in Step 2.

7. Shut down the database.

8. Use the operating system to move or rename the data file that contains your user data to a new location. In a default configuration, this is the System01.dbf data file.

9. Use RMAN to perform a time-based recovery based on the time previously noted in Step 3.

10. After recovering and opening the database, type **DESCRIBE tablea** to determine whether the table was recovered.

11. View the current contents of the V$LOG view, and compare the current sequence numbers to the sequence numbers displayed previously in Step 5. Explain any difference observed in the sequence numbers.

12. Create a new backup of the database using RMAN, and then open the database.

Assignment 9-2 Identifying the Current Online Redo Log File

In this assignment, you determine the sequence and thread number, and location, of the current online redo log file.

1. Type **SELECT * FROM V$LOG;** and determine the sequence and thread numbers of the redo log with the value CURRENT in the Status column of the view.

2. Type **SELECT * FROM V$LOGFILE;** to determine the name and location for each online redo log file.

3. Type the command **ALTER SYSTEM SWITCH LOGFILE;** to cause a log switch to occur.

4. Type **SELECT * FROM V$LOG;** and determine the sequence and thread numbers of the redo log with the value CURRENT in the Status column of the view. Has the current redo log file changed? Why or why not?

Assignment 9-3 Performing a Sequence-Based Recovery After a User Error Occurs

In this assignment, you perform a sequence-based recovery after an error is created by a user.

1. Create a backup of the database using RMAN (this step can be omitted if Step 12 of Assignment 9-1 was completed), and open the database.

2. Type **CREATE TABLE tableb (ColA NUMBER(5), ColB DATE);** to create a new table in the database.

3. Type the command **ALTER SYSTEM SWITCH LOGFILE;** to cause a log switch to occur.

4. Determine the sequence and thread numbers of the current online redo log file. Record this information for use later in this assignment.

5. Type **CREATE TABLE tablec (ColA NUMBER(5), ColB DATE);** to create a new table in the database.

6. Shut down the database.

7. Use the operating system to move the data file that contains your user data to a new location. In a default configuration, this is the System01.dbf data file.

8. Use RMAN to perform a sequence-based recovery using the sequence and thread numbers identified in Step 4.

9. Open the database and determine whether tables TABLEB and TABLEC exist. Explain your results.

10. Create a new backup of the database through RMAN.

Assignment 9-4 Performing a Sequence-Based Recovery After a Redo Log File Is Lost

In this assignment, you perform a sequence-based recovery after the loss of a redo log file.

1. Create a backup of the database using RMAN (this step can be omitted if Step 10 of Assignment 9-3 was completed), and open the database.

2. Type **CREATE TABLE tabled (ColA NUMBER(5), ColB DATE);** to create a new table in the database.

3. Determine the sequence and thread numbers of the current online redo log file.

4. Type the command **ALTER SYSTEM SWITCH LOGFILE;** to cause a log switch to occur.

5. Determine the sequence and thread numbers of the current online redo log file. Record this information for use later in this assignment.

6. Type **CREATE TABLE tablee (ColA NUMBER(5), ColB DATE);** to create a new table in the database.

7. Shut down the database.

8. Use the operating system to move one of the redo log files to a new location.

9. Use RMAN to perform a sequence-based recovery using the sequence and thread numbers identified in Step 5.

10. Open the database and determine whether tables TABLED and TABLEE exist. Explain your results.

11. Create a new backup of the database through RMAN.

Assignment 9-5 Using LogMiner to Determine the Time for DML Operations

In this assignment, you use LogMiner to determine the time of an INSERT statement. If you do not have access to LogMiner Viewer, this assignment can be completed using LogMiner through SQL*Plus.

1. Verify that a location has been assigned to the UTL_FILE_DIR parameter using either the SHOW PARAMETER command in SQL*Plus or by viewing the initialization parameters through the Enterprise Manager Console. If the parameter does not exist, add the UTL_FILE_DIR parameter to the Init.ora file to assign the destination location for the extracted data dictionary information, and open the database.

2. Type **CREATE TABLE tablef (ColA NUMBER(5), ColB DATE);** to create a new table in the database.

3. Type **INSERT INTO tablef VALUES(5, to_date('01-DEC-04', 'dd-mon-yy'));** to add a row to the table.

4. Type **COMMIT;**

5. Extract the current information contained in the data dictionary (that is, build a dictionary).

6. If you are completing this assignment using the LogMiner utility through SQL*Plus, add the online redo log files to the analysis list.

7. Instruct LogMiner to analyze the files on the analysis list.

8. Query the results of the analysis, and determine the time that the transaction resulting from the INSERT command in Step 3 was committed.

9. Terminate the LogMiner session unless you are required to complete Assignment 9-6.

Assignment 9-6 Using LogMiner to Determine the Time for DDL Operations

In this assignment, you use LogMiner to determine the time that the CREATE TABLE command is used to create a table. Complete Steps 1 through 4 only if the LogMiner session was terminated at the end of Assignment 9-5.

1. Verify that a location has been assigned to the UTL_FILE_DIR parameter using either the SHOW PARAMETER command in SQL*Plus or by viewing the initialization parameters through the Enterprise Manager Console. If the parameter does not exist, add the UTL_FILE_DIR parameter to the Init.ora file to assign the destination location for the extracted data dictionary information, and open the database.

2. Type the command **DESCRIBE tablef** to verify the table named TABLEF exists. If it does not exist in your database, type **CREATE TABLE tablef (ColA NUMBER(5), ColB DATE);** to create the table in your database.

3. Extract the current information contained in the data dictionary (that is, build a dictionary).

4. If you are completing this assignment using the LogMiner utility through SQL*Plus, add the online redo log files to the analysis list.

5. Instruct LogMiner to analyze the files on the analysis list. If completing this assignment using the LogMiner utility through SQL*Plus, use the appropriate option to include DDL commands in the analysis. If completing this assignment using LogMiner Viewer through the Enterprise Manager Console, select the appropriate option from the LogMiner Options tab to include DDL options in the analysis.

6. Query the results of the analysis and determine the time that the table named TABLEF was created. Record this time for use in Assignment 9-7.

7. Terminate the LogMiner session.

Assignment 9-7 Performing an Incomplete Recovery Based on a Time Value Obtained Through LogMiner

In this assignment, you perform a time-based recovery using the value obtained in Assignment 9-6.

1. Create a backup of the database using RMAN.

2. Use the operating system to move the data file that contains your user data to a new location. In a default configuration, this is the System01.dbf data file.

3. Perform a time-based recovery using the exact time obtained in Assignment 9-6 as the terminating parameter.

4. After the database has been recovered, determine whether the table named TABLEF exists. Explain why the table does, or does not, exist after recovery.

5. Create a new backup of the database using RMAN.

Assignment 9-8 Performing an Incomplete Recovery Based on an SCN Obtained Through LogMiner

In this assignment, first you use LogMiner to determine the SCN associated with a particular transaction, and then you use that SCN as the terminating parameter for an incomplete recovery.

1. If you did not create a backup of the database in Step 5 of Assignment 9-7, create a backup of the database using RMAN.

2. Open the database.

3. Type **CREATE TABLE tableg (colA NUMBER, colB DATE);** to create a new table in the database.

4. Type **DROP TABLE tableg;** to drop the newly created table from the database.

5. Use LogMiner to determine the SCN associated with the transaction that dropped the TABLEG table. Record this value for subsequent use.

6. End the LogMiner session.

7. Use the operating system to move the data file that contains your user data to a new location. In a default configuration, this is the System01.dbf data file.

8. Use RMAN to perform a change-based recovery using the identified SCN.

9. Verify that the TABLEG table exists after the database has been recovered.

10. Create a cold backup of the database through the operating system.

CASE PROJECTS

Case 9-1 Determining a Recovery Strategy

On Thursday afternoon, Carlos receives a call from the director of accounting. One of the employees in the Accounting Department had attempted to add data regarding new equipment into the physical assets database. However, when the user attempted to add the data, the EQUIPMENT table was not available. This table is used to store the basic information regarding equipment purchases made by Janice Credit Union. Carlos logs into the database and issues the command SELECT * FROM tab; and realizes the table is no longer contained in the database. Because the database is not accessed on a regular basis, the table could have been dropped that day or earlier in the week. Therefore, Carlos has no time frame to reference to determine when the table was dropped. However, the database is included in the regular backups that are performed each night.

Based on this information, what is the best course of action for Carlos to take to recover the EQUIPMENT table?

Case 9-2 Creating an Incomplete Recovery Procedure List

Continuing with the procedure manual created in Chapter 2, create a procedure list that can be used by the IT employees at Janice Credit Union to perform an incomplete recovery through RMAN. Make certain to include the necessary steps to:

❑ Obtain time and SCN information through LogMiner

❑ Perform a time-based recovery

❑ Perform a sequence-based recovery

❑ Perform a change-based recovery

10

RECOVERY MANAGER MAINTENANCE

After completing this chapter, you should be able to do the following:

♦ Perform a cross-check of the backup sets with the physical files
♦ Delete a backup set reference from the RMAN repository
♦ Change the status of a backup set
♦ Add operating system backups to the repository
♦ Change retention policy settings

The Recovery Manager (RMAN) repository stores the metadata regarding the target database and backup and recovery operations. Normally, the information regarding backup and recovery operations pertains to those operations performed by RMAN. However, copies of database files made by the operating system without using RMAN can also be recorded in the repository. This information is referenced by RMAN during recovery operations to determine the most appropriate backup copy to use when recovering the database.

Note that there is a downside to the wealth of information recorded by RMAN; even when a backup set is no longer available, data regarding the backup set is still contained in the repository. However, DBAs such as Carlos can use the CROSSCHECK command to identify which backup sets are currently referenced in the repository, and the status of those backup sets. In addition, he can also use the CHANGE command to change the status of a specific backup set.

In this chapter, Carlos will perform various maintenance operations on the repository. For example, he will update the repository when a backup set has been deleted, and update the repository with copies of data files that he has created through the operating system. In addition, Carlos will begin defining a retention policy for the backup sets. A retention policy can be used to specify how many backup copies of a data file must exist and defines whether a backup copy is valid.

The Current Challenge in the Janice Credit Union Database

Previously, Carlos learned how to backup and recover the database in a variety of situations. He has learned how RMAN can simplify the administrative overhead when compared to user-managed processes. He has also learned that Recovery Manager still requires the intervention of the DBA in terms of the information contained in the repository. Rather than simply tracking copies of the database files generated by the operating system, the administrative overhead is shifted to maintaining the repository.

In this chapter, Carlos is examining the maintenance requirements to determine whether there are actually any significant benefits derived from using RMAN-managed processes at Janice Credit Union. After examining the maintenance tasks he will need to perform, Carlos will then examine the options available for creating a retention policy for the backup sets created by RMAN.

Set Up Your Computer for the Chapter

Although recovery operations will not be performed in this chapter, you should create a cold backup of all data files, all copies of the control file, and any initialization files. In the event an unrecoverable error occurs while completing the examples shown in the chapter, you can use this backup to restore and recover the database.

Notes About Dual Coverage Within this Chapter

The first part of this chapter covers various maintenance operations for the RMAN repository. These tasks will be presented only using the command-line interface for the Recovery Manager utility. The latter portion of the chapter addresses changes to the retention policy, which is used to determine when a backup set is obsolete. Retention policy changes will be demonstrated using both the command-line interface and the Maintenance Wizard, which is the GUI interface for making retention policy changes through the Enterprise Manager Console. To access the Maintenance Wizard, you will be required to connect to an Oracle Management Server (OMS).

Maintaining Backup Set and Image Copy References

In the following sections, Carlos will verify that any backup sets or image copies currently referenced in the RMAN repository actually exist. If a reference is not valid, Carlos will perform the necessary maintenance to remove it from the repository. In addition, he will change the current status of a reference to prevent RMAN from incorrectly identifying it as not being valid.

Cross-Checking Backup Sets and Image Copies

Over a period of time, it is possible that the physical files generated by RMAN could accidentally be deleted. Fortunately, in such a case, RMAN still has information regarding those files in the repository, even though the files are not available for recovery purposes. Carlos can use the CROSSCHECK command to determine whether the physical files contained in the backup sets and image copies referenced by RMAN are actually available.

In the following steps, Carlos will use the CROSSCHECK command to verify the availability of the backup sets currently referenced in the repository. Carlos wants to perform this activity on a regular basis to verify that the necessary files can be accessed if database recovery ever becomes necessary.

To verify the availability of the backup sets:

1. Start the RMAN utility, and connect to the target database with a SYSDBA privileged account.

2. Type **crosscheck backup of database;**. Your results should be similar to those shown in Figure 10-1.

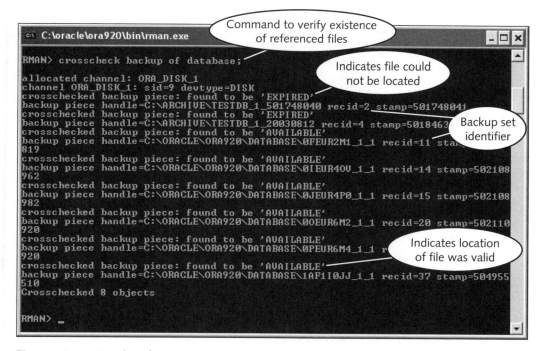

Figure 10-1 Results of CROSSCHECK BACKUP command

3. Type **crosscheck copy of database;** to check the status of all image copies that are currently referenced in the repository. RMAN displays the status of any image copies, similar to the output shown in Figure 10-2.

```
C:\oracle\ora920\bin\rman.exe                                    _ □ ×
RMAN> crosscheck copy of database;
released channel: ORA_DISK_1
allocated channel: ORA_DISK_1
channel ORA_DISK_1: sid=9 devtype=DISK
validation failed for datafile copy
datafile copy filename=C:\ARCHIVE\TOOLSIMG.DBF recid=1 stamp=501848624
Crosschecked 1 objects
```

> Indicates that the physical file could not be located

Figure 10-2   Results of CROSSCHECK COPY command

If an image copy of a data file does not already exist, you can create an image copy using the steps presented in Chapter 7.

Notice in Figure 10-1 that the first two backup pieces checked by RMAN are marked as 'EXPIRED'. This means that the physical files cannot be found in the referenced location, and the information can be removed from the repository. However, the remaining six backup pieces were successfully located by RMAN and are marked as 'AVAILABLE'. Although not shown in this example, a backup piece can also be marked as 'UNAVAILABLE' if the file was found in the expected location, but is not currently available for use (i.e., is being copied, etc.).

In Figure 10-2, the results indicate that only the image copy currently referenced in the RMAN repository could not be validated. This means that RMAN cannot use the image copy during a database recovery. Therefore, Carlos should remove this reference from the repository. If the physical copy of the file had been found in the specified location, the results would have been returned as "validation succeeded."

Deleting Repository References

Carlos can use the CHANGE command to delete references to backup sets and image copies from the repository. This is necessary whenever the results of the CROSSCHECK command indicate that a backup set is expired or if an image copy cannot be validated. However, Carlos must specify the key, or recid, if files are identified using output from the CROSSCHECK command, to specify which references should be deleted. Referencing the output previously shown in Figure 10-1, Carlos will delete the second backup piece that was marked as 'EXPIRED' from the repository.

To delete a reference to a backup set:

1. Type **change backupset** *i* **delete;**, where *i* is the key or recid of the backup piece.

2. When prompted, type **YES**, as shown in Figure 10-3.

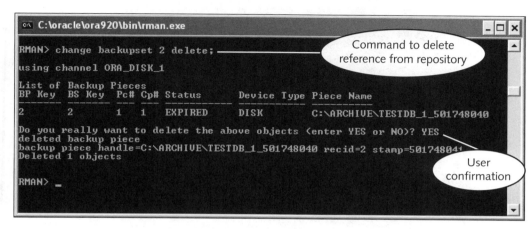

Figure 10-3   Deletion of backup set reference from repository

As shown in Figure 10-3, the name of the piece contained in the specified backup set is displayed. After Carlos confirms that the correct object is to be deleted, the reference to the backup set is removed from the repository. Carlos can also use the CHANGE command with the DATAFILECOPY keyword and the location of the data file. To delete the image copy identified in Figure 10-3, Carlos could have used **CHANGE DATAFILECOPY 'c:\archive\testdb_1_501748040' DELETE;** instead.

Changing the Status of Backup Sets and Image Copies

In addition to removing current references to a specific backup set or image copy from the RMAN repository, the CHANGE command can also be used to change the status of these files to 'AVAILABLE' or 'UNAVAILABLE'. For example, suppose Carlos needs to remove a drive containing one of the backup sets. Because the drive will be reinstalled in a few days, he does not want to delete any references to the backup set. However, he also does not want RMAN to attempt to use that backup set in the event a database recovery becomes necessary. As an alternative, Carlos can change the status of the backup set to 'UNAVAILABLE'. After the hard drive is reinstalled, the status can be changed back to 'AVAILABLE'.

In the following section, Carlos will use the LIST command to identify the backup sets currently being referenced by the RMAN repository. After locating the appropriate backup set, he will then change its status to 'UNAVAILABLE' using the CHANGE command.

To change the status of a backup set to UNAVAILABLE:

    1. Type **list backup of database;** to display the backup sets currently referenced by the RMAN repository, as shown in Figure 10-4.

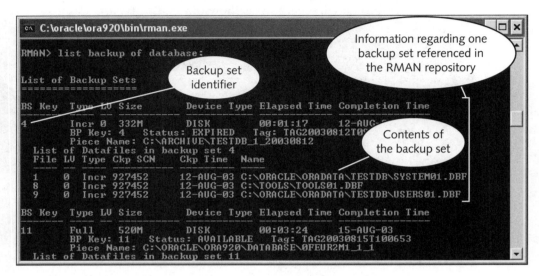

Figure 10-4 List of database backup sets (partial output shown)

    2. Type **change backupset *i* unavailable;**, where *i* is the BP (backup piece) Key of the available backup set identified in the output of Step 1, as shown in Figure 10-5.

Figure 10-5 Changing status of backup set to UNAVAILABLE

    3. Type **list backup of database;** to verify that the status of the backup set has been changed. The output should resemble Figure 10-6.

```
C:\oracle\ora920\bin\rman.exe                                    _ □ ×
     9        Full 1518749      15-AUG-03 C:\ORACLE\ORADATA\TESTDB\USERS01.DBF
    10        Full 1518749      15-AUG-03 C:\ORACLE\ORADATA\TESTDB\XDB01.DBF
    11        Full 1518749      15-AUG-03 C:\ORACLE\ORADATA\TESTDB\OEM_REPOSITORY.DBF

BS Key   Type LU Size          Device Type Elapsed Time Completion Time
---------------------------------------------------------------------
21       Full    343M          DISK        00:01:54      15-AUG-03
         BP Key: 21    Status: UNAVAILABLE   Tag: TAG20030815T111512
         Piece Name: C:\ORACLE\ORA920\DATABASE\0PEUR6M4_1_1
  List of Datafiles in backup set 21
  File LU Type Ckp SCN     Ckp Time  Name
  ---- -- ---- -------     --------  ----
     1       Full 1518754      15-AUG-03 C:\ORACLE\ORADATA\TESTDB\SYSTEM01.DBF
     3       Full 1518754      15-AUG-03 C:\ORACLE\ORADATA\TESTDB\CWMLITE01.DBF
     4       Full 1518754      15-AUG-03 C:\ORACLE\ORADATA\TESTDB\DRSYS01.DBF
     6       Full 1518754      15-AUG-03 C:\ORACLE\ORADATA\TESTDB\INDX01.DBF
     8       Full 1518754      15-AUG-03 C:\TOOLS\TOOLS01.DBF

BS Key   Type LU Size          Device Type Elapsed Time Completion Time
---------------------------------------------------------------------
37       Full    521M          DISK        00:02:48      17-SEP-03
         BP Key: 37    Status: AVAILABLE   Tag: TAG20030917T092504
         Piece Name: C:\ORACLE\ORA920\DATABASE\1AF1I0JJ_1_1
  List of Datafiles in backup set 37
  File LU Type Ckp SCN     Ckp Time  Name
```

Current status of the backup set

Figure 10-6 Current status of the backup set (partial output shown)

Now that the status of the backup set has been changed to UNAVAILABLE, Carlos can remove the hard drive without RMAN subsequently changing the status of the backup set to EXPIRED if the physical file cannot be located. How does this prevent the change? When the CROSSCHECK command is executed, any references previously marked as UNAVAILABLE are ignored; therefore, the reference to the backup set is not verified. After the hard drive is reinstalled, the status of the backup set can be changed back to AVAILABLE using the following step sequence:

To change the status of a backup set to AVAILABLE:

1. Type **change backupset *i* available;**, where *i* is the BP Key of the backup set previously changed, as shown in Figure 10-7.

```
C:\oracle\ora920\bin\rman.exe                                    _ □ ×
RMAN> change backupset 21 available;

using channel ORA_DISK_1
changed backup piece available
backup piece handle=C:\ORACLE\ORA920\DATABASE\0PEUR6M4_1_1 recid=21 stamp=502110
920
Changed 1 objects to AVAILABLE status

RMAN> _
```

Figure 10-7   Changing the status of the backup set to AVAILABLE

2. Type **list backup of database;** to verify that the status of the backup set has been changed to AVAILABLE.

10

Carlos can use the same approach to change the status of an image copy of a data file. The command to change the status of an image copy is **CHANGE DATAFILECOPY** **'location\filename' <status>;**, where *location* and *filename* are the current location and filename of the image copy, respectively, and *<status>* is either AVAILABLE or UNAVAILABLE.

ADDING OPERATING SYSTEM BACKUPS TO THE REPOSITORY

Previously, Carlos had been making backup copies of the database files through the operating system. RMAN provides the option of including operating system backups, which are the same as image copies, to the repository. Carlos would like to add these previous operating system copies to the repository so that all information regarding existing backups are contained in RMAN. Doing so makes it easier to track all copies of the database files by storing the information in a centralized location, which is the RMAN repository.

In the following step sequence, the operating system will be used to create a copy of a data file. The CATALOG command will then be used to add information about the copy to the RMAN repository.

To create a copy of a data file, and then update the repository:

1. Type **shutdown;** to shut down the database in RMAN.

2. Using the operating system, create a copy of the Users01.dbf data file.

3. Type **startup;** to start up the target database.

4. Type **catalog datafilecopy 'location\filename';**, as shown in Figure 10-8, where *location* is the location, and *filename* is the name of the file copy created in Step 2.

Figure 10-8   Adding a file copied by the operating system to the RMAN repository

5. Type **list copy of database;** to display the current image copies being referenced by the RMAN repository. As shown in Figure 10-9, this output includes the file specified in Step 4.

Note that the current status of the copy created by the operating system is AVAILABLE, as indicated by the "A" in the third column of the display which is labeled "S" for Status. The first image copy shown in Figure 10-9 contains an "X" in that column. This indicates that the image copy is EXPIRED and cannot be used in any recovery operations performed by RMAN.

```
C:\oracle\ora920\bin\rman.exe                              _ □ ×

RMAN> list copy of database;

List of Datafile Copies
Key    File S Completion Time  Ckp SCN    Ckp Time      Name
1         8  X 12-AUG-03       933859     12-AUG-03     C:\ARCHIVE\TOOLSIMG.DB
F
29        9  A 20-OCT-03       12031547   18-OCT-03     C:\BACKUPAREA\USERS01.
DBF

RMAN>
```

Figure 10-9   Image copies currently referenced by the RMAN repository

Backup copies of the archived redo log files and the control file can also be added to the repository by substituting the keywords ARCHIVELOG and CONTROLFILECOPY, respectively, for the DATAFILECOPY keyword.

DEFINING A RETENTION POLICY

Now that Carlos is familiar with the maintenance requirements of the RMAN repository, he next begins to formulate a retention policy. A retention policy identifies when Recovery Manager should consider a backup file to be obsolete and no longer required for recovery operations. Using a retention policy to its full advantage saves Carlos the administrative headache of having to determine which references should be removed from the repository every time he is performing maintenance activities.

The RETENTION POLICY parameter has different options available. These options are listed in Table 10-1.

Table 10-1   Options available with the RETENTION POLICY parameter

RETENTION POLICY Parameter Options	Description
REDUNDANCY *i*	Used to specify how many image copies or backup copies of the database files RMAN should keep
RECOVERY WINDOW OF *i* DAYS	Used to specify that the files necessary to recover the database at a stated time within the specified window cannot be considered obsolete
NONE	Used to disable the retention policy, preventing RMAN from considering any backup set or image copy as obsolete

As an example of a retention policy, suppose Carlos wants to make certain that if the database needs to be recovered to a specific point within the last week, the necessary files will be available. To indicate this policy, he simply uses **CONFIGURE RETENTION POLICY TO RECOVERY WINDOW OF 7 DAYS**. This statement is basically interpreted to mean that the database should be recoverable to any state that has occurred within the last seven days.

Therefore, any backup sets or image copies that would be required to perform the recovery cannot be marked as obsolete. This normally consists of all backup sets or image copies made within the last seven days, plus the last one made immediately before the specified time frame. Why? If the database needs to be recovered to a state it was in seven days ago, then backup files created prior the recovery point have to be restored.

In another example, suppose Carlos wants to make certain that no backup set or image copy is ever considered to be obsolete. Carlos might want to do this as a precaution until he becomes more familiar with using RMAN. In that case, he simply uses CONFIGURE RETENTION POLICY TO NONE.

Changing RMAN Retention Policy: A Command-Line Approach

For illustrative purposes, assume Carlos wants to make certain that five copies of each data file and the control file are always available through RMAN. Carlos would use RETENTION POLICY TO REDUNDANCY 5 to specify that only copies older than the five most recent backup sets or image copies can be considered obsolete. To change the current retention policy, the CONFIGURE command is executed in the RMAN utility. In the following steps, Carlos will implement this change.

To specify a retention policy:

1. Type **show all;** to display the current RMAN parameter settings, as shown in Figure 10-10.

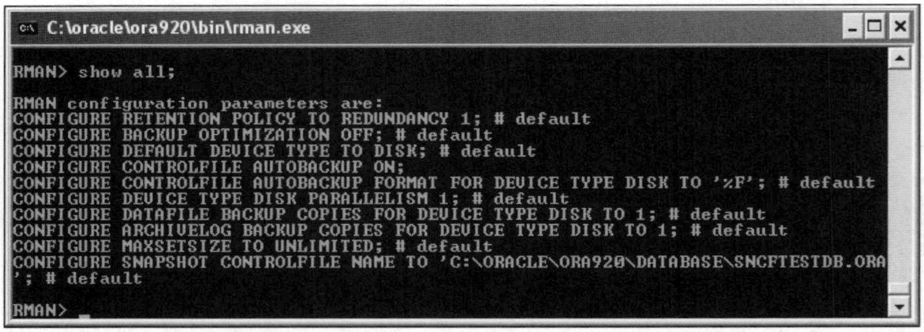

Figure 10-10 Current RMAN parameter settings

2. Type **configure retention policy to redundancy 5;**, as shown in Figure 10-11, to specify the new retention policy.

3. Type **show all;** to display the new RMAN parameter settings, as shown in Figure 10-12.

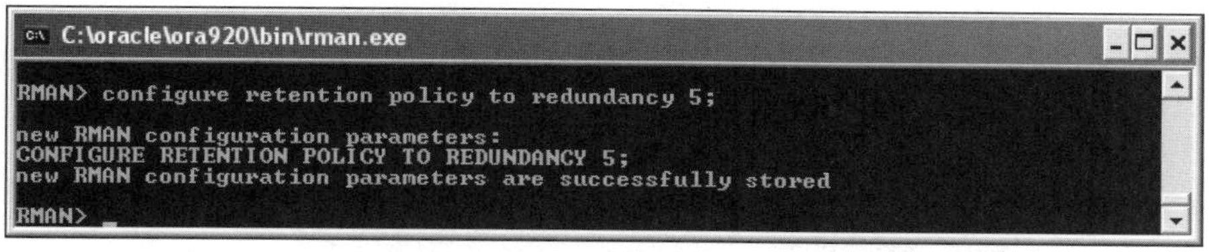

Figure 10-11   Command to change current retention policy

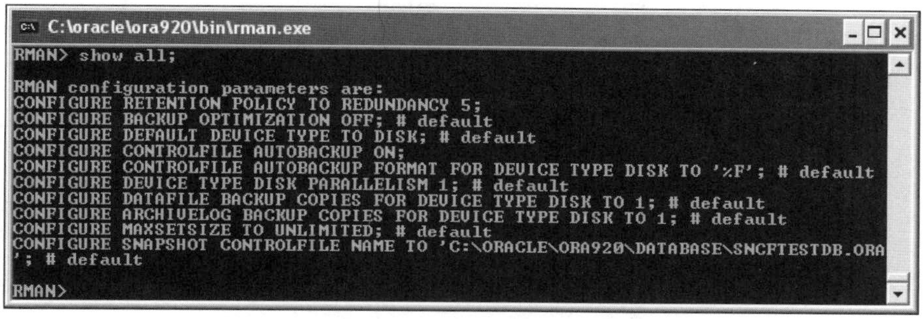

Figure 10-12   Revised RMAN parameters

10

Changing RMAN Retention Policy: A GUI Approach

In this section of the chapter, Carlos will reset the retention policy back to a redundancy of 1. A business reason for doing so might be to conserve storage space by having only one copy available (although this is not necessarily a wise decision in all cases). However, in this section, Carlos will make the change using the Maintenance Wizard for Recovery Manager through the Enterprise Manager Console.

To define a retention policy through the Maintenance Wizard:

1. Start the Enterprise Manager Console, and connect to the Oracle Management Server (OMS) with a SYSDBA privileged account.

2. Click the + (plus sign) next to the Databases folder to display the available databases.

3. Click the name of the database to identify it as the target database.

4. Click **Tools** on the menu bar, point to **Database Tools**, point to **Backup Management**, and then click **Maintenance**, as shown in Figure 10-13, to access the Maintenance Wizard.

5. If the Introduction window appears, click **Next**.

6. When the Operation choice window appears, click the first option button, if necessary, as shown in Figure 10-14. Then click **Next**.

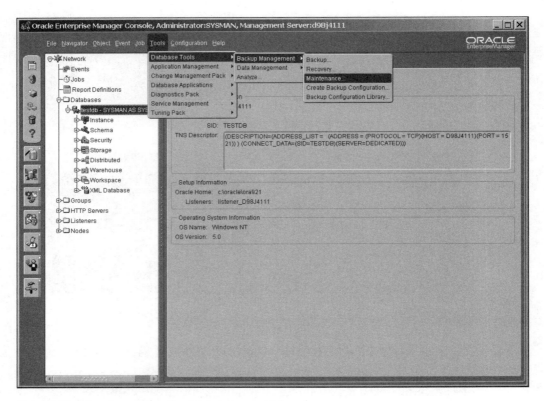

Figure 10-13   Accessing the Maintenance Wizard

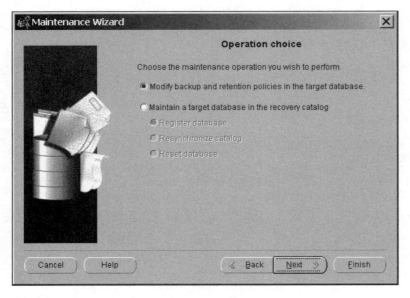

Figure 10-14   Operation choice window

7. When the Backup Policy window appears, check the first check box, if necessary, as shown in Figure 10-15. Click **Next**.

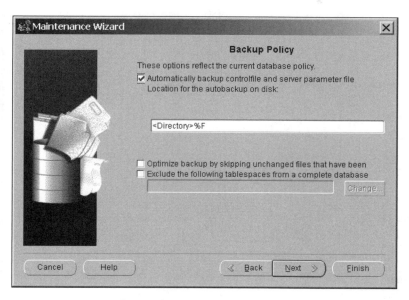

Figure 10-15   Backup Policy window

8. When the Retention Policy window shown in Figure 10-16 appears, click the **Set retention policy of backups** option button, if necessary.

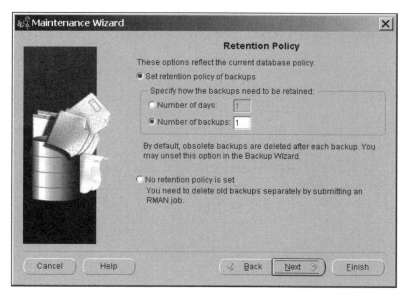

Figure 10-16   Specifying backup redundancy

10

9. If necessary, click the **Number of backups** option button to indicate that you want to set the number of backups to be retained.

10. When the text box becomes available, type **1** to set the retention policy to a redundancy of 1. Then click **Next**.

11. When the Configuration window shown in Figure 10-17 appears, click **Finish**.

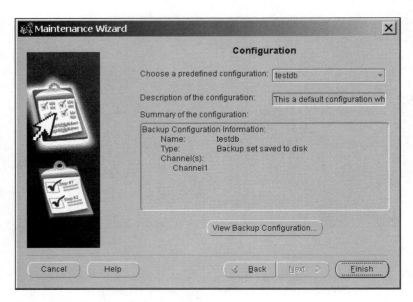

Figure 10-17   Configuration window

12. When the Summary dialog box shown in Figure 10-18 appears, click **OK**.

13. When the dialog box appears indicating that the job has been submitted successfully, click **OK**.

After the job has been executed, the retention policy is reset to a redundancy of 1. Of course, if Carlos had wanted to specify the TO WINDOW option, then he would simply have selected the first option button in Step 9, and then indicated the appropriate number of days.

At this point, Carlos has become comfortable with using some of the basic features of the Recovery Manager utility. In Chapter 11, Carlos will examine the benefits of using the recovery catalog as the repository for RMAN.

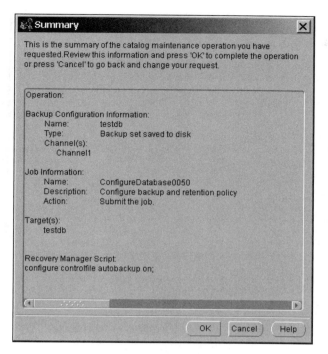

Figure 10-18   Summary dialog box

10

CHAPTER SUMMARY

❑ Backup sets and image copies referenced in the RMAN repository can be cross-checked to ensure that the physical files are available.

❑ The status of a file is considered to be EXPIRED if it cannot be located by RMAN. References to these files should be deleted if the physical files no longer exist.

❑ The DBA can change the status of a backup set or image copy using the CHANGE command.

❑ RMAN can display a list of the current backup sets or image copies referenced in the repository and the current status of each using the LIST command.

❑ The DBA can add backup copies generated by the operating system to the RMAN repository. They are classified as image copies rather than pieces in a backup set.

❑ A retention policy can be created to specify the point at which RMAN can consider a file obsolete by including the appropriate option with the CONFIGURE command.

❑ The value assigned to the retention policy can be measured by the number of days or the number of backup copies.

SYNTAX GUIDE

Command	Description	Example
CATALOG DATAFILECOPY `'location\filename'`	Used to add a file copied by the operating system to the RMAN repository	`catalog datafilecopy 'c:\backup\user01.bak';`
CHANGE BACKUPSET *i* DELETE	Used to remove any reference to the specified object from the RMAN repository	`change backupset 21 delete;`
CHANGE BACKUPSET *i* AVAILABLE|UNAVAILABLE	Used to change the status of a backup set currently referenced by the RMAN repository	`change backupset 21 available;`
CHANGE DATAFILECOPY *'location'* AVAILABLE|UNAVAILABLE	Used to change the status of an image copy currently referenced by the RMAN repository	`change datafilecopy 'c:\archive\ testdb_1_501748040' unavailable;`
CHANGE DATAFILECOPY *'location'* DELETE	Used to remove any reference to the specified image copy from the RMAN repository	`change datafilecopy 'c:\archive\ testdb_1_501748040' delete;`
CROSSCHECK BACKUP|COPY OF DATABASE	Used to check the availability of the physical files referenced in the RMAN repository	`crosscheck backup of database;`
LIST BACKUP|COPY OF DATABASE	Used to display information regarding any backup set or image copy currently referenced by the RMAN repository	`list backup of database;`
CONFIGURE...TO	Used to change parameter settings in RMAN	`configure retention policy to none;`
RETENTION POLICY Options	**Description**	**Example**
REDUNDANCY *i*	Used to specify how many copies and backups of the database files RMAN should keep	`configure retention policy to redundancy 5;`
RECOVERY WINDOW OF *i* DAYS	Used to specify that the files necessary to recover the database at a stated time frame within the specified window cannot be considered obsolete	`configure retention policy to recovery window of 7 days;`
NONE	Used to disable the retention policy, preventing RMAN from considering any backup set or image copy as obsolete	`configure retention policy to none;`

REVIEW QUESTIONS

1. Which of the following statements is used to disable the RMAN retention policy?

   a. DELETE RETENTION POLICY TO NONE;

   b. CONFIGURE RETENTION POLICY DELETE;

   c. CHANGE RETENTION POLICY TO NONE;

   d. CONFIGURE RETENTION POLICY TO NONE;

2. The RMAN repository can only reference information regarding backup copies created through RMAN. True or False?

3. If RMAN cannot locate a physical operating system file when performing a cross-check, the status of the backup piece is marked as _____.

4. The _____ keyword is used in the CHANGE BACKUPSET command to remove any reference to the specified backup set from the RMAN repository.

5. What is a retention policy?

6. Which of the following commands is used to change the current value assigned to an RMAN parameter?

   a. ALTER

   b. CHANGE

   c. CONFIGURE

   d. SET

7. Which of the following options is used with the RETENTION POLICY parameter to indicate that any backup copies created within a specific period of time cannot be considered obsolete?

   a. TO REDUNDANCY

   b. TO TIMEPERIOD

   c. TO TIME PIECE

   d. TO RECOVERY WINDOW

8. The CHANGE command can be used to add a backup copy of a data file created by the operating system to the RMAN repository. True or False?

9. Which of the following statements can be used to verify the existence of the physical files referenced by the RMAN repository?

   a. CROSS CHECK BACKUP OF DATABASE;

   b. CROSS CHECK DATABASE BACKUPS;

   c. CROSSCHECK BACKUP OF DATABASE;

   d. CROSSCHECK DATABASE BACKUPS;

10

10. Which of the following statements can be used to perform a cross-check of the information contained in the repository regarding image copies?

    a. CROSS CHECK IMAGE OF DATABASE;

    b. CROSS CHECK DATABASE IMAGES;

    c. CROSSCHECK DATAFILECOPY OF DATABASE;

    d. CROSSCHECK COPY OF DATABASE;

11. What statement can be used to display the status of the image copies currently being referenced by the RMAN repository?

12. When verifying the location of the physical files referenced in the RMAN repository, any files having a status of _____ are ignored.

13. The _____ Wizard available through the Oracle Enterprise Manager can be used to make changes to the RMAN retention policy.

14. The _____ option of the RETENTION POLICY parameter is used to specify the number of backup copies that should exist for each data file.

15. RMAN can cross-check backup sets of data files as well as image copies of data files. True or False?

16. If an image copy currently referenced by the RMAN repository cannot be located when a cross-check is performed, the status of the file is displayed as EXPIRED. True or False?

17. What statement is used to delete a repository reference to an image copy of the System01.dbf data file with the file name of System01.bak stored on drive C: in the folder named Backup?

18. When a cross-check is performed, a backup piece can be assigned the status of _____, _____, or _____ based on the results of the cross-check.

19. Information regarding the backup copy of a data file created through the operating system can be added to the repository only if the file is assigned the extension .bak. True or False?

20. Information regarding backup copies of archived redo log files can be added to the RMAN repository. True or False?

HANDS-ON ASSIGNMENTS

Assignment 10-1 Listing the Current Status of a Backup Set

In this assignment you create a backup copy of the Users01.dbf data file using RMAN, and then determine the current status of the backup set using the LIST command.

1. Open RMAN and connect to the target database using a SYSDBA privileged account.

2. Perform the necessary steps in RMAN to create a backup copy of the **Users01.dbf** data file.

3. Issue the appropriate LIST command to display the current status of all backup sets currently referenced by the RMAN repository.

4. If more than one backup set is returned in the results of Step 3, use the value displayed in the Completion Time column of the output to determine the backup set that was just created.

Assignment 10-2 Performing a Cross-Check of a Backup Set

In this assignment you delete a previously created backup set, and then perform a cross-check to determine the revised status of the backup set.

1. Use the operating system to delete the physical file created in Assignment 10-1. (*Hint*: The name and location of the file is provided in the results of the LIST command.)

2. If necessary, open RMAN and connect to the target database using a SYSDBA privileged account.

3. Perform a cross-check of the backup copies of the database files.

4. Determine the current status of the backup set associated with the file deleted in Step 1.

10

Assignment 10-3 Adding a File Copied by the Operating System to the RMAN Repository

In this assignment you create a copy of a file using the operating system, and then add that file to the RMAN repository.

1. Use the operating system to create a copy of the **Tools01.dbf** data file. Name the copy of the file **Baktls01.dbf**. Remember to shut down the database before attempting to copy the file.

2. If necessary, open RMAN and connect to the target database using a SYSDBA privileged account.

3. Use the appropriate CATALOG command to add the **Baktls01.dbf** file to the RMAN repository.

Assignment 10-4 Changing the Status of an Image Copy

In this assignment you are required to change the status of an image copy.

1. If necessary, open RMAN and connect to the target database using a SYSDBA privileged account.

2. Use the LIST command to display all the image copies currently referenced by the RMAN repository.

3. Identify the current status of the file added to the RMAN repository in Assignment 10-3.

4. Issue the appropriate CHANGE command to change the status of the file to UNAVAILABLE.

5. Reissue the LIST command to verify the revised status of the file.

Assignment 10-5 Deleting Information from the RMAN Repository

In this assignment you delete information regarding a data file from the RMAN repository, and then physically delete the file using the operating system.

1. If necessary, open RMAN and connect to the target database using a SYSDBA privileged account.

2. Use the LIST command to determine the current status of the **Baktls01.dbf** file.

3. Use the appropriate CHANGE command to delete the information regarding the file from the RMAN repository.

4. Use the operating system to physically delete the **Baktls01.dbf** file from the hard drive.

Assignment 10-6 Specifying a Retention Policy in RMAN

In this assignment you establish a window of four days as the parameter for when backup sets can be considered obsolete.

1. If necessary, open RMAN and connect to the target database using a SYSDBA privileged account.

2. Use the SHOW command to determine the current retention policy.

3. Issue the appropriate CONFIGURE command to change the retention policy so that no files required to recover the database to a previous state occurring within the last four days are considered obsolete.

CASE PROJECTS

Case 10-1 User-Managed Processes Versus Recovery Manager

Carlos has been trying to determine whether backup and recovery operations should be user-managed or managed through Recovery Manager. Based on the information you have learned in this chapter, and information from previous chapters, create a memo to present what you perceive as the advantages and disadvantages of each approach. Make certain to include your rationale for each advantage and disadvantage in the memo.

Case 10-2 Procedure Manual

Continuing with the procedure manual you began in Chapter 1, create a list of the steps or identify the command necessary to perform each of the following tasks:

❑ Cross-check the availability of backup sets and image copies

❑ Change the status of a backup set or image copy

❑ Add an operating system backup to the RMAN repository

❑ Change the RMAN retention policy

10

RECOVERY CATALOG

**After completing this chapter,
you should be able to do the following:**

♦ Create a recovery catalog
♦ Register a target database with a recovery catalog
♦ Create, store, and run an RMAN script
♦ Reset the recovery catalog
♦ Resynchronize the recovery catalog
♦ Retrieve data from recovery catalog views
♦ Identify methods of backing up and recovering a recovery catalog

The Recovery Manager (RMAN) utility uses a repository to hold information regarding previous backup and recovery operations as well as information about the target database. Previously, the repository was stored in the control file of the target database. Alternate approaches include creating a separate database that can be created with the sole purpose of serving as the repository or simply storing the repository in another existing database, as long as it is not the target database. When the repository is placed in a separate database, it includes a recovery catalog schema containing a set of tables and views that RMAN uses to store and access information.

Why should a recovery catalog be used when the repository can be stored in the control file of the target database? The most obvious answer is that if the control file of the target database becomes unavailable, then RMAN cannot access the necessary information regarding the database, or existing backups, to recover the database. However, if this information is stored in a recovery catalog database and the target database needs recovery, RMAN can still have access to the information required to perform the necessary recovery operation. In addition, a recovery catalog logs information about previous backup and recovery operations without it being overwritten, which occurs when it is stored in the target database's control file. Also, RMAN scripts can be stored in the recovery catalog and reused at a later time.

In this chapter, Carlos learns how to create a recovery catalog. In addition, Carlos learns how to update and retrieve the information stored in the recovery catalog, and how to use the recovery catalog to store scripts.

THE CURRENT CHALLENGE AT JANICE CREDIT UNION

Although Carlos has learned how to perform backup and recovery operations using RMAN, he is concerned about two potential problems. The first is that if the control file of the target database becomes unusable, then RMAN cannot recover the database. The second is that members of his staff who are not primarily responsible for backup and recovery might have a difficult time remembering the various RMAN commands in the event a recovery becomes necessary while he and other trained DBAs are on vacation or are otherwise unavailable. While reviewing the RMAN documentation regarding the recovery catalog, he realizes that the use of a recovery catalog could be a solution to both of these problems.

SET UP YOUR COMPUTER FOR THE CHAPTER

Because the content of the TESTDB database is not changed in this chapter, a cold backup is not required. However, to complete the examples, you are required to create a new database. This database will serve as the recovery catalog database. Before attempting to create the database, make certain sufficient room is available on your hard drive. As an alternative approach, the GUI examples in this chapter demonstrate using an existing database for the repository, rather than creating an additional database. Before attempting the examples in this chapter, make certain your database is in ARCHIVELOG mode.

NOTES ABOUT DUAL COVERAGE IN THIS CHAPTER

In this chapter, interaction with the recovery catalog is presented using both the command-line approach and the GUI approach. The GUI approach is available only through the Enterprise Manager Console when you are connected to an Oracle Management Server (OMS). Because an OMS is normally used in a networked environment, the GUI section is presented using the Windows 2000 Server operating system.

CREATING THE RECOVERY CATALOG AND REGISTERING THE DATABASE

To create the recovery catalog, Carlos must first create a database to store the recovery catalog schema. When initially created, the recovery catalog database is not any different from any other database that Carlos might create. However, after it is created, the RMAN account must be unlocked, the recovery catalog created, and then the target database registered with the Recovery Catalog to implement the RMAN repository.

In an Oracle9i database, the user RMAN, with the default password RMAN, is automatically created. This is the account that should be used to actually create the recovery catalog. By default, this user account is locked and the password is set as expired. In addition, it has been assigned the CONNECT, RESOURCE, and RECOVERY_CATALOG_ OWNER privileges.

To use this account, Carlos first needs to unlock it. Then, when he connects to the database as RMAN, he needs to provide a new password because the current password is set as expired. For illustrative purposes only, Carlos keeps RMAN as the password for the account. After the RMAN user account has been activated, Carlos can use the RMAN utility to create the recovery catalog.

After the recovery catalog is created, Carlos then needs to register the target database. When the target database is registered with the recovery catalog, information regarding its physical structure is stored in the catalog, and the target database can subsequently be referenced in any backup and recovery operations that are performed while connected to the catalog.

 Normally the recovery catalog database is stored on a different computer than the target database to reduce the chance of having a single point of failure. In regards to backup procedures, the recovery catalog database should be backed up regularly, like any database, in the event it becomes unavailable because of media failure.

Using the Command-Line Approach to Create the Recovery Catalog and Register a Target Database

In these steps, Carlos creates a recovery catalog using the command-line approach.

To create the recovery catalog:

1. Create a new database named RECOVCAT. Unless specified otherwise by your instructor, use the Database Configuration Assistant to create a General Purpose database.

2. Start SQL*Plus and connect to the RECOVCAT database using the SYSTEM user account and enable the SYSDBA privileges.

3. Type **alter user rman account unlock;** to unlock the user account, as shown in Figure 11-1.

4. Type **connect rman/rman** to connect to the RECOVCAT database with the RMAN user account.

5. When prompted for a new password, as shown in Figure 11-1, type **rman**. Retype the password when prompted. Then type **exit** to exit SQL*Plus.

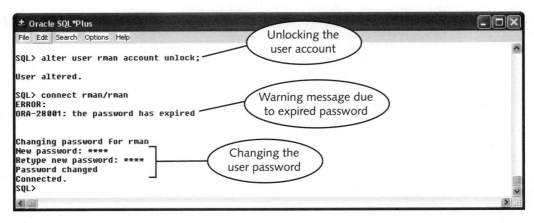

Figure 11-1   Accessing the RMAN account

Now that Carlos has created the recovery catalog database and activated the RMAN user account, he can use the RMAN utility to create the recovery catalog schema in the database.

To create the schema:

1. Click the **Start** button on the Windows XP taskbar.

2. Click **Run** from the Start menu.

3. Type **rman catalog rman/rman@recovcat**, as shown in Figure 11-2, to connect to the recovery catalog database.

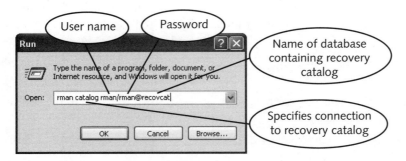

Figure 11-2   Connecting to the recovery catalog database

4. Type **create catalog tablespace data;**, as shown in Figure 11-3, to create the recovery catalog in the DATA tablespace. If the tablespace does not exist, it is created when the command is executed. The recovery catalog should never be stored in the SYSTEM tablespace. You might receive a "recovery catalog already exists" message if the catalog has previously been created.

Figure 11-3   Creating the recovery catalog

5. Type **exit** to exit RMAN.

 Rather than exiting the RMAN utility, the database can be registered at this time by first connecting to the target database and then registering the database.

After creating the recovery catalog, Carlos needs to register the target database using the following steps:

To register the database:

1. Click the **Start** button on the Windows XP taskbar.

2. Click **Run** from the Start menu.

3. Type **rman target *username/password@database***, substituting the appropriate user name, password, and name of the target database, as shown in Figure 11-4, to connect to the target database.

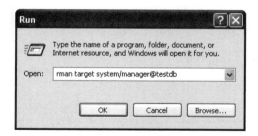

Figure 11-4   Connecting to the target database

4. Type **connect catalog rman/rman@recovcat**, as shown in Figure 11-5, to establish a connection with the recovery catalog after starting the RMAN utility.

5. Type **register database;**, as shown in Figure 11-6, to register the target database with the recovery catalog.

Figure 11-5 Connecting to the recovery catalog from the target database

```
C:\oracle\ora920\bin\rman.exe                                    _ □ ×

Recovery Manager: Release 9.2.0.1.0 - Production

Copyright (c) 1995, 2002, Oracle Corporation.  All rights reserved.

connected to target database: TESTDB (DBID=2284938461)

RMAN> connect catalog rman/rman@recovcat

connected to recovery catalog database

RMAN> register database;

database registered in recovery catalog
starting full resync of recovery catalog
full resync complete

RMAN> _
```

Registering the target database with the recovery catalog

Figure 11-6 Registering the target database with the recovery catalog

Now that the target database has been registered with the recovery catalog, Carlos can use RMAN with the recovery catalog to perform backup and recovery operations as well as store and execute RMAN scripts to perform those operations.

Using the GUI Approach to Create the Recovery Catalog and Register a Target Database

In this section, Carlos will use the Enterprise Manager Console to create the recovery catalog and register the target database. As with the command-line approach, the RMAN user account must be unlocked and the password updated. With the GUI approach the creation of the recovery catalog is performed while creating a backup configuration like the ones created in the previous chapters. When a recovery catalog is specified as part of a backup configuration and the catalog does not already exist, the user is prompted to create the catalog. In addition, if the target database is not registered in the recovery catalog, the user is also prompted to perform this task.

In the following steps, Carlos will make the RMAN user account available. However, rather than create another database, the recovery catalog will be stored in the OEMREP database created by default when the configuration of the Oracle Enterprise Manager includes the Oracle Management Server.

To make the account available:

1. Start the Enterprise Manager Console with a connection to the Oracle Management Server.

2. If necessary, click the **+** (plus sign) to the left of Databases in the objects list to expand it. Then click the **+** (plus sign) to the left of the OEMREP database.

 If the database associated with the Oracle Management Server is not listed in the Database object list, click **Navigator** on the menu bar, and then click **Refresh All Nodes**.

3. When prompted for the login information, log into the database using the SYS account with the SYSDBA privileges.

4. From the objects list for the OEMREP database, click the **+** (plus sign) next to Security, and then click the **+** (plus sign) next to Users.

5. From the users list, click **RMAN**.

6. From the General sheet displayed in the right side of the window, type **rman** in the Enter Password and Confirm Password text boxes.

7. In the Status portion of the General pane, click the **Unlocked** option button, as shown in Figure 11-7. Then click **Apply**.

8. If the dialog box shown in Figure 11-8 appears prompting you to confirm the change, click **Yes**.

9. Click **Users** from the object list in the Navigation pane on the left side of the window.

11

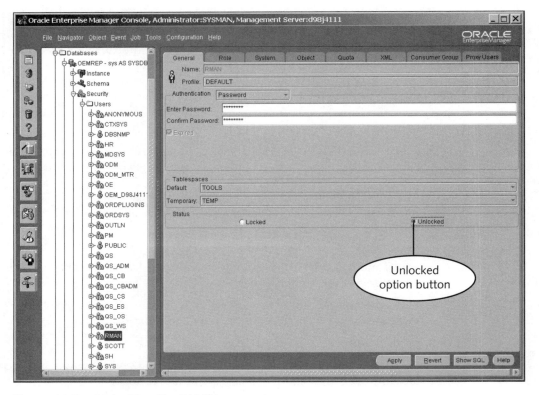

Figure 11-7 Unlocking the RMAN account

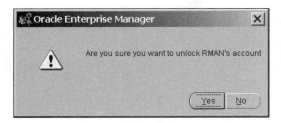

Figure 11-8 Confirming change

10. When the information regarding users appears on the right side of the window, confirm that the account status for the RMAN user account is listed as OPEN, as shown in Figure 11-9.

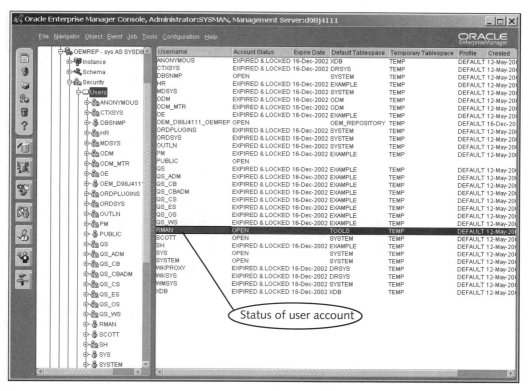

Figure 11-9   Updated account status

After the RMAN account has been unlocked, the RMAN schema is the owner of the recovery catalog. Now Carlos can create a backup configuration for the target database that results in creating the recovery catalog and registering the target database.

To create the backup configuration:

1. In the Navigation pane on the left side of the Enterprise Manager Console window, click the name of the target database.

2. On the menu bar, click **Tools**, point to **Database Tools**, click **Backup Management**, and then click **Create Backup Configuration**. See Figure 11-10.

11

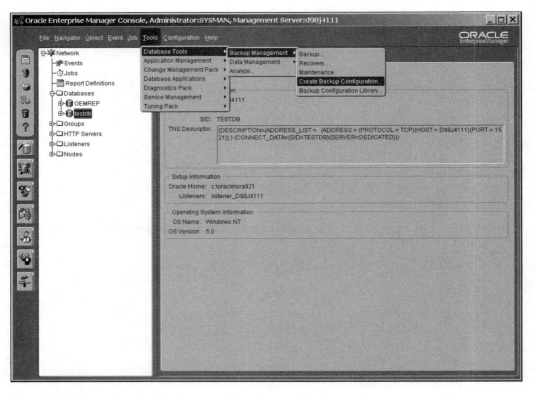

Figure 11-10 Accessing backup configuration options

3. In the General sheet displayed in the right side of the window, enter a name for the new backup configuration, as shown in Figure 11-11.

4. Click the **Channels** tab at the top of the configuration window to verify that a default channel exists, as shown in Figure 11-12.

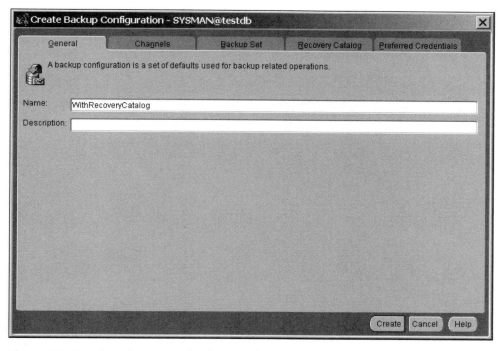

Figure 11-11   Creating a new backup configuration

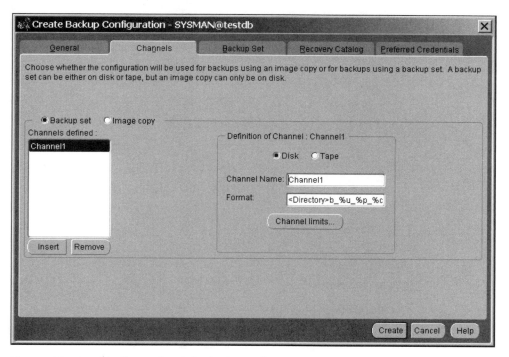

Figure 11-12   Verifying the default channel

5. Click the **Recovery Catalog** tab at the top of the configuration window.

6. Click the **In a recovery catalog** option button, and type **rman** in the Username and Password text boxes, as shown in Figure 11-13, then click **Create**.

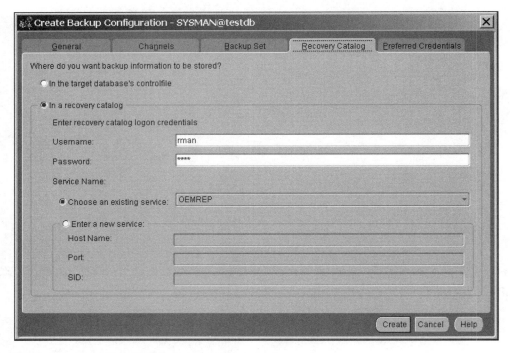

Figure 11-13 Creating the recovery catalog

7. When the dialog box shown in Figure 11-14 appears, indicating that the target database needs to be registered with the recovery catalog, click **Register**.

Figure 11-14 Registering the target database

A target database can also be registered in the recovery catalog through the Catalog Maintenance Wizard.

CREATING AND EXECUTING AN **RMAN** SCRIPT

Now that Carlos has created the recovery catalog and registered the target database, he can use the recovery catalog to store information regarding backup and recovery operations performed through RMAN, rather than the control file of the target database. Backup and recovery operations use the same commands as when the repository is stored in the control file of the target database. The only difference is that the recovery catalog is used as a repository rather than the control file of the target database.

One of the benefits of using a recovery catalog is the ability to store and execute RMAN scripts. For Janice Credit Union, this means that even if Carlos or one of the other DBAs are unavailable, another IT employee can execute stored scripts that perform routine tasks. Because Carlos is already familiar with backing up a database through RMAN, he decides to create a script that can be used to back up the target database and to store that script in the recovery catalog.

 A stored script can also be executed without user intervention by scheduling execution of the script through the Enterprise Manager Console or as a task through the operating system.

Creating and Executing an RMAN Script—Command-Line Approach

11

When working with the command-line approach, the script is created using the CREATE SCRIPT command at the RMAN prompt. The syntax for the command is CREATE SCRIPT *scriptname* {*script*}, where *scriptname* is the name of the script and *script* is the actual sequence of commands to be executed. In this section, Carlos will create a script named DBBackup that can be executed to back up all data files and all copies of the control file.

 Before attempting the following steps, make certain that the target database is in ARCHIVELOG mode.

To create the script:

1. If necessary, start RMAN by typing **rman target *username/password@ database***, using the appropriate user name, password, and database name, to connect to the target database.

2. Type **connect catalog rman/rman@recovcat**, as shown in Figure 11-15, to connect to the recovery catalog.

3. Type the following:

```
create script DBBackup {
allocate channel c1 type disk;
backup database;
}
```

This creates and stores the DBBackup script as shown in Figure 11-16.

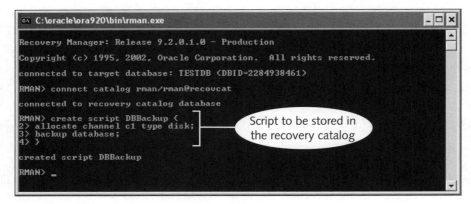

Figure 11-15 Connecting to the recovery catalog

```
C:\oracle\ora920\bin\rman.exe

Recovery Manager: Release 9.2.0.1.0 - Production
Copyright (c) 1995, 2002, Oracle Corporation.  All rights reserved.
connected to target database: TESTDB (DBID=2284938461)
RMAN> connect catalog rman/rman@recovcat
connected to recovery catalog database
RMAN> create script DBBackup {
2> allocate channel c1 type disk;
3> backup database;
4> }                                      ⟵ Script to be stored in
                                             the recovery catalog
created script DBBackup
RMAN> _
```

Figure 11-16 Creating an RMAN script

4. To execute the script, type **run {execute script DBBackup;}**, as shown in Figure 11-17. Type **exit** to exit the RMAN utility.

As demonstrated in Step 4, the EXECUTE command is used to execute an RMAN script that is stored in the recovery catalog. The syntax of the EXECUTE command is EXECUTE SCRIPT *scriptname*, where *scriptname* is the name of the script to be executed. However, this command is not a standalone command; therefore, it must be submitted as a job to be executed by the RUN command, as shown in Figure 11-17.

Comments can be included within a script by preceding the comment with a # (number sign).

Figure 11-17 Execution of a stored script

Creating and Executing an RMAN Script—GUI Approach

When creating an RMAN script through the Enterprise Manager Console, the script is actually considered to be part of a job that is stored in the job library. In the following section, Carlos will create the DBBackup script and add it to the job library. Then, he will access the job library to submit the script for execution.

To create the script:

1. If necessary, log into Oracle Management Server. Click **Job** on the menu bar, and then click **Create Job**.

2. In the Job Name text box, type **DBBackup**.

3. If necessary, click the name of the target database from the Available Targets list. Then click the **Add** button to move the target database from the Available Targets list to the Selected Targets list.

4. At the bottom of the Create Job window, click the **Add to Library** option button, as shown in Figure 11-18.

5. At the top of the Create Job window, click the Tasks tab.

6. Click **Run Rman Script** in the Available Tasks list. Click the Add button to move the task from Available Tasks to Job Tasks. Your window should be similar to the one shown in Figure 11-19 after completing this step.

11

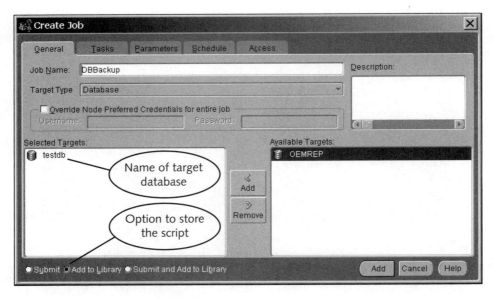

Figure 11-18   Create Job window

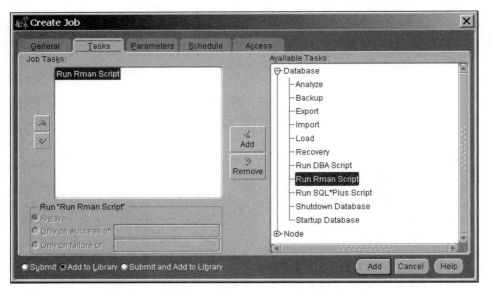

Figure 11-19   Specifying a task

7. Click the **Parameters** tab at the top of the Create Job window.

8. Type

```
{allocate channel c1 type disk;
backup database;}
```

in the Rman Script text box, as shown in Figure 11-20. Then click **Add**.

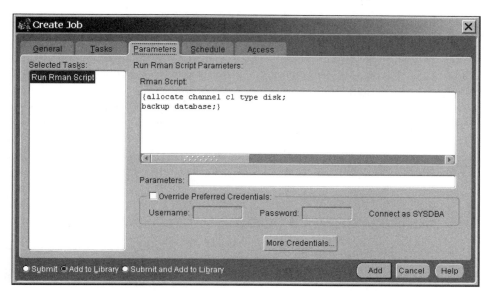

Figure 11-20   Entering an RMAN script

At this point, the RMAN script is stored in the job library. However, if Carlos had wanted to execute the script at the same time, he could simply have clicked the Execute Immediately option from the list of Schedule options, and then clicked Submit.

To execute the RMAN script after it has been stored in the job library:

1. Click **Job** on the menu bar, and then click **Job Library**.

2. If necessary, click the name of the job to be submitted from the list of displayed job names, as shown in Figure 11-21.

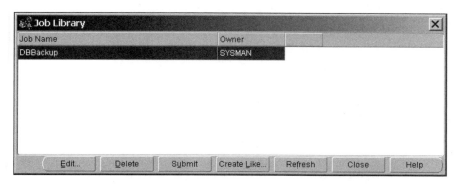

Figure 11-21   Selecting a job for submission

11

3. Click **Submit**, and then click **Close**. Then exit the Oracle Enterprise Manager Console.

As with all jobs, Carlos can verify that the job was executed by clicking Jobs from the Navigation list on the left side of the Enterprise Manager Console window. The job is executed based on the scheduling parameter identified when the script was created. Because Carlos did not explicitly indicate a scheduling parameter, it was assigned the value of Immediately by default. However, if Carlos had wanted to submit the job to be executed later that day, perhaps after the Credit Union closed, he would simply have edited the job by clicking the Edit button shown previously in Figure 11-21. Then Carlos could have clicked the Schedule tab and specified when the job needed to be executed.

UPDATING THE RECOVERY CATALOG

Recall from Chapter 10 that the contents of the RMAN repository can be updated to change the status of a backup set or to record file copies created through the operating system. This is accomplished using the CHANGE and CATALOG commands, respectively. These same operations can be performed on the recovery catalog. The only difference is that Carlos would need to make certain that he has established a connection to the recovery catalog before issuing the commands.

Other types of updates that may be required include resynchronizing the recovery catalog after changes have occurred in the target database and resetting the recovery catalog whenever an incomplete recovery is performed. Each of these processes is discussed in the following sections.

Resynchronizing the Recovery Catalog

Resynchronizing the recovery catalog is required any time there is a change in the control file of the database. Whenever a BACKUP or COPY command is executed while connected to the recovery catalog, the catalog is automatically synchronized with the contents of the target database's control file. The synchronization process includes updating the catalog with any log switches, log-archiving processes, and backup or recovery operations that occurred while not connected to the recovery catalog. There is no set rule as to how often the catalog should be resynchronized; however, it is recommended that the catalog be resynchronized every n days, where n is the value of the CONTROL_FILE_RECORD_KEEP_TIME parameter in the target database's Init.ora file.

Basically, if there are frequent structural database changes or archiving processes occurring, then the catalog needs to be updated with this information more often than if the target database was less dynamic.

Suppose Carlos adds a new tablespace to the financial database. Until the database is backed up through RMAN, with a connection to the recovery catalog, information regarding this change is not reflected in the information stored in the RMAN repository.

Therefore, he would need to resynchronize the recovery catalog. In the following sections, Carlos will resynchronize the recovery catalog by using the RESYNC command through the RMAN command-line utility executable, and then by using the Catalog Maintenance Wizard available through the Enterprise Manager Console.

Resynchronizing the Recovery Catalog—Command-Line Approach

Assuming that Carlos has added a new tablespace to the database, the recovery catalog needs to be resynchronized. Why? Because the structure of the database has changed, which subsequently changes the contents of the control file. This change needs to be updated to the recovery catalog in the event a database failure occurred.

To resynchronize the recovery catalog:

1. If necessary, start RMAN and connect to the target database with the SYSDBA privileged account. Then connect to the recovery catalog using the RMAN user account.

2. Type **resync catalog;**, as shown in Figure 11-22.

Figure 11-22   Resynchronizing the recovery catalog

When the RESYNC command is executed, the recovery catalog is updated with the relevant changes that have been made to the target database's control file.

Resynchronizing the Recovery Catalog—GUI Approach

As with the command-line approach, the recovery catalog can also be resynchronized through the Enterprise Manager Console. By responding to a series of prompts, the Maintenance Wizard creates a job that, after successful submission and execution, updates any changes contained in the target database's control file to the recovery catalog.

To resynchronize the recovery catalog:

1. If not already connected, start the Enterprise Manager Console with a connection to the Oracle Management Server.

2. Click the **+** (plus sign) to the left of Databases in the Navigation list.

3. Click the name of the target database.

4. Click **Tools** on the menu bar, and then click **Database Tools**, **Backup Management**, and **Maintenance**.

5. If the Introduction window appears, click **Next**.

6. From the Operation choice window shown in Figure 11-23, click the **Maintain a target database in the recovery catalog** option button. Once available, click the **Resynchronize catalog** option button. Then click **Next**.

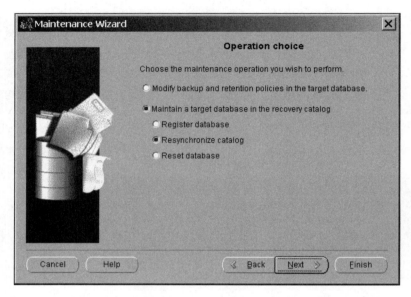

Figure 11-23   Selecting the option to resynchronize the recovery catalog

7. When the Configuration window appears, click the drop-down list to specify the WithRecoveryCatalog configuration created previously in this chapter, as shown in Figure 11-24. Then click **Next**.

8. When the Schedule window appears, click the **Immediately** option button, if necessary. Then click **Next**.

9. When the Multiple Targets window appears, verify that the name of the target database has been correctly identified in the Selected targets list, as shown in Figure 11-25, and then click **Finish**.

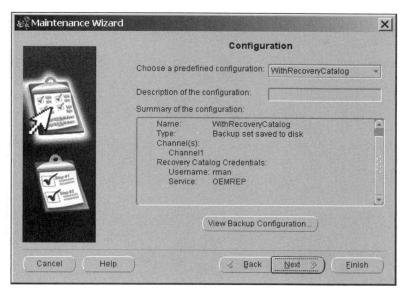

Figure 11-24   Selecting a predefined configuration

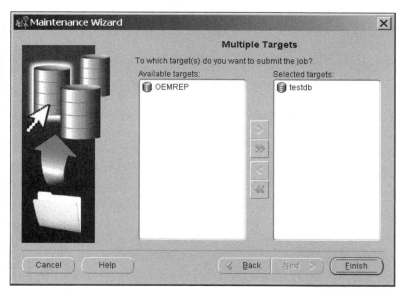

Figure 11-25   Indicating the target database

10. When the Summary window shown in Figure 11–26 appears, click **OK**.

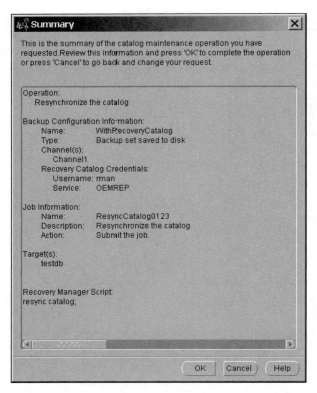

Figure 11-26   Summary window to resynchronize the recovery catalog

11. When the dialog box appears indicating that the job has been submitted successfully, click **OK**.

RESETTING THE RECOVERY CATALOG

Resetting the recovery catalog is necessary when an incomplete recovery is performed.
If Carlos opens the database with the RESETLOGS option, then all prior backup and
recovery information is then invalid because it references a different incarnation of the
current database.

What is an incarnation? An **incarnation** is simply a new reference to a target database;
it is indicated by an **incarnation number** that is assigned to each consistent version of
the database. Each backup set and image copy is associated with a particular incarnation
of the target database. The incarnation number is changed whenever the recovery catalog is reset. Why? By resetting the catalog, RMAN is prevented from attempting to
recover to the database with backup sets created before the incomplete recovery occurred.

In addition, use of the incarnation number allows Carlos to return the database to a prior state. For example, suppose Carlos has to perform an incomplete recovery. After resetting the log files, he realizes that he has recovered the database to a point after the problematic error occurred. The incarnation number could be reset to the version of the database that existed immediately prior to opening the database so that the recovery can be retried. However, note that this is not the common approach for returning the database to its prior state. Rather, the backup created before the incomplete recovery was performed should be restored and then the recovery process repeated, using a different termination point.

The following sections will demonstrate how Carlos would reset the database using the command-line and GUI approaches.

Resetting the Recovery Catalog—Command-Line Approach

In the following step sequence, Carlos will use the RESET DATABASE command to change the incarnation of the target database. The syntax for the RESET DATABASE command is RESET DATABASE [TO INCARNATION i], where i indicates a specific incarnation. If the optional TO INCARNATION clause is not included, the target database reference is simply updated to a new incarnation number. However, for illustrative purposes, in this example, Carlos will reset the database to a previous incarnation. This requires Carlos to first determine which incarnations are currently recognized by the recovery catalog and then make the necessary change.

To identify and change the incarnation of the target database:

1. At the RMAN prompt, type **list incarnation of database;**. RMAN then displays a list of recognized incarnations similar to the one shown in Figure 11-27.

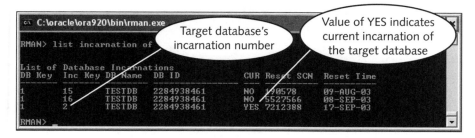

Figure 11-27   List of target database incarnations

2. Type **reset database to incarnation i;** where i is an incarnation number for a previous version of the database.

3. At the RMAN prompt, type **list incarnation of database;** again, as shown in Figure 11-28, to determine the current incarnation of the database.

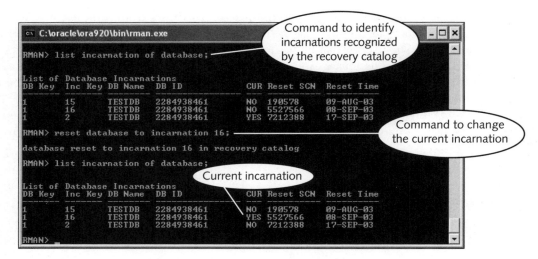

Figure 11-28   Current incarnation of the target database

4. If you will be completing the hands–on assignments for this chapter at a later time, the database incarnation needs to be reset by typing **reset database to incarnation i;**, where *i* is the current incarnation number displayed in Step 1. After pressing **Enter**, type **resync catalog;** to resynchronize the database.

Resetting the Recovery Catalog—GUI Approach

As mentioned, suppose Carlos has to perform an incomplete recovery of the financial database. After completing the recovery, the recovery catalog needs to be reset. The recovery catalog can be reset using the Enterprise Manager Console through the Catalog Maintenance Wizard.

To reset the recovery catalog:

1. If necessary, start the Enterprise Manager Console with a connection to the Oracle Management Server.

2. Click the **+** (plus sign) to the left of Databases in the Navigation list.

3. Click the name of the target database.

4. Click **Tools** on the menu bar, and then click **Database Tools**, **Backup Management**, and **Maintenance**.

5. If the Introduction window appears, click **Next**.

6. From the Operation choice window shown in Figure 11-29, click the **Maintain a target database in the recovery catalog** option button. Once available, click the **Reset database** option button. Then click **Finish**.

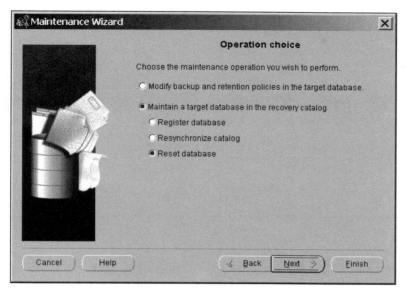

Figure 11-29   Indicating that the database incarnation is to be reset

7. When the Summary window appears, click **OK**.

8. When the dialog box appears indicating that the job has been submitted successfully, click **OK**.

Viewing the Contents of the Recovery Catalog

Carlos has several options available to view the information stored in the recovery catalog. Remember the REPORT and LIST commands introduced in Chapter 6? The LIST command can be used with various options to determine which backup sets or copies are available. Note also that it was used in the previous section to determine information regarding the database incarnation information stored in the recovery catalog.

Recall that the REPORT command can be used to provide more detailed information about the target database and what needs to be done (such as which data files need to be backed up, which data files have been recently backed up, and so on). The commands are the same as those used to obtain information from the repository when it is stored in the target database's control file, except that the RMAN utility needs to be connected to the recovery catalog.

When a recovery catalog is used to store the RMAN repository, Carlos also has the option of querying views to retrieve information. Some of the most commonly queried views include the RC_DATABASE, RC_TABLESPACE, RC_STORED_SCRIPT, and RC_STORED_SCRIPT_LINE views. These views can be referenced by Carlos when he is logged into the recovery catalog database with the RMAN user account. Each of these views will be presented in this section.

11

The RC_DATABASE view is used to identify basic information regarding each target database registered with the recovery catalog. As shown in Figure 11-30, Carlos could reference this view to quickly determine which databases have been registered with the recovery catalog, and their current incarnation numbers.

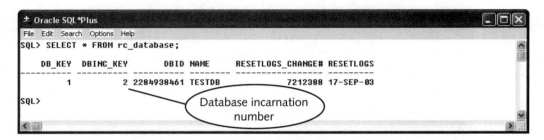

Figure 11-30   RC_DATABASE view

The RC_TABLESPACE view can be used to display information regarding the tablespaces belonging to any database currently registered with the recovery catalog, as shown in Figure 11-31. If there were tablespaces from multiple databases being displayed in the results, Carlos could use the contents of the DB_NAME column to determine the corresponding database for each tablespace listed in the output.

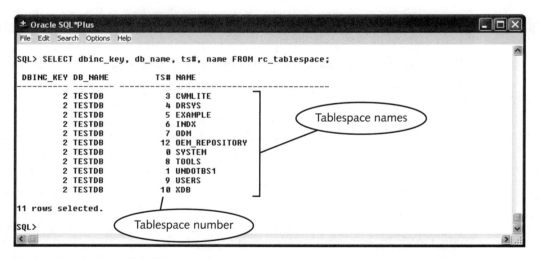

Figure 11-31   RC_TABLESPACE view

Suppose Carlos creates an RMAN script, but a few days later, he can't remember the exact name of the script. One option he has is to display all RMAN scripts currently stored in the recovery catalog using the RC_STORED_SCRIPT view, as shown in Figure 11-32.

```
± Oracle SQL*Plus                                                    _ □ X
File  Edit  Search  Options  Help

SQL> SELECT script_name FROM rc_stored_script;

SCRIPT_NAME
-------------------------------------------------------------------------
DBBackup

SQL>
```

Figure 11-32   RC_STORED_SCRIPT view

Although the output in Figure 11-32 displays only the name of one stored script, what if there had been a dozen or so scripts? If that had been the case, and Carlos needed to review the actual contents of a particular script to verify that it is the one he is looking for, he could display each line of a specific RMAN script using the RC_STORED_ SCRIPT_LINE view. Notice in Figure 11-33 that each line within a script is assigned a line number. The portion of the script that appears on each line correlates to how Carlos entered the script. If he had typed the entire DBBackup script on one line when it was created, then only one line is displayed in the results.

```
± Oracle SQL*Plus                                                    _ □ X
File  Edit  Search  Options  Help

SQL> SELECT line, text FROM rc_stored_script_line WHERE script_name = 'DBBackup';

     LINE TEXT
---------- ------------------------------------------------------------------
        1 {
        2 allocate channel c1 type disk;
        3 backup database;
        4 }

SQL>
```

Figure 11-33   RC_STORED_SCRIPT_LINE view

The contents of a script can also be viewed through RMAN using the command
PRINT SCRIPT *scriptname*; where *scriptname* is the name of stored RMAN script.

There are 27 different views that Carlos can use to retrieve information stored in the recovery catalog database. These views are easy to identify; the view name starts with RC_, for recovery catalog. Documentation on each view is available through the Oracle Technology Network at *http://otn.oracle.com*.

Backup and Recovery of the Recovery Catalog Database

As with any other database at Janice Credit Union, Carlos needs to back up the recovery catalog database. There are various strategies available when the RMAN repository

11

is stored in a separate database. The simplest method, but not necessarily the quickest, is to perform a cold backup of the database. Another option is to create a separate tablespace just for the recovery catalog, which is recommended by Oracle. Then, if the recovery catalog is operated in ARCHIVELOG mode, just the tablespace can be backed up, rather than always performing a whole database backup.

A third alternative, viable for small databases, is to export the database. If it is a large database, then just the schema containing the recovery catalog can be exported. A discussion of exporting the recovery catalog database will be presented in Chapter 12. As will also be discussed in that chapter, most database administrators export the contents of the RMAN schema on a nightly basis after all backup operations have been performed to ensure that the recovery catalog can be recovered without any data loss.

Of course, recovery of the database depends on the method Carlos selects to back up the recovery catalog database. If he uses a cold or hot backup method, and the database needs to be recovered, he can simply perform a complete recovery of the database. However, if he elects to export the recovery catalog data, then he would need to either create a new database and import the data, or import the data into another database. Again, examples of importing the recovery catalog database will be presented in Chapter 12. However, regardless of the method used, Carlos needs to resynchronize the recovery catalog with the control file of the target database as soon as he completes the recovery process.

 A cold backup of the database created during this chapter that is used to store the recovery catalog should be created before ending your session.

CHAPTER SUMMARY

- After a recovery catalog is created, the target database must be registered with the catalog before it can be used to store backup and recovery information about the target database.

- RMAN scripts can be stored in a recovery catalog.

- The recovery catalog should be resynchronized if structural changes, log switches, or archiving processes frequently occur.

- If a database is opened with the RESETLOGS option, the database must be reset in the recovery catalog.

- The contents of the recovery catalog can be displayed using the LIST or REPORT commands, or by querying RC_ prefixed views.

- The RC_STORED_SCRIPT and RC_STORED_SCRIPT_LINE views can be used to display the names and contents of a stored RMAN script, respectively.

- The recovery catalog database should be backed up, or its data exported, in the event the database needs to be recovered.

SYNTAX GUIDE

Command Syntax	Description	Example
CREATE SCRIPT *scriptname {script}*	Used to create and store an RMAN script in the recovery catalog	`CREATE SCRIPT DBBackup {` `allocate channel c1 type disk;` `backup database; }`
REGISTER DATABASE	Used to register a target database with a recovery catalog	`REGISTER DATABASE;`
RESET DATABASE [TO INCARNATION *i*]	Used to change the incarnation number of the target database; required after the target database is opened with the RESETLOGS option	`RESET DATABASE TO` `INCARNATION 2;`
RESYNC CATALOG	Used to update the recovery catalog with changes made in the target database's control file	`RESYNC CATALOG;`

LIST Command Parameter	Description	Example
INCARNATION OF DATABASE	Used to display incarnation information regarding the target database	`LIST INCARNATION OF DATABASE;`

Views	Description	Example
RC_DATABASE	Displays information regarding registered target databases	`SELECT * FROM rc_database;`
RC_STORED_SCRIPT	Displays information regarding RMAN scripts stored in the recovery catalog	`SELECT * FROM` `rc_stored_script;`
RC_STORED_SCRIPT_LINE	Displays the contents of stored RMAN scripts	`SELECT * FROM` `rc_stored_script_line;`
RC_TABLESPACE	Displays the tablespaces associated with registered target databases	`SELECT * FROM rc_tablespace;`

11

REVIEW QUESTIONS

1. A target database can only be registered with a recovery catalog using the SET command. True or False?

2. Assume a DBA is using the following steps to create a recovery catalog:

    1. Create a database.

    2. Unlock the RMAN user account of the new database through SQL*Plus.

    3. _____

    4. Issue the CREATE CATALOG command.

    What step belongs in the third spot?

    a. Connect to the target database through RMAN.

    b. Connect to the recovery catalog database through RMAN.

    c. Register the database through SQL*Plus.

    d. Register the database through RMAN.

3. Which of the following is required if the target database is opened with the RESETLOGS option?

    a. The database must be registered with the recovery catalog again.

    b. The recovery catalog must be reset.

    c. The recovery catalog must be resynchronized.

    d. All stored RMAN script files must be recreated.

4. The different versions of a target database are identified in the recovery catalog with a(n) _____ number.

5. The _____ view can be used to display the contents of a stored RMAN script.

6. Which of the following privileges, or roles, is not required by the RMAN user account?

    a. ALLOCATE

    b. CONNECT

    c. RECOVERY_CATALOG_OWNER

    d. RESOURCE

7. The CREATE CATALOG command can be executed using which of the following tools or utilities?

    a. RMAN

    b. SQL*Plus

    c. SQL*Plus Worksheet

    d. all of the above

8. The RESET command is used to update the recovery catalog with any changes contained in the control file of the target database. True or False?

9. Which of the following approaches can be used to back up a recovery catalog database?

   a. hot backup

   b. cold backup

   c. export

   d. all of the above

10. The _____ view can be used to display the current incarnation number of a target database.

11. What is the purpose of an incarnation number?

12. Which of the following events would mean that the recovery catalog should be resynchronized:

   a. when a log switch occurs in the recovery catalog database

   b. when a new user is created in the target database

   c. when a tablespace in the target database is moved to a new location

   d. when a redo log file in the recovery catalog database is archived

13. Which user account should be used to create the recovery catalog?

   a. SYS

   b. SYSTEM

   c. RMAN

   d. RCOV

14. How does a user connect to the recovery catalog after starting RMAN?

15. The default password for the RMAN user account is _____.

16. The recovery catalog is automatically resynchronized with the contents of the target database's control file if the BACKUP or _____ command is issued while connected to the recovery catalog.

17. The recovery catalog must be stored in a tablespace in the target database, or the recovery manager cannot access the target database. True or False?

18. The CHANGE and CATALOG commands used with the RMAN repository stored in a target database's control file can also be used with the recovery catalog. True or False?

19. By default, information contained in the recovery catalog views can be accessed only from the _____ user account.

11

20. From the RMAN> prompt, the EXECUTE command must be included in a
    _____ command to execute a stored RMAN script.

    a. OPEN

    b. IMMEDIATE

    c. RMAN

    d. RUN

HANDS-ON ASSIGNMENTS

Assignment 11-1 Creating and Storing an RMAN Script

In this assignment you will create an RMAN script to perform a backup of a tablespace.

1. If necessary, log into the RMAN utility with a connection to the target database.

2. If necessary, make a connection to the recovery catalog.

3. Create a script named TOOLSBKUP that creates a backup set of the TOOLS tablespace.

Assignment 11-2 Executing a Stored RMAN Script

In this assignment you will execute the RMAN script created in Assignment 11-1.

1. Shut down and mount the target database in RMAN.

2. Execute the TOOLSBKUP script created in Assignment 11-1 to back up the TOOLS tablespace of the target database.

3. Use the LIST command to display the current backup sets being referenced by the recovery catalog. Identify the backup set created in Step 2.

4. Open the target database.

Assignment 11-3 Changing the Incarnation of the Target Database

In this assignment you will reset the incarnation number of the target database.

1. Using the RMAN utility, change the incarnation of the database to a prior incarnation.

2. Log into the database using SQL*Plus with the proper user account and query the RC_TABLESPACE view to determine the tablespaces associated with the current incarnation of the target database.

3. Through the RMAN utility, execute the necessary RESET DATABASE command to change the incarnation number of the target database back to the most current incarnation.

4. From SQL*Plus, query the RC_DATABASE view to determine the new incarnation number of the target database.

5. Exit RMAN.

Assignment 11-4 Resynchronizing the Recovery Catalog

In this assignment, you will resynchronize the recovery catalog with changes from the control file of the database.

1. In SQL*Plus, type **CREATE TABLESPACE synctab DATAFILE 'location\ synctab01.dbf' SIZE 100K;**, where *location* is the current location of the target database's data files, to create a new tablespace.

2. Start RMAN with a connection to the target database.

3. Connect to the recovery catalog.

4. Resynchronize the recovery catalog with the changes that have been recorded in the control file of the target database.

5. Exit SQL*Plus.

Assignment 11-5 Viewing a Previous Incarnation

In this assignment, you will view a previous incarnation of a database and observe that the SYNCTAB tablespace created in Assignment 11-4 is not included in that previous version of the database.

1. Using the RMAN utility, change the incarnation number to a value representing the database before the SYNCTAB tablespace was created (i.e., reset to any previous incarnation).

2. Use the REPORT command to display the schema of the database. Is the new tablespace included in the schema? Why or why not?

3. Change the incarnation number to the value representing the most current state of the database.

4. Reissue the REPORT command to display the schema of the database and make certain the SYNCTAB tablespace is now included.

Assignment 11-6 Viewing the Contents of a Stored Script

In this assignment, you will view the contents of the script created in Assignment 11-1.

1. If necessary, start SQL*Plus and log into the recovery catalog database.

2. Use the RC_STORED_SCRIPT view to verify that the TOOLSBKUP script exists.

3. Use the RC_STORED_SCRIPT_LINE view to examine the contents of the TOOLSBKUP script.

CASE PROJECTS

Case 11-1 Developing RMAN Backup Scripts

Carlos has decided to begin creating scripts that other employees can use to perform backup operations of the financial database. Carlos has identified three basic scripts that should be created:

◻ A script to create a full backup of the whole database, named FullBackup

◻ A script to create a Level 0 backup of the whole database, named Level0Backup

◻ A script to create a Level 1 backup of the whole database, named Level1Backup

Assuming that the database is in ARCHIVELOG mode, and will be shutdown during the backup process, create the three scripts.

Case 11-2 Procedure Manual

To make certain that new employees know how to connect to the recovery catalog and perform basic operations through RMAN, Carlos has asked you to add the steps necessary to complete the following tasks to the procedure manual that is currently being developed at Janice Credit Union:

◻ Start RMAN with a connection to the target database and then connect to the recovery catalog.

◻ Create an RMAN script.

◻ Execute an RMAN script.

LOADING AND
TRANSPORTING DATA

**After completing this chapter,
you should be able to do the following:**

♦ Identify the purpose of the Export, Import, and SQL*Loader utilities

♦ Perform an interactive export of a table

♦ Perform a scripted export of a table

♦ Perform an interactive import of an Oracle9*i* table

♦ Perform a scripted import of an Oracle9*i* table

♦ Load external data into an Oracle9*i* database

In previous chapters, Carlos learned various strategies for performing physical backup and recovery of an Oracle9*i* database. However, in this chapter Carlos will focus on performing logical backup and recovery operations through the use of the Export and Import utilities. The **Export utility** enables Carlos to export the contents of a table or a schema to create a backup of specific objects rather than entire tablespaces. If it becomes necessary, the **Import utility** can then be used to retrieve the exported data. These utilities are also commonly used to create a backup of the Recovery Catalog.

This chapter extends the discussion of importing Oracle-generated data by also presenting the **SQL*Loader utility**. The SQL*Loader utility can be used to load data from non-Oracle databases or any other source that can generate an ASCII text file. This is useful for organizations that utilize multiple database management systems by allowing data to be transferred from one type of database to an Oracle9i database. Basically, the data is exported in a comma-delimited or fixed-width format and is then loaded into an Oracle9i database using the SQL*Loader utility. This not only allows DBAs to convert data to an Oracle database, but it also allows DBAs to use the Oracle database software as a backup or alternative to other DBMS software in the event there is a database or hardware failure with the other system.

THE CURRENT CHALLENGE IN THE JANICE CREDIT UNION DATABASE

As the size of the Janice Credit Union's financial database increases, the time required to perform backup and recovery operations also increases. Carlos is evaluating the use of various utilities as a supplement to the credit union's current backup and recovery strategy. For example, if a small static table that contains the codes of different types of accounts is accidentally deleted, Carlos can have the table manually recreated or perform an incomplete recovery. With either alternative, the process could take a long time.

However, if the table had previously been exported to an external file using the Export utility, recovery would simply consist of importing the file. Recovery time would be greatly reduced because the entire data file would not need to be restored. However, this recovery approach is viable only for static tables. Importing a previous version of a dynamic table could leave the database in an inconsistent state if its contents are linked to other data stored in the database, causing more of a headache than that resulting from simply performing an incomplete recovery.

 In this chapter, you will work only with static tables.

SET UP YOUR COMPUTER FOR THE CHAPTER

Before attempting any of the procedures demonstrated in this chapter, make certain you manually create a cold backup of all database files, including each data file, the redo log files, each copy of the control file, and any initialization files. In the event an error occurs when completing the steps within this chapter, the cold backup can be used to recover the database. In addition, make certain the database is in NOARCHIVELOG mode; otherwise, the archived redo log files also need to be backed up.

Notes About Dual Coverage Within this Chapter

In this chapter, procedures for exporting and importing data are presented using both the command-line and GUI approaches. When using the GUI approach, you access each wizard through the Enterprise Manager Console. However, the wizards are available only if you are connected to an Oracle Management Server (OMS). If an OMS is not configured for your system, you can complete only the section demonstrating use of the command-line approach.

The Export Utility

A DBA can use the Export utility for a variety of tasks. For example, it can be used to export the contents of database tables, schemas, or even an entire database. This allows a DBA, such as Carlos, to move data from one Oracle9i database to another, or to transfer data when upgrading the database to a newer version, as in upgrading from Oracle8i to Oracle9i software.

In addition, Carlos can also use the utility to help reorganize the database. Suppose a particular tablespace contains a lot of dynamic tables, which frequently results in multiple users trying to concurrently access various parts of the same tablespace. This can have a detrimental impact on the overall performance of the system. With the Export utility, tables contained in one tablespace can be exported, and then imported into another tablespace (and even another Oracle9i database, if desired).

When the data is exported, the DBA can also instruct the Export utility to compress the extents, which is a default setting for the Export utility. If a table has 15 one-megabyte extents (recall that an extent is part of the database's logical structure), compression results in one 15-megabyte extent. This reduces any fragmentation in the resulting tablespace, as well as creates a larger initial extent if a new tablespace is created. However, the DBA needs to be careful and make certain that any resulting extent is not larger than the available contiguous free space that is needed to store the tablespace and that it does not violate any size restrictions imposed by the operating system, or an error will result.

As mentioned, in regards to backup and recovery operations, the Export utility can be used to create a logical backup of database tables and schemas. Generally the logical backup is used to supplement physical backups of the database. One reason for using the Export utility is to provide quicker recovery options for static database tables. However, the utility is rarely used to export the entire database for backup and recovery purposes because of the extensive inconsistencies that can occur if the entire database is recovered. This assumes the database is recovered via import operations and new tables have been created containing linked data. It is simply easier to recover the entire database through user-managed procedures or RMAN. An import of the entire database or any part of it is very easy to do. As discussed later in the chapter, the Export and Import utilities are commonly used as a backup and recovery option for the recovery catalog of the RMAN utility.

12

A combination of physical and logical recovery techniques are also used to perform a **tablespace point-in-time recovery (TSPITR)**. Generally TSPITR is used to recover tables that have been accidentally dropped from large databases. However, the procedure is fairly complex, requiring the DBA to create and recover a clone database, recover the lost table, and then use the SYS account to export the table and then import the table back into the original database.

Exporting Data

When the Export utility is used to export data from an Oracle9*i* database, a SELECT statement is executed to retrieve all the rows in a table. The retrieved data is then dumped into a binary file, referred to as a **dump file**. By default, the dump file is named expdat.dmp. After the data is dumped into the specified file, the utility also generates the appropriate data definition language (DDL) statements to recreate the table(s) and index(es) when the data is imported through the Import utility.

There are two types of exports that can be performed: conventional-path and direct-path exports. The main difference between the two types of exports is that the conventional-path export evaluates, or verifies, the DDL statements that are stored in the dump file, while the direct-path export skips this step entirely. The direct-path approach simply takes the retrieved rows, dumps the data into the dump file, and then generates and dumps the derived DDL statements into the file. By default, the Export utility uses the conventional-path export approach. However, the DBA can specify that the direct-path approach be used by including the DIRECT=Y argument when starting the Export utility. When exporting large amounts of data, the direct-path approach can save a lot of time.

There are several options available with the Export utility. A quick way to view the available options is to call up the Help feature available within the utility. In the following step sequence, Carlos accesses the command-line version of the Export utility and includes the HELP option (-help) to display the keywords available within the utility.

To access the Export utility:

1. Click the **Start** button on the Windows XP taskbar.

2. Click **Run** from the Start menu.

3. In the Run dialog box, type **cmd** and then click **OK**.

4. When the command-line window appears, type **exp −help** at the system prompt, and press **Enter**. The resulting output displays the options and keywords available in the Export utility, as shown in Figure 12-1.

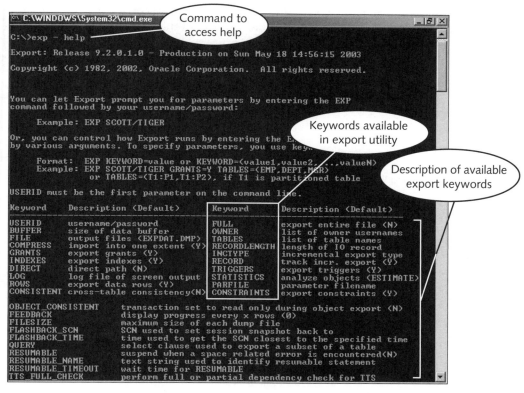

Figure 12-1   Options and keywords available in Export utility (partial output shown)

The output shown in Figure 12-1 provides the user with a brief explanation of how to use the Export utility. Basically, the Export utility is called up by entering the EXP command at the system prompt. The user has the option of including various arguments when starting the utility. In this particular example, Carlos used the HELP argument to access the Export utility and display information regarding the utility.

Notice the first example given in the display. The Export utility can also be started with the user's name and password specified. If this information is not provided when the Export utility is started, the user is required to enter it before any data can be exported from the database.

In the following section, Carlos will use some of these keywords to export data from a table. When a keyword is not specified or a prompt is not provided to input the value for a specific object, the default value shown in parentheses in Figure 12-1 is assumed.

Using the Export Utility—Command-Line Approach

The command-line approach to using the Export utility can be utilized in two ways: interactively or scripted. With the interactive method, the user is required to answer a series of prompts to export the desired data. However, the utility does not provide a prompt for all available keywords or options. As an alternative, the necessary options can be scripted, or specified, in a parameter file called a PARFILE. In this section, Carlos first creates and then exports a database table using the Export utility interactively. Then he creates a parameter file that can be used to perform the same export operation in a scripted manner.

Keywords and values are not case sensitive in the Export and Import utilities.

To practice using the Export utility, Carlos first creates a table named TRANSTABLE.

To create TRANSTABLE:

1. Log into SQL* Plus with SYSDBA privileges.

2. At the SQL> prompt, type **@*location*\ch12script01.sql**, where *location* is the path to the script file, to create TRANSTABLE, as shown in Figure 12-2.

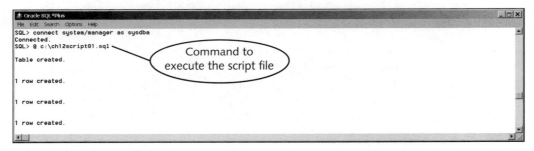

Figure 12-2   Executing the Ch12script01.sql script file

3. At the SQL> prompt, type **SELECT * FROM system.transtable;** to verify the contents of the newly created table. As shown in Figure 12-3, the table consists of ten rows.

```
± Oracle SQL*Plus                                                        _□×
File Edit Search Options Help
SQL> SELECT × FROM transtable;

    TRANS#     ACCT#
---------- ----------
    282088  822883208
    282086  842820202
    282080  622688832
    282088  840682060
    282082    2342032
    282028  823460603
    282020   22888000
    282022  828000202
    282083  884020346
    282084   88008824

10 rows selected.

SQL>
```

Figure 12-3   Contents of the TRANSTABLE table

4. Exit SQL*Plus at this time only if you are unable to complete the entire chapter during one session. SQL*Plus is used later in the Import utility example to verify that the table was imported correctly.

The Ch12script01.sql script file specifies that the table should be created in the SYSTEM schema. If your instructor requests that the schema be changed, open the script file in Notepad and change all SYSTEM references to the appropriate schema. The process can be simplified by using the Notepad Find and Replace commands on the Edit menu.

Now that the practice table has been created, Carlos can export the table using the Export utility.

To export the TRANSTABLE table:

1. If necessary, open the command-line window by clicking **Start** and then **Run** on the Start menu of the Windows XP operating system. Type **cmd** in the Run dialog box, and then click **OK**.

2. Type **exp *username/password@database*** to connect to the Export utility, substituting your user name for *username*, your password for *password*, and the name of your database for *database*, as shown in Figure 12-4.

3. When prompted for the desired size of the buffer that is used to temporarily hold the data being retrieved, press **Enter** to accept the default size of 4 KB. A larger buffer size can increase performance by reducing the number of input\output operations required. The maximum size allowed for the buffer is 64 KB.

4. Type ***location*ch12transtable.dmp**, where *location* is the path for your data disk, when the Export file prompt appears to specify the name and location of the dump file that will contain the exported data. If Enter is pressed without specifying the name of a destination file, the data is dumped into the default file named Expdat.dmp and stored in the default directory identified in the system prompt.

5. When prompted for the type of object to be exported, type **t** to specify that a table is to be exported. As an alternative, Carlos can enter a 3. If Enter is pressed without entering a choice, a user schema is specified by default.

6. When prompted as to whether to export the actual table data (rows) or just the table definition, type **y** to indicate that the table definition and the rows it contains should be exported.

12

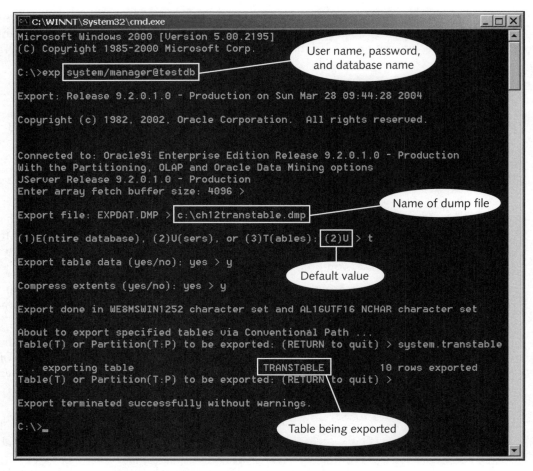

Figure 12-4 Interactive use of the Export utility

7. When prompted as to whether the extents should be compressed, type **y** to indicate yes.

8. When prompted to identify the table or partition to be exported, type **system.transtable** to specify that the TRANSTABLE table in the SYSTEM schema is to be exported. If you previously changed the name of the schema when the Ch12script01.sql script file was executed, substitute the approach schema name for SYSTEM.

9. When prompted to identify a second table or partition to be exported, press **Enter** to indicate that only the TRANSTABLE is to be exported. If there were more tables to be exported, the name of the table would be provided at this point. After pressing **Enter**, the transtable is exported, and then the Export utility terminates.

As shown in Figure 12-4, if the export process is completed successfully, the message "Export terminated successfully without warnings." is displayed. At this point, Carlos has dumped the TRANSTABLE table into a binary file which can be used at a later time to import the data into an Oracle9*i* database.

As mentioned, Carlos also has the option of automating the process by storing the necessary keywords in a parameter file. In the following example, Carlos exports the same table to a second dump file.

To create the parameter file:

1. Click the **Start** button on the Windows XP taskbar.

2. Click **Programs** on the Start menu.

3. Click **Accessories** on the Programs menu.

4. Click **Notepad** on the Accessories submenu.

5. In Notepad, type **USERID=*username*/*password*@*database***, where *username* is your user name, *password* your password, and *database* is the name of the database containing the TRANSTABLE database table. Then press **Enter**.

6. Type **TABLES=(SYSTEM.TRANSTABLE)** to identify the appropriate table to be exported and press **Enter**. Remember to replace SYSTEM with the appropriate schema name if the table was created in a different schema. If more than one table needs to be exported, then each table name is listed inside the parentheses, with each name separated by a comma.

7. Type **COMPRESS=Y** to specify that the extents should be compressed when the table is exported.

8. Click **File** on the menu bar, and then click **Save**.

9. When the Save As dialog box appears, change the location specified in the Save in text box to the appropriate location for your data disk. Then type **Ch12parfile01.pr** in the File name text box to specify the name of the parameter file. Click the **Save as type** drop-down arrow, and click **All files**. Click the **Save** button.

10. Verify that the parameter file you have created resembles the one shown in Figure 12-5, and then click **File** and then **Exit** on the menu bar to exit Notepad.

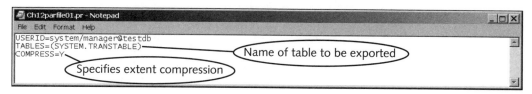

Figure 12-5   Parameter file for the Export utility

Carlos can now proceed with the actual export task.

To export the TRANSTABLE using the Ch12parfile01.pr file:

1. If necessary, open the command-line window through the operating system.

2. At the system prompt, type **exp parfile=*location*\Ch12parfile01.pr file=ch12transtable2.dmp**, substituting *location* with the approach location for the file, and press **Enter**. The location of the dump file will be stored in the default directory because a location was not specified in the FILE argument.

As shown in Figure 12-6, the Export utility is called and executed using the options specified in the Ch12parfile01.pr previously created (as specified by the PARFILE argument). The data being exported is dumped into the file named Ch12transtable2.dmp as specified by the FILE argument included with the EXP command.

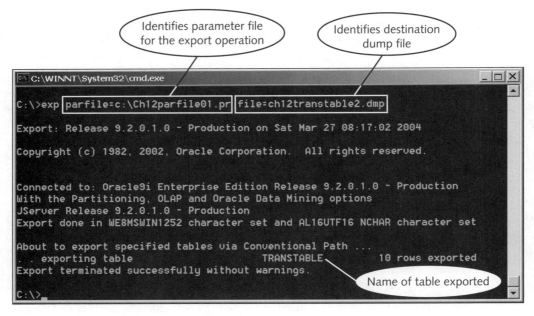

Figure 12-6   Scripted use of the Export utility

At this point, Carlos has the option of either including arguments such as TABLES and COMPRESS in the parameter file or as part of the EXP command when the utility is accessed. The rationale for determining where an argument should be referenced depends on whether the assigned value is dynamic or static. For example, values that do not change with each execution of the parameter file, such as the user information, compression, and so forth, can be placed in the parameter file. However, unless Carlos always wants the exported data written to the same file, the FILE keyword is usually included when the Export utility is accessed so that the destination file can be identified each time an export occurs. Otherwise, Carlos needs to open the parameter file before each export operation is performed and change the name of the dump file.

Notice that some of the information previously requested when the utility was run interactively is not included in the parameter file (e.g., whether the table data should be exported, etc.). If an option is not included in the parameter file, or it is not specified when the Export utility is accessed, the default value for that option is assumed.

Using the Export Utility—GUI Approach

In this section, Carlos uses the Export Wizard available through the Enterprise Manager Console to export the TRANSTABLE to a dump file. As with previous wizards, Carlos is required to respond to a series of prompts. Basically, the wizard creates a parameter file that is submitted as a job to be executed by the Enterprise Manager. Therefore, you can consider the use of the wizard as the interactive counterpart of using the command-line approach and the resulting job as the parameter file counterpart.

To export the TRANSTABLE using the Export Wizard:

1. Launch the Enterprise Manager Console and connect to the Oracle Management Server with a SYSDBA privileged account.

2. Click the **+** (plus sign) next to Databases in the object list on the left side of the console.

3. Click the name of the database from the Database list to identify which database contains the table to be exported.

4. Click **Tools** on the menu bar. On the Tools menu, click **Database Tools**, **Data Management**, and then **Export**, as shown in Figure 12-7.

5. If the Introduction screen appears, click **Next**.

6. When the Export File screen appears, click **Next** to indicate that the default dump file name should be used for the destination of the exported data, as shown in Figure 12-8.

12

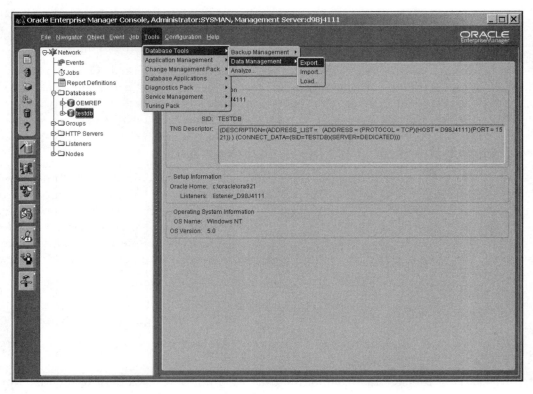

Figure 12-7   Accessing the Export Wizard

Figure 12-8   Specifying the export's dump file name

7. When the Export Type window shown in Figure 12-9 appears, click the **Table** option button to indicate that a table is to be exported rather than the entire database or a particular schema. Then click **Next**.

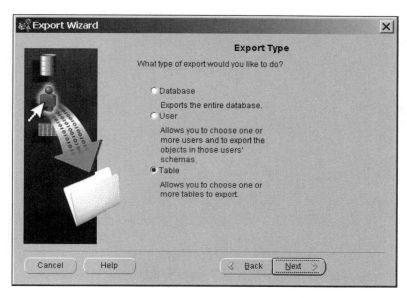

Figure 12-9   Specifying the type of object to be exported

8. When the Table Selection window appears, use the scroll bar in the Available Tables portion of the window to locate the name of the schema containing the TRANSTABLE table. This should be the SYSTEM schema unless your instructor previously specified a different schema. Click the **+** (plus sign) next to the appropriate schema name to display the list of available tables.

9. After the schema's objects are displayed, use the scroll bar to locate the TRANSTABLE table, and click the table's name. Click the right arrow located between the Available Tables and Selected Tables lists to display the table name in the right pane of the window as shown in Figure 12-10. Then click **Next** to display the other options available through the Export Wizard.

10. When the Associated Objects window appears, verify that the check box for the Rows of table data object is checked, as shown in Figure 12-11, to indicate that the data and the table definition are to be exported. Then, click the **Next** button.

11. When the Schedule window appears, click **Next** to indicate that that job is to be submitted and executed immediately.

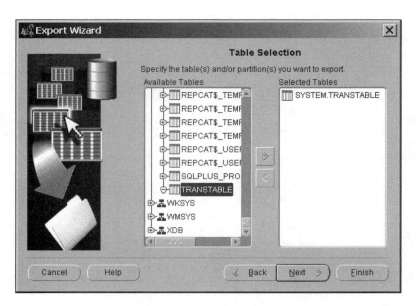

Figure 12-10   Selecting the table to be exported

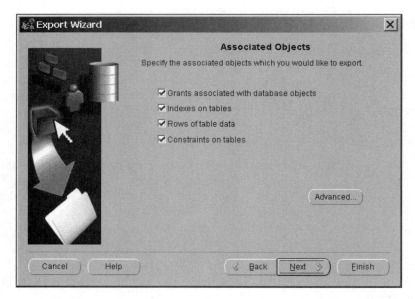

Figure 12-11   Specifying the associated objects to be exported with the
TRANSTABLE table

12. When the Job Information window appears, click **Finish**.

13. When the Summary Window shown in Figure 12-12 appears displaying the
job to be submitted and the contents of the parameter file, click **OK**.

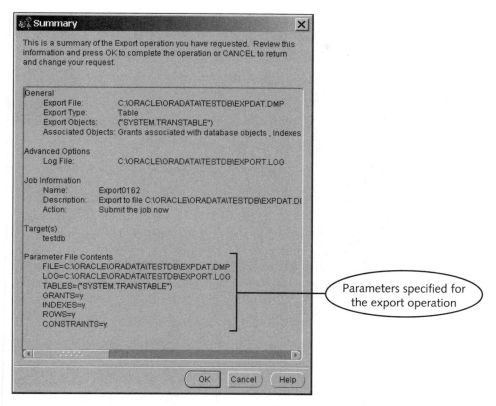

Figure 12-12   Summary information regarding export operation to be performed

14. When the dialog box appears indicating that the job has been successfully submitted, click **OK**.

As always, you can verify that the job has been executed successfully by viewing the history portion of the Jobs window available within the Enterprise Manager console.

Now that Carlos has exported the TRANSTABLE table, he can practice using the Import utility.

THE IMPORT UTILITY

As mentioned, Carlos can use the Export utility to create a logical backup of a database. So what happens if a table is accidentally dropped and needs to be recovered from the logical backup? Carlos can simply use the Import utility to restore the table. The Import utility is used to perform a logical recovery of an Oracle9*i* database. Basically, the utility reads the DDL statements stored in the binary dump file previously created by the Export utility. If the table does not already exist, the table is re-created. After the table is re-created, the data contained in the dump file is then inserted back into the table.

The Import utility is called by entering the command IMP at the system prompt. As with the Export utility, you can also view the available options using the –HELP argument, as shown in Figure 12-13.

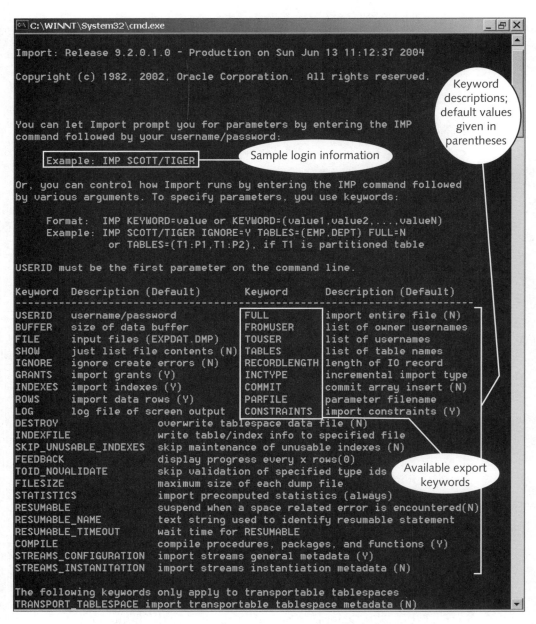

Figure 12-13 Available Import utility options (partial output shown)

Similar to the Export utility, the Import utility can be used in an automated or scripted mode. However, before Carlos can practice using either mode of the Import utility, he first needs to drop the TRANSTABLE table within the database through SQL*Plus, and then he can perform the logical recovery of the table using the Import utility.

To drop the TRANSTABLE table:

1. If necessary, start SQL*Plus and log into the database with SYSDBA privileges enabled.

2. At the SQL> prompt, type **DROP TABLE system.transtable;** to drop the specified table as shown in Figure 12-14. If the table was created in a different schema, substitute the appropriate schema when entering the DROP TABLE statement.

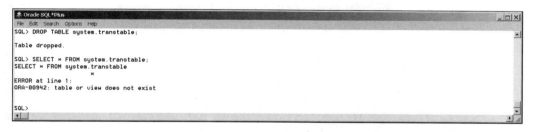

Figure 12-14   Dropping the database table

3. Type **SELECT * FROM system.transtable;** at the SQL> prompt, as shown in Figure 12-14, to verify that the table no longer exists. Remember to specify the correct schema name if it was previously changed.

Now that the table has been dropped, Carlos can use the Import utility to perform a logical recovery of the table.

Using the Import Utility—Command-Line Approach

As mentioned, Carlos has two options when using the command-line approach to access the Import utility. He can either use the utility interactively or with a parameter file. In the first sequence of steps in this part of the chapter, Carlos lets the Import utility prompt him for the information necessary to logically recover the TRANSTABLE table. In the second sequence of steps in this part of the chapter, he creates a parameter file and uses that file to specify the options necessary to recover the table.

To recover the database table interactively through the Import utility:

1. If necessary, open the command-line window through the operating system.

2. At the system prompt, type **imp system/manager@testdb**. As in the previous section, change the user name, password, or database name if appropriate.

3. As shown in Figure 12-15, when prompted for the name of the file to be imported, type **c:\ch12transtable.dmp**. If necessary, modify the location of the dump file to the location previously specified when the file was created.

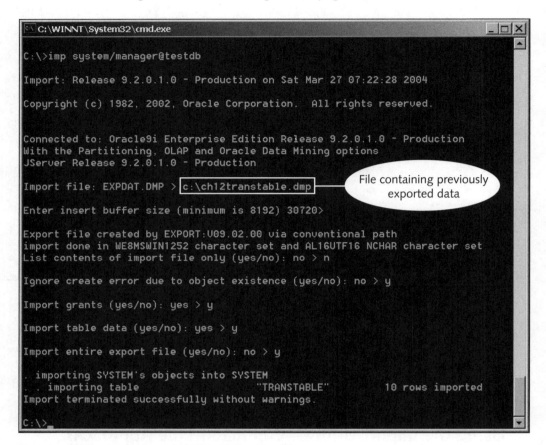

Figure 12-15   Importing the TRANSTABLE table

4. When prompted for the buffer size to be used when inserting data into the table, press **Enter** to accept the default value of 30 KB.

5. When prompted as to whether the contents should only be displayed or actually recovered, type **n** to indicate that the data should actually be imported into the database.

6. When prompted as to whether an error should be ignored if the object already exists, type **y** to indicate that the error message should be ignored.

7. When prompted as to whether any exported grants for the objects should be imported, type **y** to indicate that the grants should be imported.

8. When prompted as to whether or not the table data should be imported, type **y** to indicate that both the table definition and the table data should be imported.

9. When prompted as to whether or not the entire contents of the export (dump) file should be imported, type **y** because the file only contains the table that is to be imported.

10. After the Import utility returns the message "Import terminated successfully without warnings.", type **/** in SQL*Plus to verify that the table was actually recovered, as shown in Figure 12-16.

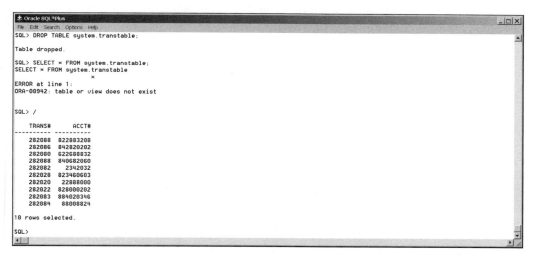

Figure 12-16   Recovered contents of the TRANSTABLE table

Carlos can also create a parameter file to import the TRANSTABLE database table. In the following example, Carlos again drops the table, and then creates a parameter file and performs a logical recovery using the Import utility along with the parameter file.

To drop the database table a second time:

1. At the SQL> prompt, type **DROP TABLE system.transtable;** to drop the specified table. Remember, if the table was created in a different schema, substitute the appropriate schema when entering the DROP TABLE statement.

2. Type **SELECT * FROM system.transtable;** at the SQL> prompt to verify that the table no longer exists.

Now that the table has been dropped, Carlos creates the parameter file that is used to specify the options to be included in the logical recovery.

To create the file:

1. Click the **Start** button on the Windows XP taskbar.

2. Click **Programs** on the Start menu.

3. Click **Accessories** on the Programs menu.

4. Click **Notepad** on the Accessories submenu.

5. In Notepad, type **USERID=***username/password@database*. As before, remember to substitute the appropriate user name, password, and name of the database.

6. Type **TABLES=(TRANSTABLE)** to identify the appropriate table to be imported. When importing a database table, the schema name cannot be specified with the TABLES option; it must be specified using the FROMUSER option.

7. Type **IGNORE=Y** to specify that any error message generated because the object already exists should be ignored.

8. Type **ROWS=Y** to specify that the data, as well as the table definition, is to be imported.

9. Click **File** on the menu bar, and then click **Save**.

10. When the Save As dialog box appears, change the location specified in the Save in text box to the location of your data disk. Then type **Ch12parfile02.pr** in the File name text box to specify the name of the parameter file. Click the **Save as type** drop-down arrow, and click **All files**. Click the **Save** button.

11. Verify that the parameter file you have created contains the same information as the one shown in Figure 12-17, except for the user name, password, and database name, then click **File** and **Exit** on the menu bar to exit Notepad.

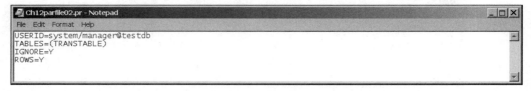

Figure 12-17   Parameter file to import data through the Import utility

At this point Carlos is ready to recover the dropped database table. Remember the previous export example that used the parameter file to export the TRANSTABLE table? In that example, the table was exported to the Ch12transtable2.dmp dump file. This file will be used as the source dump file for this particular recovery example.

To recover the TRANSTABLE database table using the parameter file:

1. If necessary, open the command-line window through the Windows XP operating system.

2. At the system prompt, type **imp parfile=*location*\Ch12parfile02.pr file=*location*\ch12transtable2.dmp**, where *location* is the path for your data disk, as shown in Figure 12-18, where the PARFILE argument specifies the location and name of the parameter file, and the FILE argument specifies the name of the source dump file containing the data to be imported, and press **Enter**.

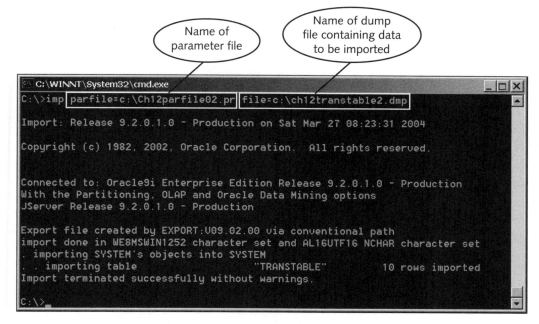

Figure 12-18 Automated import of the TRANSTABLE database table

The message "Import terminated successfully without warnings." indicates that no errors were encountered while the database table was being imported. As before, Carlos can verify that the table was imported correctly by reissuing the SELECT statement to view the contents of the table.

Using the Import Utility—GUI Approach

In this section, Carlos uses the Import Wizard to import a dropped table. Although the wizard provides a point-and-click approach to importing data, the data could have been previously exported using either the command-line or GUI approach. The resulting dump file will be the same. The decision to use a particular approach lies with the individual DBA and his or her "comfort zone." Some individuals prefer to interact directly with the utility while others are more comfortable in a GUI environment.

The Import Wizard is available through the Enterprise Manager Console when connected to the Oracle Management Server (OMS). As with the Export Wizard, Carlos is required to respond to a series of prompts provided by the Import Wizard. In essence, these responses are used to create a parameter file that is referenced during the actual importing of the specified data. For inexperienced DBAs, the information provided in the resulting parameter can be referenced as the individual becomes more familiar with the various keywords available through the Import utility.

Before using the Import Wizard, Carlos needs to drop the TRANSTABLE database table. The following steps can be performed through either SQL*Plus or the SQL*Plus Worksheet to remove the table from the database.

To drop the TRANSTABLE database table, perform the following steps:

1. Type **DROP TABLE system.transtable;** to drop the specified table. However, if the table was created in a different schema, you need to substitute the appropriate schema when entering the DROP TABLE statement.

2. Type **SELECT * FROM system.transtable;** at the SQL> prompt to verify that the table no longer exists.

Because the table has been dropped, Carlos can now use the Import Wizard to recover the TRANSTABLE table.

To recover the table:

1. Start the Enterprise Manager Console, and connect to the Oracle Management Server with a SYSDBA privileged account.

2. Click the **+** (plus sign) next to Databases in the object list on the left side of the console.

3. Click the name of the database from the Database list to identify which database contains the table to be exported.

4. Click **Tools** on the menu bar. On the Tools menu, click **Database Tools**, **Data Management,** and then **Import** from the submenus that subsequently appear, as shown in Figure 12-19.

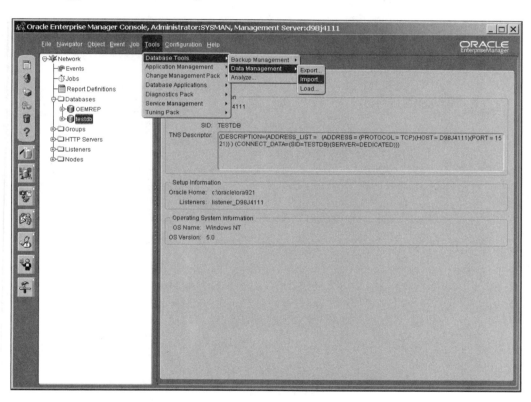

Figure 12-19   Accessing the Import Wizard

5. If the Introduction window appears, click **Next**.

6. When the Import File window shown in Figure 12-20 appears, click the first option button if necessary, and then click **Next** to import the contents of the default dump file.

Figure 12-20   Specifying the dump file containing the data to be imported

7. When the Import Type window shown in Figure 12-21 appears, click the **Table** option button. In the Users text box, type **SYSTEM**, or the appropriate name of the schema containing the table to be imported. In the Tables text box, type **TRANSTABLE**, and then click **Next**.

8. When the User Mapping window shown in Figure 12-22 appears, simply click **Next** because the table is being imported into the same schema from which it was previously exported.

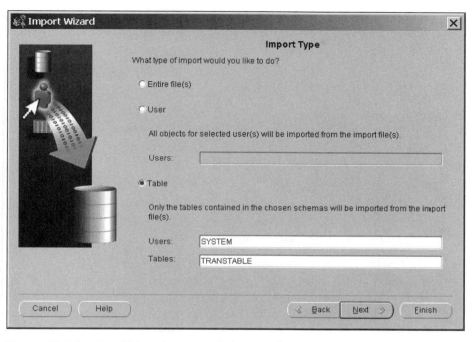

Figure 12-21 Specifying the type of object to be imported

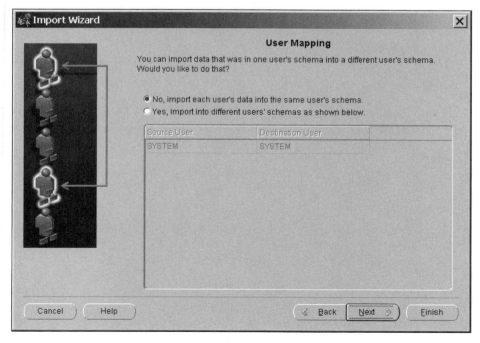

Figure 12-22 Option to import the object into a different schema

9. When the Associated Objects window shown in Figure 12-23 appears, verify that at least the box for the Rows of table data object contains a check mark, and then click **Next**.

10. When the Schedule window appears, click **Next** to indicate that the job should be executed immediately.

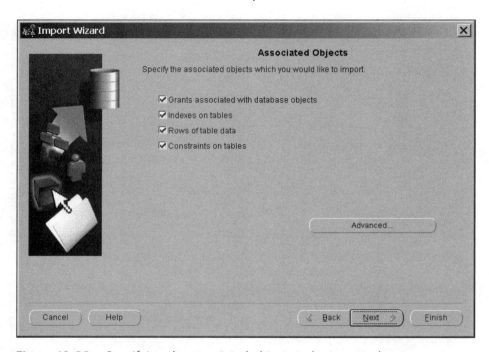

Figure 12-23   Specifying the associated objects to be imported

11. When the Job Information window shown in Figure 12-24 appears, click **Finish** to indicate that the job should be submitted now rather than have the parameter file added to the job library.

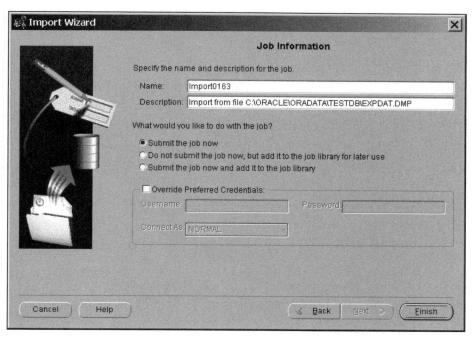

Figure 12-24   Submission information regarding the import job

12. When the Summary window appears, click **OK**. As shown in Figure 12-25, the lower portion of the window provides the options and their values that are referenced when the import operation is performed.

13. When the dialog box appears indicating that the job has been successfully submitted, click **OK**.

To verify that the import was successfully executed, Carlos can issue a SELECT statement referencing the imported table to retrieve the ten imported rows. After verifying that the import was successful, any modifications should be made to ensure the table is up to date and ready for use.

12

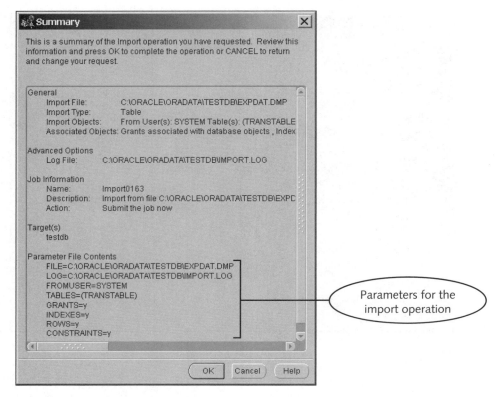

Figure 12-25 Summary information

BACKUP AND RECOVERY OF THE RECOVERY CATALOG

Oracle recommends keeping the Recovery Catalog in a separate database from the target database. When the Recovery Catalog is created, it is stored in the RMAN schema. For backup and recovery purposes of the catalog, Carlos has three choices: (1) create physical backups of the database containing the Recovery Catalog, (2) create logical backups of the RMAN schema, or (3) a combination of both.

Oracle recommends that a strategy consisting of both physical and logical backups be used to protect against loss of the Recovery Catalog. In addition to creating a physical backup of the actual database containing the catalog, Carlos can also perform nightly exports of the RMAN schema after RMAN backups of all the other databases have been created. In the event a database failure occurs with the database containing the Recovery Catalog, either the entire database can be recovered from the physical backup or the schema can simply be imported into a new database. Because the Recovery Catalog is not highly dynamic and its contents change only when RMAN performs backup or recovery operations, it is an ideal candidate for a logical backup strategy. The following is an example of the EXP command and arguments that can be used to export the Recovery Catalog:

```
exp rman/rman@database_name
file=dumpfilename.dmp owner=rman
```

Inclusion of the OWNER argument indicates that the entire RMAN schema should be exported, rather than just a table. Execution of this command on a nightly basis provides extra protection against loss of all information regarding the backup and recovery operations that have been performed by the Recovery Manager utility. If the information stored in the catalog becomes lost or corrupt, the RMAN schema could simply be imported.

SQL*LOADER

Janice Credit Union has several departmental databases that use a simplified DBMS to accommodate inexperienced Oracle users. However, Carlos needs to be able to extract the data from these databases to update the contents of the Oracle9i database. Because the Import utility is designed to work only with the data previously exported from an Oracle9i database, this presents a problem. How can Carlos import the data generated by a different type of database? Fortunately, he can do so by using the SQL*Loader utility. The SQL*Loader utility available in the Oracle9i database is designed to transfer or load data from non-Oracle databases that have been stored in a comma-delimited text file or fixed-width records file. The utility uses three types of files during an import operation: a control file, an input file, and log files. Each are discussed in turn.

The Control File

The control file specifies what data should be loaded, the format of that data, and what procedures to use during the loading process—in essence, the parameter file for the load operation. The data to be loaded into the database can be included in the control file or stored separately in an input file.

The Input File

An input file, also called a data file, is used to store the data to be loaded into the Oracle9i database if that data is not included in the control file. Most commonly, the data is stored in an external file in a comma-delimited format, which can easily be generated by various database and spreadsheet programs. However, this is not required because the DBA can specify what type of symbol is used to separate the fields. By default, the SQL* Loader utility requires that each record be placed on a separate line in the input file.

The Log Files

The log files consist of three types. A **bad log** is created during the loading process to store any data that was rejected due to a violation of a table constraint or that have the wrong data type. A **discard file** may also be created if a criteria was specified for the data and certain rows failed to meet the criteria. A **general log file** is also created to provide information regarding when the load process was started and ended, how many records or rows were read, how many were discarded or bad, and how many were actually loaded into the table.

Bad Log Files

As mentioned, there are three types of log files. The bad log file stores any rows rejected because they violate a constraint (e.g., PRIMARY KEY, NOT NULL, etc.) or because they are not consistent with the column definition. In Figure 12-26, an entry for one of the columns exceeded the defined column width. When Carlos attempts to load the data into the database table, that row is rejected and written to the bad log file.

Figure 12-26   Contents of the bad log file

Notice that the format of the bad log file is basically the same as the input file. Why is this important? This allows the DBA to make any necessary corrections to the data and then simply change the extension of the file and subsequently use it to load the modified data into the database table. (As an alternative, Carlos could simply specify a different file name for the bad log file when loading the modified data into the database.)

Discard Log File

In this particular example, no criterion is specified for selecting which records should be loaded into the database. Therefore, no discarded rows are generated during the loading process. However, if a discard log file were created by SQL*Loader, it would have the same format as the original input file and bad log file to allow the DBA to make any modifications to the criteria and subsequently load any missed data into the database table—or perhaps into a different database table.

Suppose Carlos only needs to load data from transactions affecting accounts with account numbers higher than 600000. How would he specify this criterion? Carlos can simply modify the control file to include a WHEN clause followed by the criteria, as shown in the following modified control file:

```
LOAD DATA
INFILE 'c:\Ch12data.dat'
BADFILE 'c:\Ch12data.bad'
DISCARDFILE 'c:\Ch12data.dsc'
APPEND INTO TABLE system.transtable
WHEN acct# > 600000
FIELDS TERMINATED BY ','
```

```
(trans#, acct#)
```
Any records not meeting the specified criterion of acct# > 600000 are written to the discard log file.

General Log File

The final log file to be discussed is the general log file, sometimes simply referred to as a log file. As shown in Figure 12-27, the log file generated during the loading process provides information regarding the files referenced, parameter values used (both specified and default values), rows rejected and rationale for the rejection, as well as the number of rows processed. In essence, the general log file provides Carlos with an overview of the loading process and lets him know whether any problems occurred (e.g., rejected rows). If necessary, he can consult the appropriate bad or discard log file to determine the root of any problems, and thus make any necessary changes.

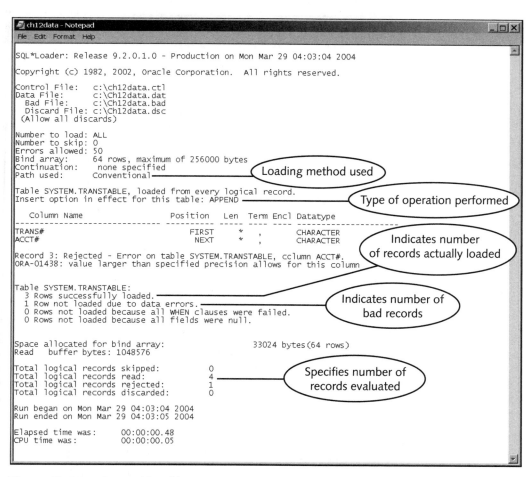

Figure 12-27   General log file

The Import Operation

Now that you have an idea of the types of files that can be referenced during a load operation, you should examine the actual contents of the files. Figure 12-28 shows the contents of the control file that Carlos uses to load data into the TRANSTABLE database table. The first line of the Ch12data.ctrl control file contains the statement LOAD DATA. This statements instructs SQL*Loader that this is the beginning of a data load. This statement is a requirement.

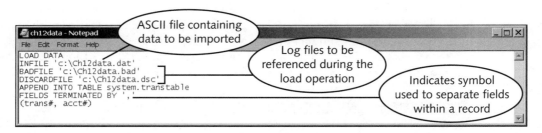

Figure 12-28   Control file for a data load process through SQL* Loader

The second line in Figure 12-28 specifies the INFILE parameter, which identifies the external file containing the data to be loaded. If the data has to be included in the control file itself, the parameter uses an asterisk (*) rather than a file name—for example, INFILE *. In addition, a section has to be added to the end of the control file, preceded by the keyword BEGINDATA, to list the data to be loaded into the database.

The third and fourth lines identify the names of the bad and discard log files, respectively. If the names for these files are not included as parameters in the control file and they become necessary because some records are rejected, they are created by default. The default name for the bad log file is the name of the control file itself, with the extension of .bad. The default file name for the discard log file is the name of the control file followed by an extension of .dsc, while the general log file would have the .log. If Carlos decides to specify the name of the general log file in the control file itself, he uses the parameter LOG followed by the name of the file in single quotation marks.

The fifth line of the control file in Figure 12-28 identifies whether the data being loaded into the database table should replace the existing data (REPLACE), be added to the data already contained in the table (APPEND), or be entered into an empty table (INSERT). In this example, the data is added to the existing database table because Carlos has specified the APPEND keyword. The keywords INTO TABLE followed by the table name are used to identify into which table the data is loaded.

The subsequent two lines provide more information regarding the data and the receiving table. The keywords FIELDS TERMINATED BY are used to specify what symbol is used to separate the individual fields within a record. Because some databases export non-numeric data within quotation marks or other symbols, Carlos can modify this parameter to include an OPTIONALLY ENCLOSED BY clause to indicate that some of the columns may be enclosed within special symbols. For example, if alphanumeric data were enclosed in double quotation marks when it was exported from the non-Oracle database, the line would have been modified and displayed as such:

```
FIELDS TERMINATED BY ',' OPTIONALLY ENCLOSED BY ' " '
```

The last line of the control file identifies the names of the columns that are to receive the data being loaded. The column names must be enclosed in parentheses. The order in which the column names are listed in the control file indicates the order of the data within the input file, not the order of the columns within the destination table. As an option, SQL*Loader allows the DBA to specify the format of the data being loaded (the format of dates, for example) and the type of data being loaded—characters, and so forth. However, because the data being loaded is in the same format as the table columns, the formats have been omitted.

If the data to be loaded is not stored in the control file, it must be provided in an external input file. Figure 12-29 shows the input file that Carlos uses to add data to the TRANSTABLE database table. The file is in a typical comma-delimited format. This means that each column or field of data is separated by a comma. As mentioned, any symbol can be used as long as it is specified by the FIELDS TERMINATED BY parameter in the control file.

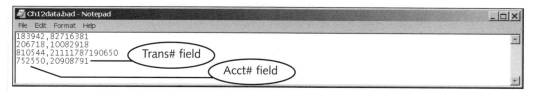

Figure 12-29   Example input file

Notice that the third line of data to be loaded in the database contains an unusually long entry. When Carlos attempts to load this data into the database table, the length of the field exceeds the maximum width defined for the ACCT#, causing this row to be rejected. Because it violates the defined data type, or more specifically the data width, this row is written to the bad log file generated by the SQL*Loader utility.

LOADING METHODS

Notice the Path used entry in the general log file previously displayed in Figure 12-27. This entry identifies the type of loading method used by SQL*Loader during the loading process. There are three loading methods that can be used by SQL*Loader: conventional-path, direct-path, and parallel direct-path.

When the conventional-path method is used to load data into a database table, Oracle9*i* basically uses an INSERT statement to insert each row of data. As each row is processed, any triggers are fired and referential integrity is checked. Although this is the default method for the SQL*Loader utility, it is slower than the alternative methods because the data is processed through the buffer cache, just like a traditional, or conventional, SQL INSERT statement.

The direct-path method bypasses the buffer cache and writes directly to the data file. In this case, the table and any indexes associated with the table are locked during the loading process, thus preventing other users from making changes to the table. Any associated triggers and associated referential integrity constraints are disabled when the direct-path method is used. In addition, only rows that violate a NOT NULL constraint are rejected. If a row violates a PRIMARY KEY or UNIQUE constraint, the row is added to the table, however the respective index is marked as unusable. The offending data needs to be corrected before the constraints can be re-enabled.

The parallel direct-path method is simply a variation of the direct-path method. It uses multiple load sessions to allow multiple data segments to be written at the same time. However, this approach uses more storage space during the loading process and requires the use of parallel processes. Of the three methods, the parallel direct-path method is the fastest, while the conventional-path method is the slowest.

 In addition, the direct-path method also differs from the conventional-path method because it loads the data above the high water mark (HWM). After the loading process is complete, the HWM is reset and the locks on the database table are released.

Loading Data with SQL*Loader—Command-Line Approach

In this section Carlos loads the records to be added to the TRANSTABLE table using the SQL*Loader utility. Although this example only includes four records, the procedure is the same even if Carlos were loading thousands of records into the table (it would just take longer).

The control and input files to be referenced during this example are provided in the Chapter12 folder of your data disk. You need these files to complete this example.

To load the data into the TRANSTABLE database table:

1. Log into SQL*Plus with SYSDBA privileges.

2. At the SQL> prompt, type **@ _drive_\Chapter12\ ch12script01.sql** to re-create the TRANSTABLE table, where _drive_ is the disk drive containing your data disk.

3. At the SQL> prompt, type **SELECT * FROM system.transtable;** to verify that the table consists of ten rows.

4. Click the **Start** button on the Windows XP taskbar, and then click **Run**.

5. In the Run dialog box, type **cmd** and then click **OK**.

6. At the system prompt, type **sqlldr *username/password@database* control=*drive*\Chapter12\ch12data.ctl**, as shown in Figure 12-30, to perform the load operation. Substitute the correct information for the *username*, *password*, and *database* name for the user ID and the correct *drive* letter for the location of the control file.

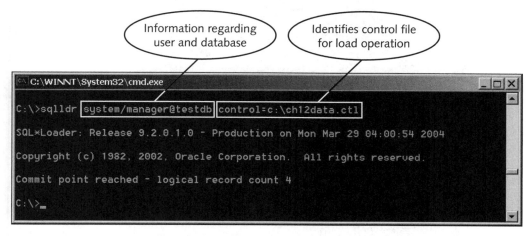

Figure 12-30   Using SQL*Loader to load data into an Oracle9*i* database table

In Figure 12-30, SQLLDR is the command that calls the SQL*Loader. As with the Export and Import utilities, the command can include various arguments. In this case, the user's name, password, and database name are provided, as well as the name and location of the control file as identified by the CONTROL parameter. If the CONTROL parameter is omitted, SQL*Loader prompts the user for the name of the control file to be referenced during the load procedure. However, as shown in Figure 12-31, if just the command to start the SQL*Loader is entered, then a list of available keywords and options are displayed, just as if the –HELP argument had been included.

12

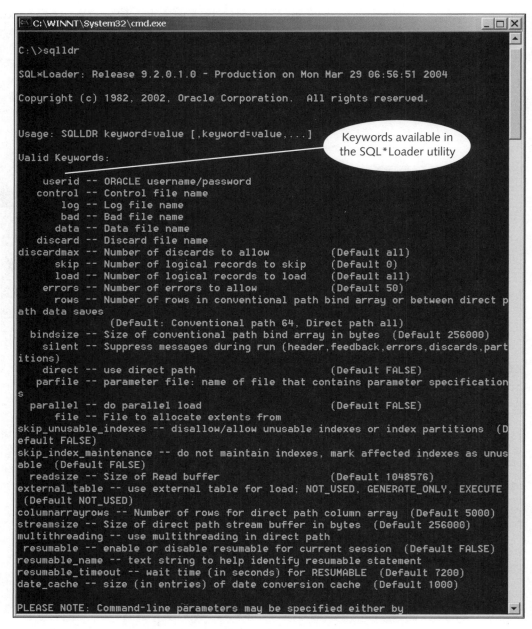

Figure 12-31 Keywords and options available with SQL*Loader

Now that Carlos has loaded the external data into the database, he can verify the modified contents of the TRANSTABLE table by issuing a SELECT statement through SQL*Plus. In Figure 12-32, the SELECT statement is issued before and after the data load has occurred. However, notice that after the data is loaded, the table contains only 13 rows rather than 14 (the original 10 rows plus the four rows contained in the input file). Why? Recall that one of the rows contained an entry that violated the defined width for the ACCT# column. Therefore, that row was rejected. If you use Notepad to open the bad log file, you can find the missing row.

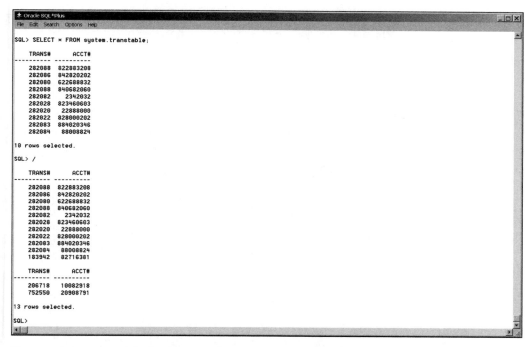

Figure 12-32   Modified contents of the TRANSTABLE database table

Loading Data with SQL*Loader—GUI Approach

In this section Carlos loads the new records into the TRANSTABLE database table using the Load Wizard available through the Enterprise Manager Console. This wizard is available only if you are connected to an Oracle Management Server (OMS).

To load the data through the Load Wizard:

1. Run the Ch12script01.sql script file to drop and re-create the TRANSTABLE database table. The script can be run through SQL*Plus or the SQL*Plus Worksheet.

2. If necessary, start the Enterprise Manager Console and log into the OMS with a SYSDBA privileged account.

3. If necessary, click the + (plus sign) next to Databases in the object list on the left side of the console.

4. If necessary, click the name of the database in the Database list to identify the target database for the load operation.

5. Click **Tools** on the menu bar. On the Tools menu, click **Database Tools**, **Data Management**, and then **Load** on the submenus that subsequently appear, as shown in Figure 12-33.

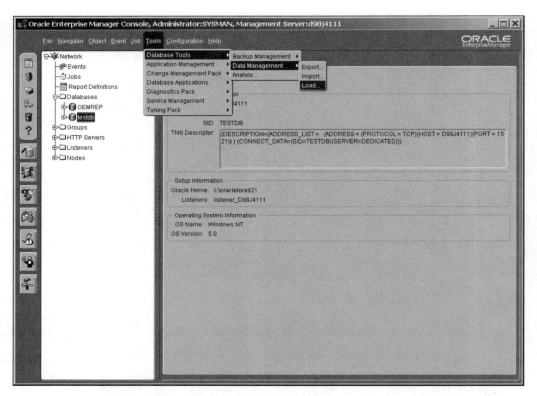

Figure 12-33 Accessing the Import Wizard

6. If the Introduction window appears, click **Next**.

7. When the Control File window shown in Figure 12-34 appears, enter the appropriate name and location of the control file in the text box, and click **Next**.

8. When the Data File window shown in Figure 12-35 appears, click the second option button and enter the appropriate name and location of the input file into the text box when it becomes available. Then click **Next**.

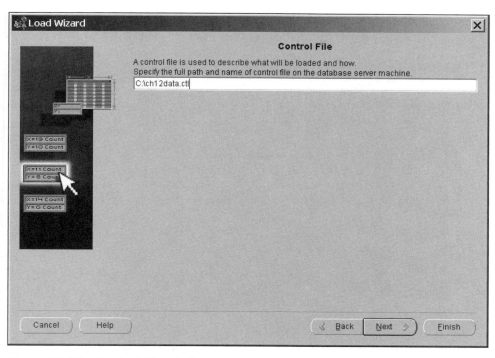

Figure 12-34 Control File window

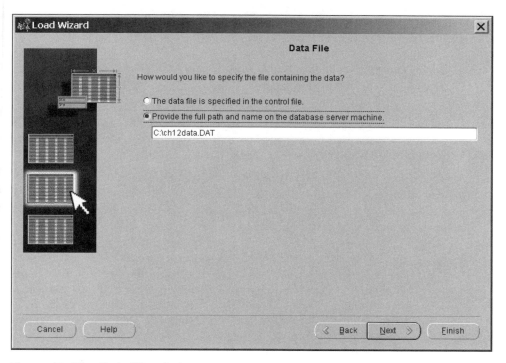

Figure 12-35 Data File window

9. When the Load Method window shown in Figure 12-36 appears, verify that the **Conventional Path** option button is selected, and then click **Next**.

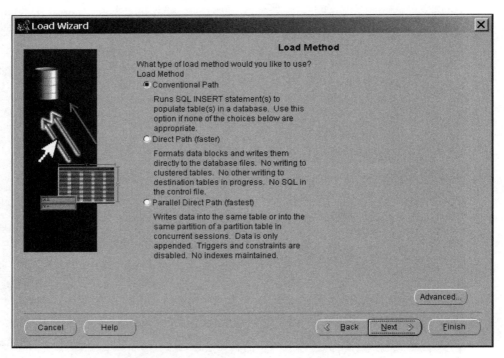

Figure 12-36　Load Method window

10. When the Schedule window appears, click **Next**.

11. When the Job Information window appears, click **Finish**.

12. When the Summary window shown in Figure 12-37 appears, click **OK**.

13. When the dialog box appears indicating that the job has been submitted, click **OK**.

At this point, Carlos can query the TRANSTABLE database table to verify that the rows have been added. As in the previous section, only three rows have actually been loaded into the table. Again, the row that violates the defined width for the ACCT# column has been written to the bad log file.

The Export, Import, and SQL*Loader utilities have numerous options available to make the utilities more flexible based on the needs of the database administrator. This chapter has covered only the basic points with the intention of familiarizing the student with the basic operations of these utilities. More detailed information can be obtained by searching for "Database Utilities" at *http://otn.oracle.com*.

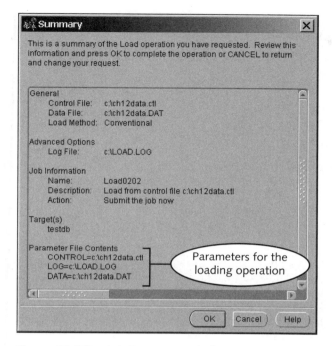

Figure 12-37   Job Summary window

At this point, Carlos has reviewed the basic information regarding backup and recovery operations in Oracle9*i*. The remaining chapters of this textbook focus on what network administrators require to ensure that a database is accessible to users across a network.

 If you are not completing the hands-on assignments at this time, the database should be placed back in NOARCHIVELOG mode after completing the steps demonstrated in this chapter.

CHAPTER SUMMARY

- The Export utility is used to export data from an existing Oracle9*i* database table.

- The entire database, a user's schema, or simply a database table can be exported using the Export utility.

- Exported data is written to a binary file, known as a dump file.

- The Export utility can be operated in an interactive or scripted mode.

- If a value is not specified, the Export utility assumes the default value for the parameter.

- The Import utility is used to import data that was previously exported to a dump file.

▫ The Export and Import utilities can be used along with a physical backup strategy to create a backup of the Recovery Catalog.

▫ The SQL*Loader utility is used to load data from text files.

▫ SQL*Loader utilizes three types of files: a control file, log files, and input files.

▫ The control file specifies the loading parameters and the structure of the data.

▫ The log files are used to provide information about the loading process and identify any bad or discarded records.

▫ The input file, also called a data file, contains the records to be loaded into the database; otherwise, the data must be contained in the control file.

SYNTAX GUIDE

Command	Description	Example
EXP	Used at the system prompt to start the Export utility	`exp system/ manager@testdb`
IMP	Used at the system prompt to start the Import utility	`imp system/ manager@testdb`
SQLLDR	Used at the system prompt to start the SQL*Loader utility	`sqlldr system/ manager@testdb`

REVIEW QUESTIONS

1. Which of the following utilities is used to transfer data between existing Oracle9*i* databases?

   a. Export and Import utilities

   b. SQL*Loader and SQL*Unloader utilities

   c. Dump and Undump utilities

   d. none of the above

2. Which Oracle9*i* utility is used to import or load externally generated data into an Oracle9*i* database?

3. When the Export utility is used to export data from a database table, the parameters must be specified in a control file. True or False?

4. The INSERT keyword must be used to instruct the SQL*Loader to add the loaded data to the rows already contained within the specified table. True or False?

5. Which of the following log files contain the rows that do not meet the criteria specified during a load operation?

   a. bad log file

   b. discard log file

   c. general log file

   d. exempt log file

6. What is the default load method for the SQL*Loader utility?

7. The default name for the binary file used to store the data exported by the Export utility is _____.

   a. Dump.exp

   b. Export.dmp

   c. Output.dmp

   d. Expdat.dmp

8. Which SQL*Loading loading method is similar to using a series of INSERT statements?

9. When exporting data, which of the following approaches is considered the quickest?

   a. direct-path

   b. direct-dump

   c. conventional-path

   d. parallel direct-path

10. Which of the following is used to perform a logical recovery of a database object?

    a. SQL*Loader

    b. RMAN

    c. Import utility

    d. Export utility

11. The _____ keyword is used to instruct the SQL*Loader utility to load the data into an empty table.

12. When using SQL*Loader to load data into a table, the data must be stored in an input or data file or in the _____ file.

13. The minimum buffer size when importing data is:

    a. 8 KB

    b. 8192 KB

    c. 64 KB

    d. none of the above

12

14. The input file for a load operation is identified using the INFILE parameter. True or False?

15. The general log file contains a copy of all the data loaded into the database. True or False?

16. If the values assigned to the ROWS parameter is specified as N, or No, when importing data, what will occur?

17. What is the difference between running the Export utility interactively as compared to using a scripted mode?

18. How do you indicate that an export operation should use the direct-path approach?

19. If a highly dynamic database table becomes inaccessible, why is physical recovery of the table preferred versus a logical recovery?

20. When importing data for a database table, which parameter identifies the name of the destination table?

    a. TAB

    b. DEST

    c. TABLES

    d. INTO

HANDS-ON ASSIGNMENTS

Assignment 12-1 Interactively Exporting Data Using the Direct-Path Approach

In this assignment you export the contents of the TRANSTABLE table interactively using the direct-path method.

1. Access the Export utility and have the contents of the TRANSTABLE dumped into a file named Ch12HandsOn1.dmp without the use of a parameter file. However, make certain that the direct-path approach is used when the data is dumped into the file. (*Hint*: Remember that the path is not one of the prompts provided when the utility is used interactively.) What other option is available to provide information to the utility if a parameter file is not being used?

2. Use Windows Explorer to verify that the file named Ch12HandsOn1.dmp actually exists.

Assignment 12-2 Interactively Importing Data Using the Direct-Path Approach

In this assignment you import the contents of the TRANSTABLE table using the dump file created in the Assignment 12-1.

1. Type **RENAME transtable TO transtable2;** in SQL*Plus to rename the TRANSTABLE database table to TRANSTABLE2.

2. Access the Import utility and import the contents of the Ch12HandsOn1.dmp file to re-create the TRANSTABLE database table.

3. After the import operation is completed, look at the screen display provided by the Import utility and determine the importing method used. Why did it use the direct-path approach?

4. Verify that the TRANSTABLE database table was created and contains the same information as the TRANSTABLE2 database table.

Assignment 12-3 Using a Parameter File to Export Data Using the Export Utility

In this assignment you export the contents of the TRANSTABLE database table through the use of a parameter file using the direct-path method.

1. Open Notepad and create a parameter file named Ch12HandsOn3.pr that exports the contents of the TRANSTABLE database table into a file named Ch12HandsOn3.dmp. Make the necessary entry into the file to ensure the export operation uses the direct-path approach when the data is exported.

2. Access the Export utility and use the parameter file to export the specified data.

3. Use Windows Explorer or My Computer to verify that the file named Ch12HandsOn3.dmp actually exists.

Assignment 12-4 Using a Parameter File to Import Data

In this assignment you import a table definition from a dump file through the use of a parameter file.

1. Open Notepad and create a parameter file named Ch12HandsOn4.pr that imports *only* the definition of the table previously exported to the file named Ch12HandsOn3.dmp. The table definition should be used to create a table named TRANSTABLE3.

2. Access the Import utility and use the parameter file to create the table named TRANSTABLE3. After the operation has been completed, use the DESCRIBE command in SQL*Plus to verify the existence of the table.

12

Assignment 12-5 Correcting Data Load Errors

In this assignment you perform the steps necessary to correct and load data previously rejected by the SQL*Loader utility. If you did not complete the command-line example for the SQL*Loader utility presented in the chapter, you must do so before attempting this assignment.

1. Use Notepad to open the Ch12data.bad file created previously in this chapter.

2. Modify the second column of data by truncating enough digits from the right side of the data so it can be entered into the ACCT# column of the TRANSTABLE without causing an error. Save the changed file with its original file name.

3. Open the Ch12data.ctl file in Notepad and make the necessary changes so that the Ch12data.bad file is used as the input file for the load operation. Specify a new file name for the bad log file. Save the modified control file as Ch12data2.ctl.

4. Access SQL*Loader without specifying the name of the control file to be used during the load operation.

5. When prompted, enter the location and name of the Ch12data2.ctl file created in Step 3.

6. After the load operation is complete, verify that the row is now included in the TRANSTABLE table.

Assignment 12-6 Replacing Data Using SQL*Loader

In this assignment you change the APPEND parameter in the Ch12data2.ctl control file so that the original contents of the database table are removed when the load operation is performed.

1. If necessary, start SQL*Plus and determine how many rows are currently in the TRANSTABLE database table.

2. Open the Ch12data2.ctl control file created in the previous assignment in Notepad.

3. Locate the APPEND keyword and change the keyword to REPLACE. Save the modified file as Ch12data3.ctl.

4. Access the SQL*Loader utility and specify Ch12data3.ctl as the control file to be used during the load operation.

5. After the loading has been completed, view the contents of the TRANSTABLE database file. How many rows currently exist in the database table? What caused this result?

CASE PROJECTS

Case 12-1 Developing a Logical and Physical Backup Strategy for the Recovery Catalog

As mentioned in the chapter, the Export and Import utilities can be used to supplement a physical backup strategy. Currently, Carlos is creating a cold backup of the database containing the Recovery Catalog each night after all RMAN backup operations have been completed. Using the information provided in this chapter, develop a combination of physical and logical backup strategies that Carlos can use to ensure the Recovery Catalog can always be recovered in the event it becomes inaccessible. Make certain you provide your rationale as to when each type of backup (physical and logical) is appropriate and how the Recovery Catalog can be recovered in the event of a database failure.

Case 12-2 Procedure Manual

Continuing with the procedure manual created in Chapter 1, create lists that identify the steps necessary to perform the following tasks:

- Export the contents of a database table interactively.

- Import the contents of a database table interactively.

- Load data generated by a non-Oracle database.

Recall that the purpose of the procedure manual is to provide a reference to new employees. Therefore, the focus should be placed on the meaning of the prompts that are provided by the utilities so the individual can understand the implication of the responses he or she provides. In addition, it should also instruct the user on how to obtain help for each utility.

12

13

NETWORK ADMINISTRATION AND SERVER-SIDE CONFIGURATION

> **After completing this chapter,**
> **you should be able to do the following:**
>
> ♦ Identify the different types of architecture available in an Oracle network environment
> ♦ Identify the components available within Oracle Net Services
> ♦ Identify features of the Oracle Connection Manager
> ♦ Specify the purpose of each layer in the Oracle communications stack
> ♦ State the purpose of a listener
> ♦ Create, configure, and delete a listener

In the previous chapters, various techniques for ensuring the availability of an Oracle9i database have been examined. However, what about accessibility to the database? Accessibility is just as important as availability, because even if the database is started at Janice Credit Union, it is insufficient if one of the tellers is unable to connect to the database.

In today's business environment, the majority of data is accessed through networks. Part of a DBA's responsibility is to ensure that a properly authorized client or end user can send a request for services to a database and receive an appropriate response. This statement implies that the client is able to contact the Oracle9i database server and that the server is able to receive the transmission, preferably in a manner that is transparent to the user. To ensure that accessibility, the DBA needs to make certain that both the server and the client have been configured properly.

This chapter focuses on the behind-the-scenes activities and services necessary to ensure that, if the network is functioning properly, the database is accessible to clients.

THE CURRENT CHALLENGE IN THE JANICE CREDIT UNION DATABASE

Although Janice Credit Union employs a full-time network administrator, that person's responsibility is limited to the network and does not include configuring the Oracle9*i* software for network communication. That responsibility belongs to the database administrators, including Carlos.

When Carlos first started working for Janice Credit Union, the Oracle9*i* networking components had already been configured and were working properly. However, the credit union is planning on adding more computers to the network and they need to have access to various databases. In addition, Carlos needs to make certain that any new database servers added to the existing system are accessible via the network.

In this chapter, Carlos learns the basic architecture for an Oracle network. In addition, Carlos learns the configuration requirements for a server to detect and respond to a user request. In the subsequent chapters, he learns how to configure the Oracle9*i* client and methods for processing the expected increased volume with limited resources.

SET UP YOUR COMPUTER FOR THE CHAPTER

While completing this chapter, you will be required to create a listener and make changes to an existing listener. These activities result in changes being written to the Listener.ora file, the configuration file for the database listener. This file is located in the \network\admin folder of the Oracle9*i* home directory. Before attempting the examples given in this chapter, you should create a copy of the Listener.ora file using your operating system. In the event a listener failure occurs and you cannot connect to the database, the copy of the original file should be restored using the computer's operating system.

NOTES ABOUT DUAL COVERAGE WITHIN THIS CHAPTER

During the server configuration examples given in this chapter, you will be using both Net Manager, a GUI interface for creating and managing a listener, and the listener control utility, a command-line interface for managing a listener. Both programs can be accessed using the Windows XP and Windows 2000 Server operating system. In this chapter, the GUI approach primarily is used to create and delete the listener because this option is not available through the listener control utility. The listener control utility is used when the configuration of the listener needs to be changed or if the listener needs to be stopped or started.

Another GUI interface is also available for creating a listener: the Net Configuration Assistant. The GUI is basically a wizard that creates and configures the listener based on the responses provided by the user. The Net Configuration Assistant can also be used to configure the client and test the client's ability to connect to the listener. Use of the interface will be demonstrated in Chapter 14.

NETWORKING OVERVIEW

Although a database administrator is not responsible for the actual network used to connect the client, or user, with the database server, the DBA should have at least an elementary understanding of how a network operates. A **network** is a group of computers and other devices that are connected together through a transmission medium. This connection allows users to share electronic data, as well as hardware resources.

As shown in Figure 13-1, a basic network consists of two devices and a transmission medium. The two devices may consist of two computers, such as a sender and a receiver of data, or a computer and a printer or other type of hardware. Each device is connected to the network through communication hardware, such as a modem or a network interface card (NIC). The exact communication hardware required to connect to the network is dictated by the transmission medium. The transmission medium used to connect the devices can be copper wiring, such as the telephone wiring found in most homes; coaxial cable, the type of cable used by cable companies to provide television services; fiber optics which transmits data through pulses of light rather than through electrical pulses; or wireless, using microwave or laser signals.

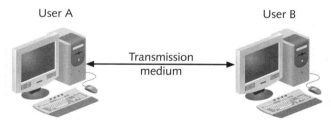

Figure 13-1   Basic network

In addition to the basic hardware requirement, each computer or device must also be configured to function in a network environment. At a minimum, each computer or device must have the necessary software or instructions to be able to send or receive electronic data being transmitted across the network. For a device such as a printer, these instructions are included in the communication hardware, such as a network interface card. However, for a networked computer, additional software configuration may be required.

Typically the operating system of the computer is used to identify what type of data can be shared across the network. In addition, communication software can be installed to specify how data is to be transmitted across the network, which would include the type of protocol to be used for the transmission. A **protocol** is a set of rules that is used for electronic communications. It specifies how the data is organized for the transmission and the procedures for how the computers or devices communicate with each other. In essence, it ensures that the sender and receiver are both speaking the same dialect of an electronic language.

In more complex networks, such as those composed of more than eight nodes or connecting devices (i.e., not a peer-to-peer network), administration and maintenance of the network can become difficult. Network operating systems, as well as a number of utility programs, are available to simplify the tasks of the network administrator. In addition, other services can be made available through the network to improve the performance. For example, rather than have each computer on the network keep a file containing the name and other identifying information (IP address, etc.) regarding each node on the network, there can be a centralized repository to house this data. Then each computer needs to store information only about the location of the repository. When a user needs to contact another computer, the identifying information is retrieved from the repository and then the user can transmit the data with the appropriate recipient address. As you will see in Chapter 14, Carlos can use this same approach to provide the necessary information to connect a client with an Oracle9*i* database.

Individuals who have not had previous training or experience in networking should consider reviewing introductory networking materials.

LAYERED ARCHITECTURE

There are three basic types of network configurations available in an Oracle9*i* network environment: single-tier architecture, two-tier architecture, and *n*-tier architecture.

Single-Tier Architecture

A **single-tier architecture** is the oldest form of network configuration and consists of a mainframe computer with dumb terminal connections. The term "dumb terminal" refers to the fact that the terminals used to interface with the mainframe computer do not have processing capabilities. All processing operations are actually performed on the mainframe computer.

As shown in Figure 13-2, each terminal has a direct cable link to the mainframe computer. A user enters his or her requests at one of the terminals and the request is sent via the cable directly to the mainframe computer. The request is processed and any response is then displayed on the user's terminal screen. The problem with this type of configuration is that it has limited **scalability**. In other words, there are a finite number of physical connections to the mainframe computer. Furthermore, as more connections are made to the computer, the overall response time of the central computer is increased.

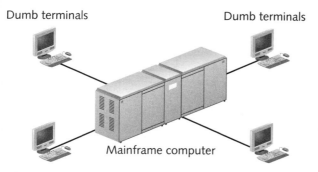

Figure 13-2   Single-tier architecture

Two-Tier Architecture

As the concept of networking evolved, another layer was added to the Oracle9*i* network environment. With the **two–tier**, or client/server, **architecture**, clients connect to the server over a network, as shown in Figure 13–3. The clients that connect to the network have individual computing capabilities that allow some of processing to occur on the individual workstations before requests are sent to the Oracle9*i* database server.

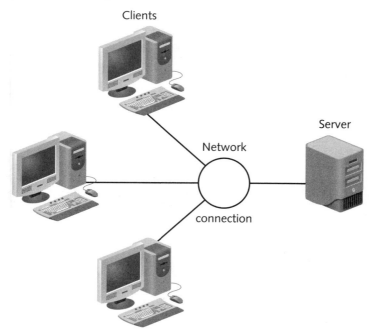

Figure 13-3   Two-tier architecture

13

Although this type of architecture can reduce the processing burden placed on the database server, there is still a limitation on the number of users that can concurrently access the database without decreasing database performance, i.e., creating an increase in response time. Furthermore, the two-tier architecture is more complex than the single-tier architecture due to the addition of other network components such as routers, switches, and so forth.

N-Tier Architecture

In a two-tier architecture, the work is basically handled by two machines—the client and the server. However, with an ***n-tier architecture***, different parts of a task can be handled by various computers. Think of the Internet. Should Carlos really allow one of the credit union's customers to be able to have a direct connection to the Oracle9*i* database with the purpose of viewing the balance of his or her account? Perhaps a better approach would be to allow the user to connect to another server, such as a Web server, and allow that server to retrieve the required data from the database server on behalf of the client, as shown in Figure 13-4. The Web server could then be responsible for services such as verifying the identity of the user, as well as how the data is presented. This reduces the processing burden placed on the database server and enables the server to handle requests from more users within the same amount of time.

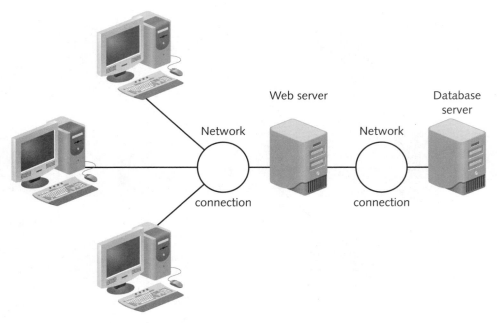

Figure 13-4   *N*-tier architecture

ARCHITECTURE IMPLEMENTATION

Viable implementation of any of the aforementioned network architectures requires Carlos to address four issues:

- **Connectivity**: The client's ability to request and receive information from the Oracle9i database

- **Manageability**: The ease with which Carlos can configure, monitor, and administer the Oracle network environment

- **Scalability**: The addition of equipment and resources without degrading performance

- **Security**: Preventing unauthorized access to data

Fortunately, Oracle Net Services and Oracle Advanced Security are available to address these issues. Each is discussed in the subsequent sections.

ORACLE NET SERVICES

Oracle Net Services is a set of components available with the Oracle9i Database software that is designed to address the issues of connectivity, manageability, and scalability. Oracle Net Services is composed of Oracle Net, Oracle Net Listener, Oracle Connection Manager, and Networking Tools. Each of these components will be discussed in turn.

Oracle Net

Oracle Net is a set of software that is used to create a connection either between a client and an Oracle server, or between two Oracle servers. A copy of the software must reside on the client and the server for a connection to be established. Oracle Net is also responsible for ensuring that the connection is maintained during the session. The software is designed to use common protocols such as TCP/IP to allow the computers to communicate.

Oracle Net Listener

How does the database software on the Oracle9i server know that a client is requesting information? The server must have a listener configured. A **listener** is a server process that accepts and processes client requests for connections to the database. The listener is configured to "listen" for client requests on a particular port.

A **port** is an address or location on the computer where a particular program or service can be accessed. You can think of a port as a door. When a service is requested from a computer, the request must be sent to a preassigned port (i.e., the client must knock on the correct door). Of course, this requires that the client be configured with the correct port address for the listener. After the server's listener verifies that it is a valid request for database services, the listener then provides the necessary information for the client to establish the connection with the Oracle9i server.

Oracle Connection Manager

The Oracle Connection Manager is a software program stored in the middle layer of an *n*-tiered Oracle network, and it is sometimes called middleware. It can serve three basic functions: multiplexing, access control, and protocol conversion.

Multiplexing

In a single-tier or two-tier environment, each client session is required to have a separate connection to the Oracle9i server. Rather than require the database server to manage multiple client sessions, the clients can first connect to the Oracle Connection Manager, which has a multiplexed connection to the database server. The Oracle Connection Manager bundles the incoming client requests into one session and forwards them to the Oracle9i database server, as shown in Figure 13-5. The responses are sent back to the Oracle Connection Manager, the software sorts out which client generates each response, and then the response is transmitted to the appropriate client. This feature supports the scalability of the Oracle9i network environment because it allows Carlos to add more users without overwhelming the database server.

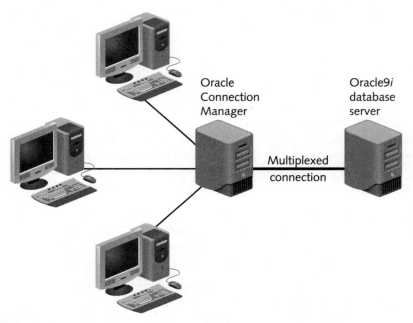

Figure 13-5 Oracle Connection Manager

Access Control

Because the Oracle Connection Manager receives the client connection requests, the software can be used to restrict access to the database server. This allows the middle layer to act as a firewall by determining whether access to the database server should be allowed based on the IP address of the client. Carlos can use this feature to prevent users who are outside the credit union's intranet from accessing certain database servers, such as the one that stores the credit union's payroll information.

Protocol Conversion

The final feature of the Oracle Connection Manager software is protocol conversion. Recall that a protocol is a set of communication rules that must be followed to transport data across a network. However, there are several different protocols in existence; the most common being TCP/IP, which is required for access to the Internet. If one of the departmental networks within Janice Credit Union uses a protocol different from the one used by the Oracle9*i* database server, a client in that department would not be able to communicate with the server. However, if the user first connects to the Oracle Connection Manager using TCP/IP, TCP/IP with SSL, Named Pipes, LU6.2, or VI, the client can retrieve data from the database.

Oracle Net does not support the IPX/SPX protocol used with Novell networks.

Networking Tools

Oracle Net Services also provides a set of tools that can be used to configure and manage different network components and services, such as testing for connectivity and so forth. These tools are provided in both a command-line and GUI environment. Carlos will be working with the tools when he begins configuring the new clients that are being added to the Oracle database network.

13

ORACLE ADVANCED SECURITY

Oracle Advanced Security is a set of features designed to provide enhanced security for the Oracle9*i* database. Some of the more commonly used features include data encryption, user authentication, and single sign-on.

Janice Credit Union is not currently using any of these features because the software is not included with the Oracle9*i* DBMS and must be purchased separately. But given the security benefits, Carlos is definitely considering making the expenditure in the near future.

Data Encryption

When data is transmitted across a network, unauthorized individuals can intercept the data through a variety of methods. However, if that data has been encrypted, the data may get intercepted, but it is not in a readable form unless the individual also knows the algorithm necessary to decode the data and the key needed for decryption. Oracle Advanced Security supports Standard DES, Triple DES, and RSA RC4 encryption algorithms.

User Authentication

One component of security is having the individual prove his or her identity. A basic authentication method is to have the user supply a valid account name and password. However, this may not be a sufficient level of protection for sensitive data—especially if the user has a password that can easily be guessed, or it is on a piece of paper tucked under the keyboard. To increase the level of protection against unauthorized access, Oracle Advanced Security also supports several third-party authentication methods. These methods include smart cards, biometrics (fingerprints, retina scans, etc.), and token cards, and industry-standards such as SSL, RADIUS, CyberSafe, Kerberos, and DCE.

Single Sign-On

Another benefit of Oracle Advanced Security is that the user information, such as user names and passwords, is stored in a centralized location enabling single sign-on. With single sign-on a user can log into the system and access multiple accounts and applications without having to remember a different password for each one, as long as the user has been granted appropriate access. This feature can also save time from having to log into numerous accounts and applications.

COMMUNICATION ARCHITECTURE

Have you ever received a word-processing document or a spreadsheet file and not been able to open it because it is not in a recognized file format or was generated by a different version of the software that you have on your computer? Imagine trying to transmit data across different types of networks, to and from different types of computers, and so on, without some type of communication standard or way of transmitting data. You would find that everyone either needed to buy the same equipment and software or be restricted to sharing data with a narrow group of users.

Oracle Communications Stack

To resolve the problems that could result from the use of various platforms or technologies, development began in the early 1980s on what would become the Open Systems

Interconnection (OSI) model. The purpose of the model was to develop a set of specifications for how data could be transmitted from one computer to another in a network environment. The model consists of several layers, each layer having a different function.

Oracle has developed a similar communication model, commonly referred to as the Oracle communications stack or the client/server stack, specifically addressing how data is transmitted between clients and servers in an Oracle network environment. This model is shown in Figure 13-6.

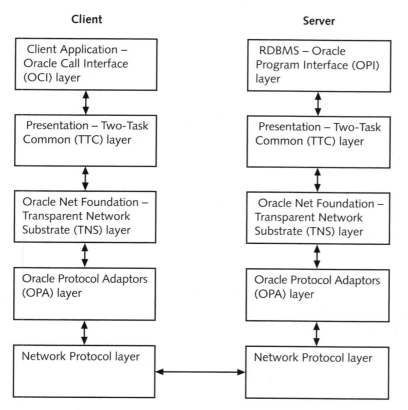

Figure 13-6   Oracle communications stack

Client Requests and the Oracle Communications Stack

As shown in Figure 13-6, each layer of the communications stack can interact only with the layer immediately above or below it. On the client side, a client makes a request through the Application layer using the Oracle Call Interface (OCI). The request can be made through a program like SQL*Plus, a Web browser using HTTP, or even using XML through the Internet Inter-ORB Protocol (IIOP).

The OCI is responsible for all SQL processing, such as managing cursors and fetching rows. The request is passed down to the Two-Task Common (TTC) layer where any character set or data type differences that may exist between the client and server are resolved. The TTC layer then passes the data to the Oracle Net Foundation layer that determines the location of the destination computer, the most appropriate path to reach the destination computer, and what type of protocol is required for transmission. The data is subsequently passed to the Oracle Protocol Adapters (OPA) layer, which is responsible for making certain that the format of the data is compatible with the protocol actually used by the network.

The lowest layer of the communication stack is the Network Protocol layer. This layer represents the actual network that is used to transmit data between a client and a server. When the server receives the transmission, the process is repeated in reverse until the request has been reconstructed at the Oracle Program Interface (OPI) layer. At that point, the database software processes the request and a response can be sent to the client, again transversing the different layers of the communications stack.

At this point, why should Carlos even care about the Oracle communications stack? Well, it serves as a framework that he can reference to ensure that data is properly transmitted across the network. Different Oracle components operate at or interact with different layers within the stack. For example, remember Oracle Net, the software that creates and maintains the connection between the client and the server? It operates at the Oracle Net Foundation and Oracle Protocol Adapters layers. If Oracle Net returns an error message, Carlos would be able to narrow down the cause of the problem to a specific portion of the communication stack, eliminating the network protocol or an application program as the source of the problem.

In some instances, the communications stack utilized may be a modification of the one previously presented in Figure 13-6. For example, to support the connection of a client to an Oracle9i database via the Internet, only the layers shown in Figure 13-7 are needed. Because the client is accessing the database via the Internet, the client does not need to determine the most appropriate path to the database (this is handled by mechanisms already included in the infrastructure of the Internet), and a constant connection does not need to be maintained. Therefore, the Oracle Net Foundation and Oracle Protocol Adapters layers can be eliminated on the client side of the stack.

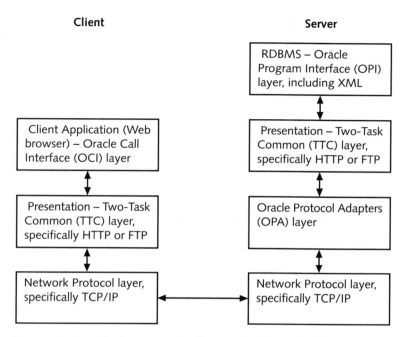

Figure 13-7   Stack communication model for Web client connections

SERVER CONFIGURATION

Now that Carlos has a basic overview of the Oracle networking environment, he can begin to examine the configuration of the database server and the clients. Specifically, this includes learning how to configure a listener on a server and using various naming methods to allow a client to locate and access an Oracle9*i* server. In the following sections, Carlos performs the steps necessary to configure the server to listen for client requests. In Chapter 14, client configuration will be presented.

Regardless of whether the database server resides on the same machine or across a network, the server must have a listener configured to identify a client's request for services. The listener is an application that operates at the Oracle Net Foundation layer of the stack communications model. Think of a listener as an ear. When the listener hears a client requesting a connection to the database, it passes the request to the appropriate service handler for the database and then listens for requests from other clients.

LISTENER CONFIGURATION

Each database server must have at least one listener configured to respond to client requests. Carlos has three options for configuring a listener: during creation of the database, through Net Manager, or through Net Configuration Assistant after the database is created. Regardless of the method used, the default name of the first listener for the database is LISTENER.

A database can have multiple listeners, and a listener can process requests for multiple databases as long as the databases are on the same server as the listener. Why would Carlos want, or even need, multiple listeners for a database? A listener can process only one request at a time. If several clients are trying to access the database at the same time, this can cause a delay in the response time from the listener. Therefore, Carlos can configure multiple listeners as a means of balancing the traffic or volume handled by each listener.

When a listener is configured, Oracle Net Manager creates three files, if they do not already exist: Listener.ora, Tnsnames.ora, and Sqlnet.ora. Listener.ora is discussed in the subsequent sections. The latter two files will be discussed in Chapter 14 because they relate to the client side of the connection process.

The Listener.ora File

The Listener.ora file is used to store configuration information for all listeners. This file is stored in the \network\admin folder of the Oracle9i home directory. An example of the contents of a Listener.ora file is shown in Figure 13-8.

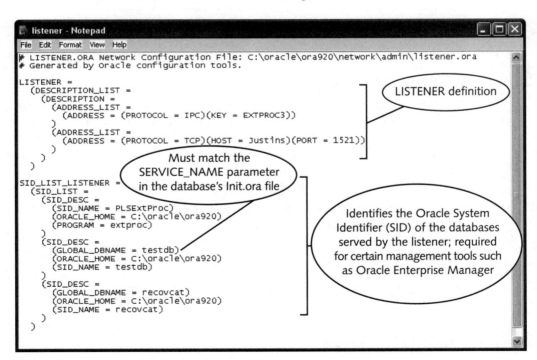

Figure 13-8   Listener.ora file

There is one entry for each configured listener. In this particular example, there is only one listener, named LISTENER. Following the name of the listener is a description of the listening location. This listener is configured to listen for requests made through TCP or an IPC protocol. An IPC protocol is used by client applications that are on the same machine as the database itself. TCP is commonly used in networked environments, including the Internet. The first listening address is for any external procedure calls. The second listening address is for any requests for services from the Oracle9i database located on the machine named Justins. In addition, the listener is configured to listen on port 1521. This is the default listening port for all Oracle9i listeners.

The SID_LIST_LISTENER Section of Listener.ora

The SID_LIST_LISTENER section of the Listener.ora file is used to specify the list of Oracle9i services for which the listener is to listen. In Figure 13-8, the listener was configured to listen for any external procedure calls, and requests for connections to the TESTDB and RECOVCAT databases. The GLOBAL_DBNAME listed for the two databases must match the database name assigned to the SERVICE_NAMES parameter in the database's Init.ora file. The ORACLE_HOME parameter identifies the location of the RDBMS software and the SID_NAME is the SID for the Oracle9i instance (the same name as the GLOBAL_DBNAME if the network is not using domain names).

Although the Listener.ora file can be created manually, a syntax error can leave the database inaccessible. Therefore, it is always recommended that any changes be made through the Net Manager or the Net Configuration Assistant. In the following section, Carlos creates a listener for the Oracle9i database and then reviews the changes made to the Listener.ora file previously shown in Figure 13-8.

13

CONFIGURING A LISTENER THROUGH NET MANAGER

Before configuring a listener, Carlos needs the following information: a name for the listener, the protocol, the name of the host machine where the database is stored, and the port on which the listener needs to listen. Before attempting the steps provided in this section, make certain you have this information available or request the necessary information from your instructor.

To configure a listener through Net Manager:

1. Click the **Start** button on the Windows XP taskbar.

2. Click **All Programs**, and then click **Oracle–OraHome92** on the submenu.

3. On the Oracle submenu, click **Configuration and Migration Tools**, and then click **Net Manager**.

4. When the Net Manager window appears, click the + (plus sign) next to Local in the left pane of the window.

5. After the Local list is expanded, double-click the word **Listeners**.

6. Click **Edit** on the menu bar, and then click **Create,** as shown in Figure 13-9.

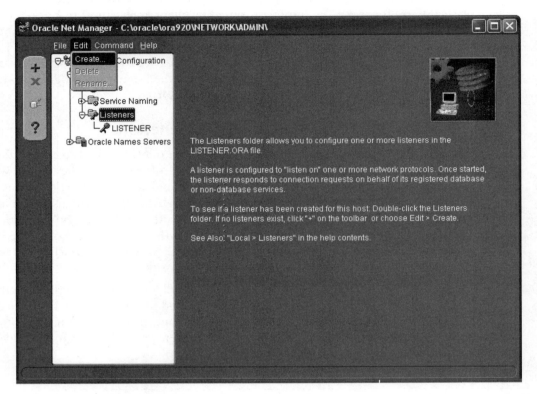

Figure 13-9   Creating a new listener

7. When the Choose Listener Name dialog box appears, type **LISTENER1** for the name of the new listener, if necessary, as shown in Figure 13-10, and click **OK**.

Figure 13-10   Entering a listener name

8. Click the **Add Address** button on the right side of the Net Manager window.

9. Select the appropriate protocol for your network from the protocol drop-down list.

10. Enter the name of the computer in the Host text box, if necessary.

11. Enter a port address for the listener that is assigned as shown in Figure 13-11. Use 1681 as the port address unless your instructor provides a different address.

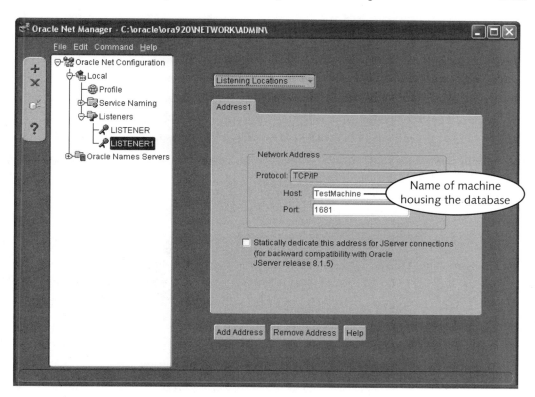

Figure 13-11 Configuration information entered for a new listener

12. After verifying the information entered for the new listener, click **File** on the menu bar, then click **Save Network Configuration** to save the changes made in this section.

If the listener is listening for client requests made via the Internet and it is possible that a pre-Oracle8i instance might attempt to connect to the server, the check box for the Statically dedicate this address for JServer connections should be selected because dynamic service registration is not supported for certain options. In addition, the selected protocol should be either TCP/IP with an assigned port of 2481 or TCP/IP with SSL with 2482 as the listener's assigned port.

MAKING CHANGES TO AN EXISTING LISTENER WITH NET MANAGER

Carlos can also use Net Manager to make changes to an existing listener. If he made a mistake when the listener was created and needed to make changes to the listener named LISTENER, he could click the name of the listener from the left pane of the window. From the right pane of the window he could then change the protocol, host name, and so forth, for the listener. In addition, other options are available from the drop-down list at the top of the right pane.

To view the general parameters for the LISTENER1 listener:

1. Click **LISTENER1** from the listener list.

2. Click the **down** arrow at the top of the right pane of the window.

3. Select **General Parameters** and review the current settings by clicking the **General**, **Logging & Tracing**, and **Authentication** tabs.

In Figure 13-12, Carlos has selected the General Parameters, which allow him to change basic parameters, such as whether listener changes should be automatically saved when the database is shut down and whether logging and tracing options for troubleshooting should be enabled.

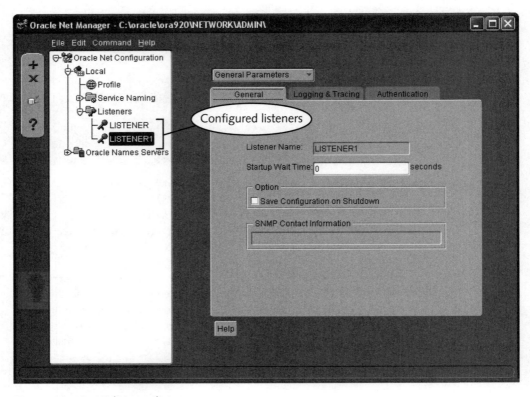

Figure 13-12   Editing a listener

After the listener has been created or any changes made, the changes are updated to the Listener.ora file. As shown in Figure 13-13, after the LISTENER1 listener has been created, the configuration is added to the existing Listener.ora file.

```
listener - Notepad
File  Edit  Format  View  Help
# LISTENER.ORA Network Configuration File: C:\oracle\ora920\NETWORK\ADMIN\listener.ora
# Generated by Oracle configuration tools.
LISTENER1 =
  (DESCRIPTION =
    (ADDRESS = (PROTOCOL = TCP)(HOST = TestMachine)(PORT = 1681))
  )
LISTENER =
  (DESCRIPTION_LIST =
    (DESCRIPTION =
      (ADDRESS = (PROTOCOL = IPC)(KEY = EXTPROC3))
    )
    (DESCRIPTION =
      (ADDRESS = (PROTOCOL = TCP)(HOST = Justins)(PORT = 1521))
    )
  )
SID_LIST_LISTENER =
  (SID_LIST =
    (SID_DESC =
      (SID_NAME = PLSExtProc)
      (ORACLE_HOME = C:\oracle\ora920)
      (PROGRAM = extproc)
    )
    (SID_DESC =
      (GLOBAL_DBNAME = testdb)
      (ORACLE_HOME = C:\oracle\ora920)
      (SID_NAME = testdb)
    )
    (SID_DESC =
      (GLOBAL_DBNAME = recovcat)
      (ORACLE_HOME = C:\oracle\ora920)
      (SID_NAME = recovcat)
    )
  )
```

Definition for new listener

Figure 13-13   Modified Listener.ora file

13

CONTROLLING THE LISTENER THROUGH THE LISTENER CONTROL UTILITY

Although the Net Manager can be used to control some of the listener operations, the Listener control utility supports more options. For example, the listener can be started or stopped and reloaded without stopping if changes are made. The status of the listener and the services it supports can also be viewed. The Listener control utility is a command-line program that is executed through the operating system. The HELP option of the utility displays the available commands. When a command is entered at the utility prompt, the action is applied to the listener named LISTENER unless the command is followed by the name of another listener.

In most Oracle9i documentation, the Listener control utility is simply referred to as lsnrctl; LSNRCTL is the command used to launch the utility.

Listener Operation Controls

Carlos performs some of the basic control operations through the Listener control utility. The utility can be accessed from the operating system using the LSNRCTL command at the operating system prompt. After the utility is launched, commands are simply entered at the utility prompt.

As an alternative, Carlos can execute utility commands from the operating system prompt without first launching the utility by using the command syntax of LSNRCTL *command listenername*. By first typing lsnrctl at the operating system prompt, Carlos is specifying that the subsequent command is to be executed through the Listener control utility. The syntax of the command is the same as when actually typed in the Listener control utility. In addition, if the name of the listener is not provided, then the listener named LISTENER is assumed by default.

To control the operation of the default listener:

1. Click **Start** from the Windows XP taskbar, then click **Run**. Type **cmd** to open the command line window.

2. Type **lsnrctl** at the operating system prompt to start the Listener control utility.

3. At the prompt, type **help** to display the list of available commands.

4. Type **stop** at the prompt to stop the listener as shown in Figure 13-14. If a client requests services for the database at this time, the client cannot connect to the database unless another listener is available.

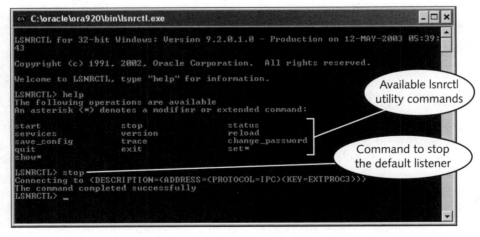

Figure 13-14 Stopping the listener named LISTENER

5. Type **start** at the utility prompt to restart the listener. As shown in Figure 13-15, when the listener is started, information such as the services, listening configuration, and so forth, are displayed.

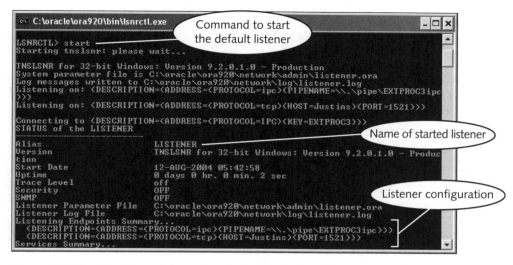

Figure 13-15   Starting a listener (partial output shown)

6. Type **status** at the utility prompt to determine the status of the listener, as shown in Figure 13-16.

Figure 13-16   Status of the listener

7. Type **services** at the utility prompt to determine to which databases the listener has been assigned and the protocols for which the listener is configured, as shown in Figure 13-17.

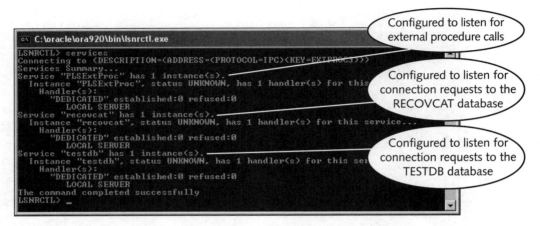

Figure 13-17 Determining for which services the listener is listening

As shown in Figure 13-17, after the listener is started, it listens on behalf of three services: PLSExtProc (external procedures) and the RECOVCAT and TESTDB databases. If Carlos needs to control the operations of a different listener, he can simply include the listener name in the command.

8. To verify the services for which the LISTENER1 listener is configured, type **services listener1** at the utility prompt.

Parameter Changes Through the Listener Control Utility

The operational changes Carlos made in the previous section (start, stop, etc.) did not require any entry into the listener configuration file (Listener.ora) because the actual configuration of the listener was not altered. However, there are some listener parameters that can be changed that subsequently record changes in the configuration file. For example, Carlos may decide to enable tracing or logging for troubleshooting purposes. This change can be made through either the Net Manager or through the Listener control utility. If tracing is enabled, then information regarding the listener's activities is recorded into a trace file.

To view the commands available through the Listener control utility, Carlos can type "show" at the utility prompt. As shown in Figure 13-18, the utility responds with a list of all available parameters.

Figure 13-18   Available configuration parameters

A summary of each of these parameters is provided in Table 13-1.

Table 13-1   Summary of Listener Parameters

Listener Parameter	Description
CURRENT_LISTENER	Shows the name of the current listener and can change the default listener
DISPLAYMODE	Allows display to be changed to COMPACT, NORMAL, RAW, or VERBOSE
LOG_DIRECTORY	Identifies the log directory location
LOG_FILE	Identifies the name of the listener log file
LOG_STATUS	Identifies listener's logging status as on or off
RAWMODE	Indicates the amount of detail to be displayed for STATUS and SERVICES commands; can be set to ON or OFF
SAVE_CONFIG_ON_STOP	Indicates that all changes are to be saved when the utility is exited
SNMP_VISIBLE	Indicates whether listener can respond to queries from an SNMP-based network management system
STARTUP_WAITTIME	Specifies the number of seconds a listener waits to respond to a START command
TRC_DIRECTORY	Identifies the name of the directory containing the listener's trace file
TRC_FILE	Identifies the name of the file containing trace information regarding the listener
TRC_LEVEL	Specifies the level of tracing as OFF, USER, ADMIN, or SUPPORT

13

The current value assigned to a listener parameter can be displayed by entering the SHOW command followed by the name of the parameter. Carlos has the option of changing the value of a parameter by either using the SET command in the listener control utility, or through Net Manager. In the following steps, Carlos verifies the logging mode of the listener using the SHOW command and then changes the current logging mode using the SET command.

To verify and change the current logging mode of the default listener:

    1. Type **show log_status** at the utility prompt to display the current logging status of the listener, as shown in Figure 13-19.

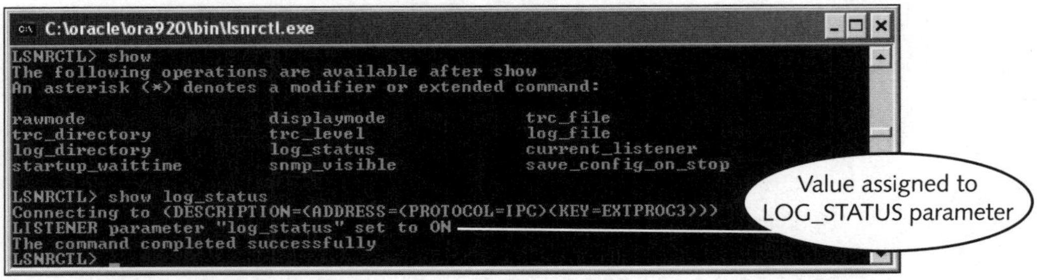

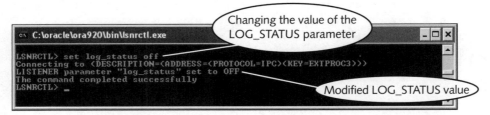

Figure 13-19   Current listener logging status

    2. Type **set log_status off** at the utility prompt, as shown in Figure 13-20, to disable logging for the listener.

Figure 13-20   Changing logging status for a listener

    3. Type **exit** to close the Listener control utility.

As mentioned, Carlos can also change the status of most of the listener parameters through the Net Manager. As shown in Figure 13-21, after Carlos selects the appropriate listener name, he can change, or simply view, the current logging setting by clicking the Logging & Tracing tab and entering the appropriate information.

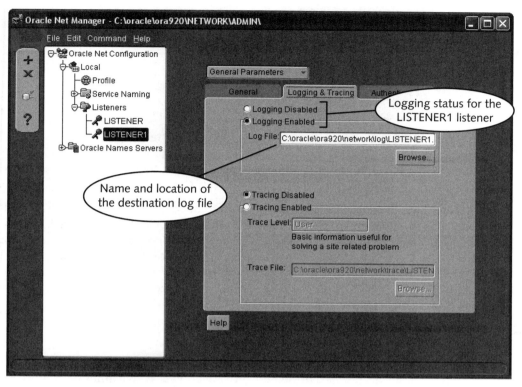

Figure 13-21   Current logging and tracing information

Deleting a Listener

Carlos can also use the Net Manager to delete an existing listener. For example, Carlos may need to make so many configuration changes to a listener that he feels it would be easier to delete a listener and then re-create it. When the listener is deleted, any reference to that listener is deleted from the Listener.ora file.

To delete the LISTENER1 listener:

1. From the Net Manager interface, click the **+** (plus sign) next to Local in the Oracle Net Configuration list to expand it, if necessary.

2. If necessary, click the **+** (plus sign) next to Listeners in the Local list.

3. Click the name of the listener previously created in this chapter.

4. Click **Edit** on the menu bar, and then click **Delete** on the submenu.

5. When the Delete dialog box shown in Figure 13-22 appears, click **Yes** to delete the listener.

Figure 13-22   Dialog box to confirm deletion of a listener

6. Click **File** on the menu bar, and then click **Save Network Configuration** to save the change. Click **File**, and then **Exit** to close the Oracle Net Manager.

Now that Carlos has deleted the listener, the default listener named LISTENER handles all client requests made for the database until another listener is created. If there is a subsequent lag time when the default listener is responding to client requests, Carlos may need to create an additional listener. The additional listener would allow load balancing and also provide an option for automatic failover if there is a problem with the default listener. In addition, clients would be able to contact other listeners that are available.

DYNAMIC REGISTRATION OF SERVICES

When a database instance is started, the instance's PMON can automatically register with an existing listener, through a process known as dynamic registration of services. This process allows Carlos to take advantage of the load balancing and automatic failover options available when using multiple listeners. To allow an instance to automatically register with a listener, the listener must be configured as the default listener, or identified by the LOCAL_LISTENER parameter in the Init.ora file.

Note that the INSTANCE_NAME and SERVICE_NAMES parameters within the Init.ora file must also be set to allow automatic registration. The INSTANCE_NAME parameter identifies the name of the instance that should be registered with the listener. The SERVICE_NAMES parameter is a combination of any domain name specified by the DB_DOMAIN parameter of the Init.ora file and the instance name. For example, if the value assigned to the DB_DOMAIN parameter is JCU.COM and the value of the INSTANCE_NAME parameter is OPERATION, then the value that Carlos assigns to the SERVICE_NAMES parameter is OPERATION.JCU.COM.

Now that Carlos knows how to configure a server to respond to client requests, he can begin configuring the clients with the necessary information to make contact with the Oracle9*i* database server. This will be the topic of Chapter 14.

If you are not completing the hands-on assignments at this time, the original copy of the Listener.ora file should be restored to the \network\admin folder.

CHAPTER SUMMARY

- ❏ A single-tier architecture consists of a dumb terminal directly connected to a mainframe computer.

- ❏ A two-tier architecture consists of clients connecting to a server over a network with the processing burden being shared between the client and server.

- ❏ An *n*-tier architecture includes middleware between the client and database server tiers, such as a server that can be used to multiplex client connections or function as a Web server.

- ❏ Components of Oracle Net Services address the connectivity, manageability, and scalability of the Oracle9*i* network environment.

- ❏ Oracle Net Services includes Oracle Net, Oracle Net Listener, Oracle Connection Manager, and Networking Tools.

- ❏ Oracle Connection Manager can be used to multiplex client connections and to block or allow access from different clients and protocol conversion.

- ❏ Oracle Advanced Security is a separately licensed product that provides support for data encryption and user authentication.

- ❏ The Oracle communications stack is based on the OSI model and provides the specifications for transmitting data across an Oracle network environment.

- ❏ Each database must have at least one listener configured to respond to client requests.

- ❏ A database listener can be configured when the database is created through Net Manager or through the Net Configuration Assistant.

- ❏ Operations such as starting, stopping, and reloading a listener can be performed through the Listener control utility (lsnrctl).

- ❏ Net Manager can be used to change the configuration for a listener, delete a listener, or add a listener.

13

SYNTAX GUIDE

Command	Description	Example
LSNRCTL	Command to access the Listener control utility	`C:\> lsnrctl`
RELOAD	Used to reload a listener after configuration changes have been made without stopping the listener	`LSNRCTL> reload listener1`
SERVICES	Displays a summary of services and details on the number of connections	`LSNRCTL> services listener1`
SET	Used to assign a value to a listener parameter	`LSNRCTL> set log_status off`
SHOW	Used to display the current value of a listener parameter	`LSNRCTL> show log_status`
START	Used to start a listener	`LSNRCTL> start listener1`
STATUS	Shows the current status of a listener	`LSNRCTL> status listener1`
STOP	Used to stop a listener	`LSNRCTL> stop listener1`

REVIEW QUESTIONS

1. The Two-Task Common layer of the Oracle communications stack is responsible for determining the address of the specified database server. True or False?

2. A dumb terminal connected to a mainframe computer is representative of a(n) _____-tier architecture.

3. What type of services can be provided by the Oracle Connection Manager?

4. The ease with which an Oracle9i network can be administered is known as

   _____.

   a. manageability

   b. scalability

   c. connectivity

   d. security

5. In the Oracle communications stack, the OPI is found on the server machine. True or False?

6. The default name for a database listener is _____.

7. A(n) _____ is a process that is contacted for client requests to connect to the database.

   a. protocol

   b. listener

   c. OCI

   d. security agent

8. In comparison to other Oracle9*i* network architectures, a two-tier architecture allows thousands of clients to be connected to the Oracle9*i* database server without overwhelming the server. True or False?

9. The Network Protocol layer of the client directly interacts with the _____ layer of the server in the Oracle communication stack.

10. VI and LU6.2 are examples of _____.

    a. network protocols

    b. encryption algorithms

    c. token cards

    d. none of the above

11. A user requests data from the Oracle9*i* database through a Web server. This configuration is an example of:

    a. single-tier architecture

    b. two-tier architecture

    c. *n*-tier architecture

    d. a sneakernet

12. The Oracle communication stack is based on the _____ model.

    a. API

    b. OCI

    c. TTC

    d. OSI

13. _____ refers to the ability to include additional hardware or nodes to an existing network.

14. _____ can be used to restrict client access to the Oracle9*i* database server based on the client's IP address.

15. Identify three ways to create a listener.

16. What is the purpose of a communication model?

17. Oracle Net operates at the _____ layer(s) of the Oracle communication stack.

13

18. Oracle Net needs to be installed only on the client side, not the server side of an Oracle network connection. True or False?

19. The Oracle communication stack for a client/server connection consists of _____ layers.

    a. seven

    b. six

    c. five

    d. four

20. Which protocol is required if a client connects to an Oracle9*i* database via the Internet?

HANDS-ON ASSIGNMENTS

Assignment 13-1 Creating a Listener

In this assignment you create a new listener for your database.

1. Start the Net Manager.
2. Click the + (plus sign) next to Local in the Oracle Net Configuration list.
3. Click the word **Listeners** from the Local list.
4. Click **Edit** on the menu bar, and then click **Create** on the submenu.
5. Create a listener named LISTENER5 that will listen for requests on port 25921 from clients using a TCP/IP protocol for the TESTDB database located on your computer.

Assignment 13-2 Stopping and Starting a Listener

In this assignment you will perform the necessary tasks to stop and restart a listener.

1. Start the Listener control utility.
2. Stop the LISTENER5 listener.
3. Issue the command necessary to verify the status of the LISTENER5 listener.
4. Start the LISTENER5 listener.
5. Re-issue the command from Step 3 to verify the current status of the LISTENER5 listener.

Assignment 13-3 Changing the Configuration of a Listener

In this assignment you change the value of the STARTUP_WAITTIME parameter of a listener.

1. If necessary, launch the Listener control utility.

2. Type **show startup_waittime** to display the current value assigned to the STARTUP_WAITTIME parameter.

3. Issue the appropriate SET command to change the value assigned to the STARTUP_WAITTIME to 10 seconds.

4. Redisplay the value currently assigned to the STARTUP_WAITTIME parameter.

5. Reset the STARTUP_WAITTIME parameter back to its previous value using the SET command.

Assignment 13-4 Enabling Logging Through the Listener Control Utility

In this assignment you perform the steps necessary to enable logging through the Listener control utility.

1. Using your operating system, create a folder named **Logging**.

2. If necessary, start the Listener control utility.

3. Use the SET command to change the LOG_DIRECTORY parameter for the LISTENER5 listener and specify the Logging folder as the log directory.

4. Use the SET command to assign the file name of Logging1 as the name of the listener log file.

5. Issue the appropriate command through the listener control utility to turn logging on for the LISTENER5 listener.

6. Perform the necessary steps to stop all listeners except LISTENER5.

7. Use SQL*Plus to log into your database.

8. Using My Computer from the operating system, locate the Logging1 file and open it using Notepad. Determine whether the connection request you made in Step 7 has been recorded to the file.

9. Close the Logging1 file and then exit SQL*Plus.

Assignment 13-5 Changing the Logging Status Using Net Manager

In this assignment you disable the logging status for the LISTENER5 listener. Make certain you complete Assignment 13-4 before attempting this assignment.

1. If necessary, start Net Manager and click the + (plus sign) next to Local in the Oracle Net Configuration list.

2. If necessary, click the + (plus sign) next to Listeners in the Local list to display the name of all configured listeners.

13

3. Click **LISTENER5** in the Listeners list.

4. From the Logging & Tracing page in the right pane of the window, make the change(s) necessary to disable logging for the listener.

5. Save the change and exit Net Manager.

6. Use SQL*Plus to log into your database.

7. Using My Computer from the operating system, locate the Logging1 file and open it using Notepad. Determine whether the connection request you made in Step 6 has been recorded to the file.

8. Close the Logging1 file and then exit SQL*Plus.

9. Delete the Logging1 file and remove the Logging folder created in Assignment 13-4.

Assignment 13-6 Deleting a Listener

In this assignment, you delete a listener.

1. If necessary, start Net Manager.

2. From the left pane of the Net Manager window, locate and delete the LISTENER5 listener previously created in Assignment 13-1.

3. Save the change and exit Net Manager.

CASE PROJECTS

Case 13-1 Identifying Appropriate Network Architecture

Currently the tellers access customer information directly from the Oracle9*i* database through an application program stored on their individual computers. However, within the next few months, more tellers will be hired to handle the increasing workload and to decrease customer wait time. In addition, the board of directors has decided that customers should also have access to their accounts via the Internet and that this project will begin within the next six months.

Part of Carlos' responsibility is to determine how to meet the increased demand on the Oracle9*i* server that houses the financial database. His initial response was to simply get a more powerful machine. However, this would only be a temporary solution because at some point, the new machine would also become overburdened with processing requests. Carlos is beginning to wonder if reconfiguring the Oracle9*i* network environment would be more beneficial.

Create a memo and provide Carlos with suggestions for reconfiguring the Oracle9*i* network to minimize the impact that a significant increase in volume will have on the database's performance.

Case 13-2 Increasing Security

The security of data at Janice Credit Union has not been a big concern. At the present time, employees of the credit union can access the information contained within the Oracle9*i* financial database from computers located on the work premises. All external transactions, such as ATMS, and so forth, have been handled through a clearinghouse that also has a connection to the database. Tellers at Janice Credit Union are required only to provide a user name and password to access customer account and transaction information, and the clearinghouse has a dedicated connection to the database. However, once Web connections are permitted, security will become a major concern.

Carlos needs to determine what type of changes needs to be made to ensure the security and integrity of the data contained in the financial database. What suggestions would you make to reasonably assure database security? Make certain you explain the rationale for each suggestion.

13

14

CLIENT-SIDE CONFIGURATION

**After completing this chapter,
you should be able to do the following:**

♦ List the different name resolution methods available

♦ Identify the purpose of the Sqlnet.ora and Tnsnames.ora files

♦ Modify the contents of the Sqlnet.ora file

♦ Modify the contents of the Tnsnames.ora file

♦ Perform basic troubleshooting operations

In Chapter 13, the configuration necessary for an Oracle9*i* server to listen and respond to clients' requests was presented. However, for the Oracle9*i* network environment to be complete, the client must be provided with the information necessary to locate the listener. The listener is integral to the process because it is used to accept connection requests from the client. Therefore, if the client cannot contact the listener, the client is not able to connect to the server.

In this chapter, the common naming resolutions available in Oracle will be identified. When appropriate, any client-side configuration requirement for the naming resolution is also included. The client uses a naming resolution to determine how to contact the database server; the client-side configuration provides the client with this information. In addition, common troubleshooting techniques used when a client is unable to connect to a server are demonstrated.

THE CURRENT CHALLENGE IN THE JANICE CREDIT UNION DATABASE

Recall from Chapter 13 that Carlos is preparing to expand the Oracle9*i* network environment by including several new clients and possibly another Oracle9*i* server. In that chapter, Carlos examined networking concepts and how to configure a server to listen for client requests.

In this chapter, Carlos will examine the Tnsnames.ora file, which is used to provide clients with connection information to Oracle9*i* servers. In addition, he will also review other name resolution options available in an Oracle network environment and how to specify the sequence for the options in a Sqlnet.ora file. After he masters the information and has configured the client, the client will then be able to contact the Oracle9*i* server.

SET UP YOUR COMPUTER FOR THE CHAPTER

The examples in this chapter will change the settings in two files on your computer: Sqlnet.ora and Tnsnames.ora. By default, these files are located in the \network\admin folder of your ORACLE_HOME directory. A copy of these files should be created through your operating system in the event an error preventing you from connecting to the Oracle9*i* server occurs.

 The ORACLE_HOME directory is the location where you installed the Oracle9*i* database software. For example, if you installed the software in the Ora92 folder of the Oracle subdirectory on the C: drive of your computer, then ORACLE_HOME is C:\oracle\ora92.

NOTES ABOUT DUAL COVERAGE WITHIN THIS CHAPTER

The tasks performed in this chapter will be completed using the Net Configuration Assistant and Net Manager. These tools are available through the Oracle9*i* Database software, regardless of whether you are using the Windows XP or Windows 2000 (Server or Professional) operating systems. Therefore, the examples in this chapter will be presented only in the Windows XP environment.

NAME RESOLUTION METHODS

A client must supply three pieces of information to be able to connect to a database: the user name, password, and the net service name (which represents the database service on the server). Carlos previously configured the Oracle9*i* server to listen for client requests; however, he has to ensure that the clients can locate the database server for a connection to occur. Otherwise, the client is not able to connect to the database, even if the two are on the same network.

For the client to connect to the database server, the database's net service name must be resolved, or translated, into the appropriate connect descriptor. The **net service name** is a simple name, such as a host name without network location information, used to identify a database. The **connect descriptor** identifies the name of the desired service and network route information, such as the listener address that the client uses to contact the database server. Figure 14-1 depicts this resolution process.

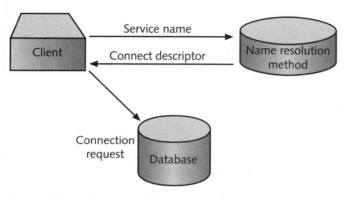

Figure 14-1   Name resolution procedure

Although Oracle9*i* supports external naming methods such as Network Information Services (NIS) and Cell Directory Services (CDS), typically DBAs use host naming, directory naming, or local naming to resolve the net service name. Each naming resolution method is discussed in turn in the following sections.

Oracle9*i* also includes support for the Oracle Names Server, a server used to store network addresses. However, it is being replaced by the Oracle Internet Directory (directory naming) and will not be covered in this textbook.

14

Host Naming

The **host-naming method** for name resolution is probably the simplest approach because there is no configuration requirement for the client. If the client requests a connection to a database service, the client supplies the user ID, password, and the host name (name of the machine) where the database is stored. This information is necessary to authenticate the client and to determine to which database the client is attempting to connect.

In the next step sequence, the host name is resolved into the connect descriptor information through either a HOSTS file or an external naming service that already exists on the network. (Examples of external naming services include NIS and DNS, or Domain Name System.) In essence, the burden of ensuring the clients can connect to the database server is shifted from the DBA to the network administrator. That is, the DBA is responsible only for ensuring that the database server and client machines are configured to operate across the network, and the network administrator is responsible for providing the information necessary to resolve the name and location of the database server.

There are, however, some requirements that must be met before Carlos can use the host-naming method:

- The network protocol must be TCP/IP.
- He cannot use any advanced networking features, such as Oracle Connection Manager.
- The GLOBAL_DBNAME parameter must be assigned the name of the host machine so that the connect descriptor contains the correct information to identify the computer.
- A HOSTS file or external naming service must be available to the client.

Although the use of the host-naming method may seem simple because there is no extra client-side configuration required, this method is not recommended for a very large network or if there are several Oracle9*i* servers. Why? In a large network, any user trying to transmit data across the network needs to reference the HOSTS file or the external naming service; use of the HOSTS file is not limited to just database users. The result is that in a network with heavy traffic, a bottleneck can occur because all users are required to use the same naming service. Therefore, this approach is recommended for small networks and those with only a few Oracle9*i* servers.

Directory Naming

The **directory naming** approach uses a centralized directory server to store name resolution information. Clients can use the Lightweight Directory Access Protocol (LDAP) to request name resolution information from the centralized directory server. LDAP is an implementation of the International Standardization Organization (ISO) X.500 standard. LDAP simplifies management of directory services by allowing users of various applications to reference a single interface to obtain directory information.

A directory server allows for centralized administration of database services. This name resolution approach is similar to the use of a DNS. However, this approach requires that the directory server's configuration be LDAP compliant. Examples of an LDAP-compliant directory server include Oracle Internet Directory or Microsoft Active Directory. If Carlos does not have this type of configuration available, then directory naming is not an option for the Janice Credit Union's Oracle network.

Local Naming

The most widely used approach for name resolution is **local naming**. It is widely used due to its ease of implementation and the reduction in network traffic because the name resolution occurs locally. With this approach, Carlos configures a Tnsnames.ora file that resides on each client. This file is used to map a net service name to the appropriate connect descriptor enabling the client to locate the Oracle9*i* database server.

Although the Tnsnames.ora file must reside on each client, there is a way to simplify the process of Carlos configuring each one individually. After the file has been configured correctly on one client, a copy of the same file can be placed on all clients without modification. The creation and syntax for the Tnsnames.ora file is presented later in this chapter.

NAME RESOLUTION ORDER

Carlos does not have to choose only one name resolution approach. In fact, he can utilize all the available approaches if he has access to the necessary resources. By having more than one name resolution method available, this avoids having a single point of failure for a client in the Oracle9i network. If Carlos chooses to use only the local naming method, and the client's Tnsnames.ora file became corrupt, that client would be unable to connect to the Oracle9i server until the problem is corrected because another naming resolution method is not available.

The order for naming resolution is provided in the Sqlnet.ora file. How does Carlos decide what order should be used? He can look at the advantages and disadvantages of each method and then decide which approach is the most appropriate. The advantage and disadvantage of each method is provided in Table 14-1.

Table 14-1 Naming resolution methods advantages and disadvantages

Resolution Naming Method	Advantage	Disadvantage
Host naming	No client-side configuration; uses standard HOSTS file available on the network	Increased network traffic required for client to access the HOSTS file; certain conditions must be met to use this method
Directory naming	Does not require client-side configuration; directory server can be used by multiple applications	Increased network traffic
Local naming	Reduced network traffic	Requires configuration of the Tnsnames.ora file on each client

The following sections will discuss how Carlos can use the Sqlnet.ora file to specify the name resolution order to be used when clients access the Oracle9i database server.

Sqlnet.ora File Contents

How does Carlos instruct the client which naming method to use, and in what order if multiple naming methods are available? The resolution order is provided in the Sqlnet.ora file in the \network\admin folder of the client's ORACLE_HOME directory. An example of the contents of this file is shown in Figure 14-2.

14

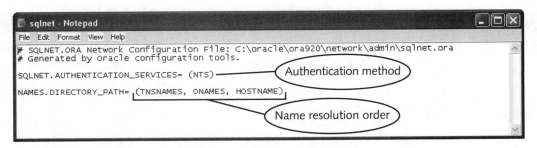

Figure 14-2   Example Sqlnet.ora file

The Sqlnet.ora file displayed in Figure14-2 contains four lines of entries. The first two lines are comments indicating the location of the file and how the file was generated. These lines are not required; they are simply for documentation purposes. The first parameter listed in the file, SQLNET.AUTHENTICATION_SERVICES, is an entry to support an Oracle Advanced Security authentication feature. In this example, the value NTS specifies that the Windows NT Native Authentication method can be used.

The last entry in the file, the NAMES.DIRECTORY_PATH parameter, correlates the naming methods to be used and the order that is to be used for name resolution. In Figure 14-2, this client first locates the local Tnsnames.ora file and attempts to resolve the service name. If the name cannot be resolved using this approach, or if the file cannot be located, the Oracle Names Server is consulted next.

 Remember that the Oracle Names Server is no longer being developed and will eventually be replaced by the directory naming approach. The last option to resolving a service name is to consult a HOSTS file. If none of these methods are successful, the client displays an ORA-12154 error message indicating that the name could not be resolved.

Modifying the Sqlnet.ora File

The Sqlnet.ora file is created when the client software is installed. By default, Oracle9*i* assigns the following resolution sequence: local naming, Oracle Names Server, and host naming. However, Carlos can override this at any time using the Oracle Net Manager. In the following step sequence, Carlos uses Net Manager to change the sequence so that host naming is always the first approach used for name resolution.

To change the name resolution order:

1. On the Start menu of the operating system, click **All Programs**, click **Oracle-OraHome92**, click **Configuration and Migration Tools**, and click **Net Manager**.

2. When the Net Manager window shown in Figure 14-3 appears, click the **+** (plus sign) next to Local.

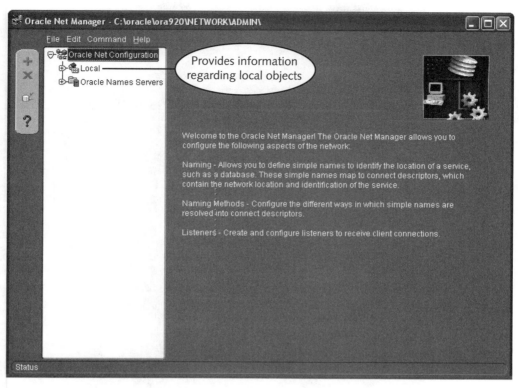

Figure 14-3   Net Manager window

3. After Local objects appear, click the word **Profile** from the list.

4. When the Profile window shown in Figure 14-4 appears, click **HOSTNAME** from the Selected Methods list.

5. Click the **Promote** button twice to move HOSTNAME to the top of the list. The sequence of the naming methods in the Selected Methods list specifies the order for resolving service names.

6. Click **File**, and then click **Save Network Configuration** to save the change.

7. Click **File**, and then click **Exit** to exit Net Manager.

When the change is saved, Net Manager automatically updates the Sqlnet.ora file.

14

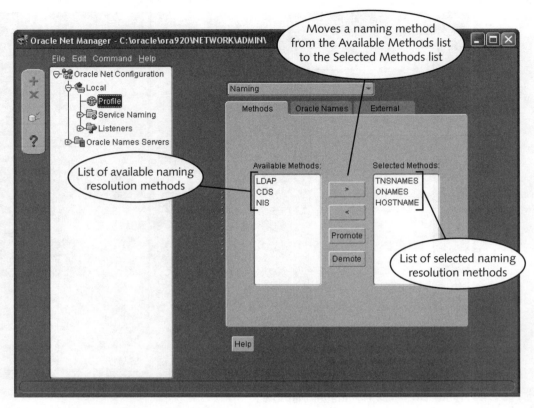

Figure 14-4   Name resolution methods

To view the Sqlnet.ora file on your computer:

1. On the Start menu, click **My Computer**.

2. Navigate to the appropriate drive by clicking the folder icons necessary to reach the \network\admin subdirectory. In a default installation the path would be: C:\oracle*ORA_HOME*\network\admin, where *ORA_HOME* is the name of the folder where the Oracle9*i* database software was installed.

3. Double-click the icon for the **Sqlnet.ora** file to display its contents.

 If the file is not associated with a text editor, it will not open and you must select a text editor such as Notepad from the list of available programs displayed by your computer. Do not use a word processor such as WordPad or Word because the saved file contains formatting codes that cannot be processed by the Oracle software.

Your Sqlnet.ora file should resemble the one shown in Figure 14-5, with HOSTNAME being listed first in the NAMES.DIRECTORY_PATH parameter.

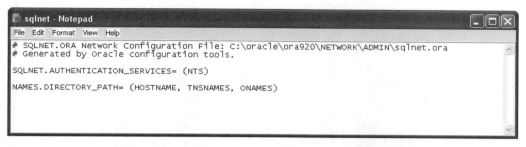

Figure 14-5   Modified Sqlnet.ora file

In addition to changing the resolution order, Carlos can also use the Profile pane to delete a method from the Selected Methods list by selecting the method to be removed, and then clicking the left arrow button. Carlos can add additional methods by selecting the name of the method from the Available Methods list and clicking the right arrow button.

CONFIGURING THE CLIENT FOR LOCAL NAMING

The Tnsnames.ora file must reside on each client that uses the local name resolution method. This file is typically created when the client software is installed on the machine. However, it can also be created, or modified, using Net Manager or the Net Configuration Assistant, or by using any text editor. In the following sections, Carlos will modify the Tnsnames.ora file on an existing machine and then examine its contents.

Modifying a Tnsnames.ora File

In this section, Carlos will modify a Tnsnames.ora file for a client that is being added to the Oracle9*i* network. Remember, as an alternative, Carlos could also have copied a valid file from another client. However, in this case, a Tnsnames.ora file already exists on the client to provide connection information to another database and it simply needs to be updated to be able to access the TESTDB database.

To add connection information to the Tnsnames.ora file for the TESTDB database:

1. On the **Start** menu, point to **All Programs**, click **Oracle-OraHome92**, click **Configuration and Migration Tools**, and then click **Net Configuration Assistant**.

2. When the Welcome window for the Oracle Net Configuration Assistant appears, click the **Local Net Service Name configuration** option button, as shown in Figure 14-6, and click **Next**.

14

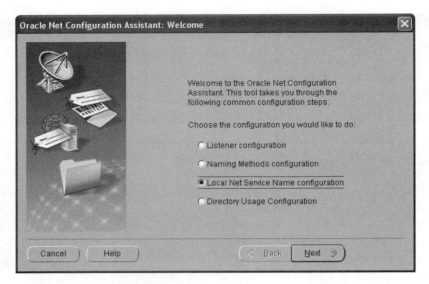

Figure 14-6   Oracle Net Configuration Assistant: Welcome window

3. If necessary, click the **Add** option button in the Net Service Name Configuration window shown in Figure 14-7 to indicate that a new service is to be added to the Tnsnames.ora file. Then click **Next**.

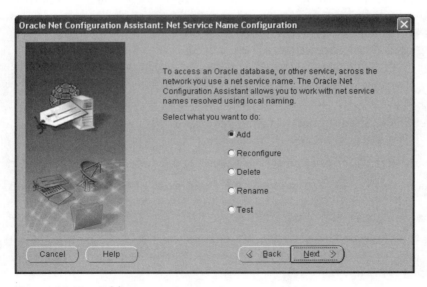

Figure 14-7   Adding a service name

4. When the Database Version window appears, click the first option button, if necessary, to indicate that the service is for a database that is an Oracle8*i* or later version. Then click **Next**.

5. When the Service Name window shown in Figure 14-8 appears, enter the name of your database in the text box, and then click **Next**. If your database has been configured with a global database name, your instructor will provide you with the appropriate information.

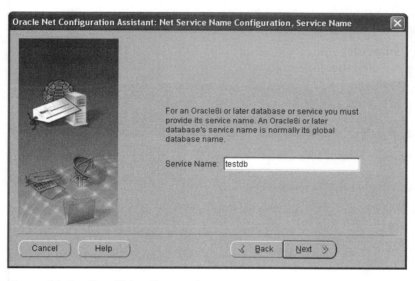

For an Oracle8i or later database or service you must provide its service name. An Oracle8i or later database's service name is normally its global database name.

Service Name: testdb

Figure 14-8 Specifying the service name

6. When the Select Protocols window appears, click the appropriate protocol (normally TCP), and click **Next**.

7. If the TCP/IP protocol window appears, or another protocol window if a different one was selected in the previous step, enter the name of the host machine where the database is located, and then enter the appropriate port for its listener. Other information may be required if you selected a different protocol. Then click **Next**.

8. When the Test window appears, click the **Yes, perform a test** option button, and then click **Next**.

9. When the Connecting window shown in Figure 14-9 appears indicating that the test was successful, click **Next**. If an error message was returned, click the **Change Login** button to change the user ID and password used to connect to the database and then reattempt the connection.

14

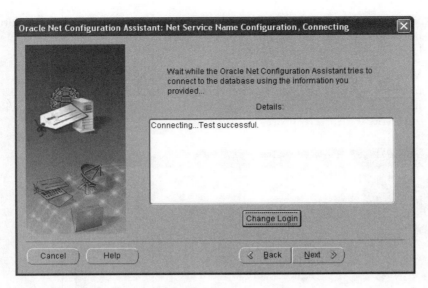

Figure 14-9   Successful connection test

10. When the Net Service Name window appears, click **Next**.

11. After the Another Net Service Name window appears, click the **No** option button, and then click **Next**.

12. When the Net Service Name Configuration Done window appears, click **Next**.

13. When the Welcome window reappears, click **Finish** to complete the addition of the new service name.

At this point, the client's Tnsnames.ora file is updated with the connection information to allow the client to locate the TESTDB database. As an alternative, Carlos could have made this addition, or made modifications to an existing service, using Net Manager. As shown in Figure 14-10, Service Naming can be expanded in the Oracle Net Configuration list to identify existing services. After an existing service is selected, relevant connection information is displayed in the right pane of the window. At that time, Carlos has the option of making any necessary changes.

The basic difference between the two approaches is that with the Net Configuration Assistant, Carlos provides the necessary information through his responses to the wizard. In the case of Net Manager, Carlos simply enters the data without receiving a series of prompts.

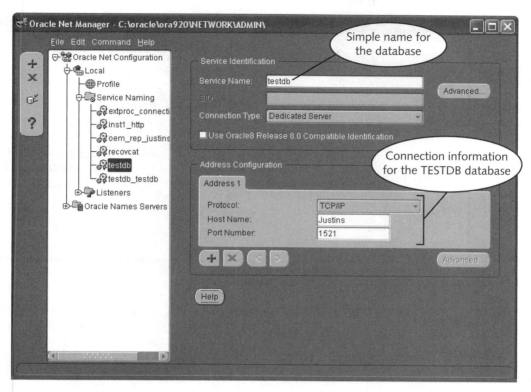

Figure 14-10   Service name information

Tnsnames.ora File Syntax

Now that Carlos has added the TESTDB connection information to the new client's Tnsnames.ora file, he can review the file's syntax and how this information is stored. Figure 14-11 is an example of the file updated with the connection information. The Tnsnames.ora file on your computer may look different depending on the file's previous contents.

Do some of the parameters in the file look familiar? You may have seen some of these parameters in the Listener.ora file examined in Chapter 13. In this example, information for two net service names is included in the file. Each connect descriptor section is identified by the parameter DESCRIPTION. The ADDRESS_LIST parameter begins the descriptor address information. The address begins with identification of the appropriate protocol, name of the machine housing the database, and the port for the database's listener.

14

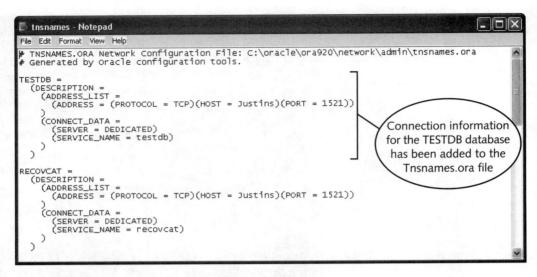

Figure 14-11 Contents of the Tnsnames.ora file

 Although the example given in Figure 14-11 lists the subparameters for the ADDRESS parameter in the order of PROTOCOL, HOST, and then PORT, the subparameters can be listed in any order. For example, stating the ADDRESS parameter as (ADDRESS=(HOST=Justins) (PROTOCOL=TCP) (PORT=1521) would be equivalent to the parameter given in Figure 14-11.

The CONNECT_DATA section identifies the type of Oracle9i server (typically Dedicated or Shared, which will be discussed in Chapter 15) and the name or global name of the database. Another example of a Tnsnames.ora file is provided in Figure 14-12. Notice that the service names are actually the global names for the databases, not just the name of the database. In this case, the databases are located in the CPA domain of the network, so "CPA" is appended to the database name.

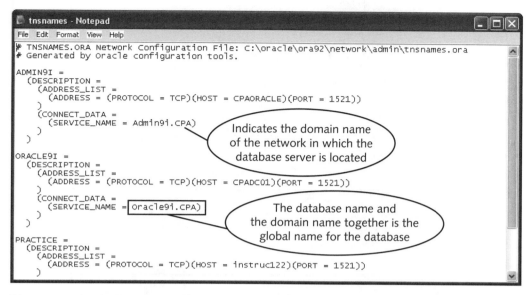

Figure 14-12   Tnsnames.ora file containing global database names (partial output shown)

CONFIGURING THE CLIENT TO USE HOST NAMING

Although Carlos does not need to create any special configuration files if he elects to use the host-naming approach for name resolution, he would still need to instruct the client which approach to use. The Net Configuration Assistant can be used to specify the desired name resolution. When the process is completed, the Sqlnet.ora file is updated with the information, just as when Carlos previously changed the resolution sequence.

To specify that a HOSTS file should be used for name resolution:

1. On the Start menu, point to **All Programs**, click **Oracle-OraHome92**, click **Configuration and Migration Tools**, and then click **Net Configuration Assistant**.

2. When the Net Configuration Assistant Welcome window appears, click the **Naming Methods configuration** option button, and then click **Next**.

3. If more than one naming method is listed in the Selected Naming Methods list of the Select Naming Methods window, select and remove all but the Host Name method. Click **Next**.

If naming resolution had not previously been configured for the client, the Selected Naming Methods list would be empty and you would simply need to add the host name to the Selected Naming Methods list.

14

4. When the Host Name window shown in Figure 14-13 appears, click **Next**.

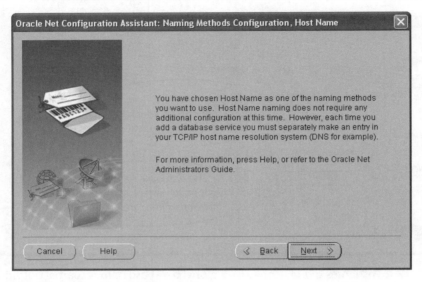

Oracle Net Configuration Assistant: Naming Methods Configuration, Host Name

You have chosen Host Name as one of the naming methods you want to use. Host Name naming does not require any additional configuration at this time. However, each time you add a database service you must separately make an entry in your TCP/IP host name resolution system (DNS for example).

For more information, press Help, or refer to the Oracle Net Administrators Guide.

Cancel   Help   ≪ Back   Next ≫

Figure 14-13   Host Name window

5. After the Naming Methods Configuration Done window appears, click **Next**.

6. When the Welcome window appears, click **Finish** to indicate that no other configuration needs to be performed.

In addition, Carlos could have added any of the other name resolution methods using the same procedure. However, as previously shown, if the local naming method is used, additional information is required.

TROUBLESHOOTING A DATABASE NETWORK CONNECTION

After Carlos has performed the necessary server-side and client-side configurations, the user should be able to connect to the database, provided the network is operating properly. However, if a problem should occur, there are some basic steps Carlos can perform to narrow down the source of the problem.

The Oracle Net Configuration Assistant provides Carlos with a method of testing whether the local naming method has been configured properly on the client. As shown in Figure 14-14, the Test option is available after choosing Local Net Service Name configuration from the Welcome window. However, this procedure tests only the local naming method, and may not be sufficient if the problem is not caused by the Tnsnames.ora file.

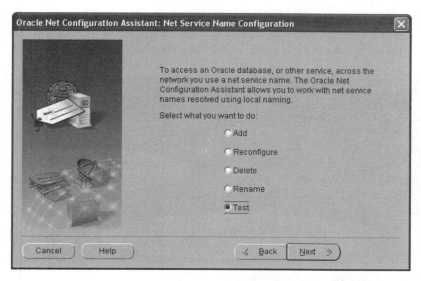

Figure 14-14 Test option from the Oracle Net Configuration Assistant

When Carlos is troubleshooting a problem with the Oracle Net Configuration Assistant, his best strategy is to identify the general area causing the problem and then focus on that area until it is narrowed down to the actual cause. If a client is unable to connect to the database, there are three broad areas that Carlos needs to examine: the server, the client, and the network. The typical procedure is to verify that the network connection is established and that the client can locate the listener. This would indicate that the network is operational and that the client can correctly locate the database server. If the client can locate the listener, then Carlos needs to verify that the listener is started and running using the listener utility, as demonstrated in Chapter 13.

Another utility provided by Oracle9i to verify that the client can locate the listener is the TNSPING utility. The database can be referenced using either the service name or the IP address of the machine housing the database. If the utility locates the listener using the service name, then Carlos would know that not only is the network operational and that the name is being resolved correctly, but that the listener can be located. However, it does not actually confirm that the client can connect to the database itself. In the following step sequence, Carlos will determine whether the listener can be located.

To locate the listener:

1. On the Start menu, click **Run**.

2. Type **cmd** in the Run dialog box and press Enter to open a command-line window.

3. At the system prompt, type **tnsping testdb** to perform a ping of the listener for the TESTDB database. To ping a different database, simply substitute the appropriate database name.

4. As shown in Figure 14-15, the name resolution method used is displayed. If the listener is located, the results include the response of OK and the number of seconds it took to receive a response from the listener.

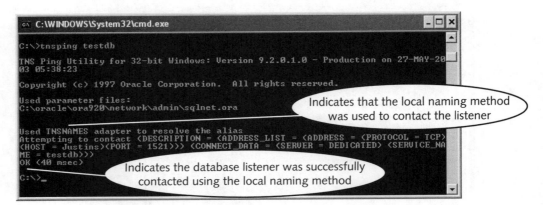

Figure 14-15 Using the TNSPING utility to locate a listener

If the utility had been unable to locate the listener and it is a TCP/IP-based network, Carlos could have used the operating system's PING utility to verify that the machine housing the database can be reached. To use the PING utility, Carlos would simply type ping *host_name* or ping *IP address* at the operating system's prompt. If the network is operational, the utility displays the amount of time it took to actually receive a response from the computer. If an error occurs, Carlos could have the client ping itself using the IP address 127.0.0.1, as shown in Figure 14-16. An error at this point would indicate that there could be a problem with the network card or the machine's configuration.

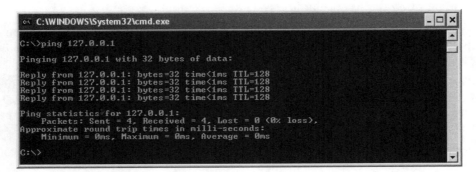

Figure 14-16 PING utility

If Carlos is able to locate the listener for the database using the TNSPING utility, it does not mean that the client was actually able to connect to the listener. After verifying that the client and network are operating correctly, Carlos would need to verify that the listener is running. Of course, Carlos should also double-check that there are no other problems occurring with the database. For example, if several clients were having problems connecting to the database, this would indicate that he would need to focus his efforts on verifying that the network is operational and that there are no problems with the server. In other words, there is no absolute troubleshooting procedure; it simply depends on what type of information is available to Carlos.

Now that Carlos has learned the configuration procedures for the servers and clients in an Oracle network environment, in Chapter 15 he will begin examining the use of an Oracle Shared Server. With a shared server configuration many users can share server processes. Typically in this type of configuration, Carlos needs fewer resources to handle more clients. With the credit union expanding the type of services being provided to its clients, this could be a viable option considering Carlos' limited budget.

After completing the examples in this chapter, restore the original copies of the Sqlnet.ora and Tnsnames.ora files.

CHAPTER SUMMARY

- ❏ After the listener for an Oracle9*i* server has been configured, the client still needs to be provided with the necessary information to contact the listener.

- ❏ Common name resolution methods include host naming, directory naming, and local naming.

- ❏ Host naming uses a HOSTS file or a Domain Name System (DNS).

- ❏ Directory naming references an LDAP-compliant directory server.

- ❏ Local naming requires the configuration of a Tnsnames.ora file on the client.

- ❏ The Tnsnames.ora file can be created (and edited) manually by the DBA or through Net Manager or the Net Configuration Assistant.

- ❏ The advantage to using local naming is that there is no network traffic generated during name resolution.

- ❏ The advantage to using host or directory naming methods is that the information needed to resolve the name of a database server is stored in a centralized location on the network.

- ❏ The name resolution order is specified in the Sqlnet.ora file.

- ❏ The TNSPING utility is used to locate a listener on a specific Oracle9*i* server.

- ❏ The PING utility is used to verify that the machine housing the database can be reached.

14

Syntax Guide

Sqlnet.ora Parameter	Description	Example
NAMES. DIRECTORY_PATH	Identifies the name resolution method and sequence	`NAMES.DIRECTORY_PATH = (TNSNAMES, ONAMES, HOSTNAME)`
SQLNET. AUTHENTICATION_SERVICES	Used with Advanced Oracle Security to identify the supported authentication methods	`SQLNET. AUTHENTICATION_SERVICES = (NTS)`

Review Questions

1. Which of the following configuration files must be included on the client if the local naming approach is used for name resolution?

   a. Init.ora

   b. Tnsnames.ora

   c. Listener.ora

   d. Host.ora

2. If the client is attempting to connect to a database using the local naming approach for name resolution, the network must use the TCP/IP protocol. True or False?

3. What is the default name resolution sequence in Oracle9*i*?

4. Which name resolution method(s) does not require special configuration on the client?

5. If you elect to use the host-naming method, what protocol is required?

6. Which of the following is *not* an Oracle9*i* utility?

   a. TNSPING

   b. PING

   c. LSNRCTL

   d. all of the above

7. What is the purpose of the HOST parameter in the Sqlnet.ora file?

8. Which file is used to specify the sequence for the available name resolution methods?

   a. Init.ora

   b. Tnsnames.ora

   c. Listener.ora

   d. Sqlnet.ora

9. The ADDRESS_LIST parameter of the Tnsnames.ora file is used to identify the name of the database to which the client is attempting to connect. True or False?

10. What protocol is required to use an Oracle Internet Directory in an Oracle network environment?

11. By default, the path to the Sqlnet.ora file is _____.

12. During name resolution, the net service name is resolved to its appropriate connect descriptor. True or False?

13. Which option from the Oracle Net Configuration Assistant is used to create or modify the client's Tnsnames.ora file?

   a. Listener configuration

   b. Naming Methods configuration

   c. Local Net Service Name configuration

   d. Directory Usage configuration

14. To change the order of name resolution through the Net Manager, select _____ from the Oracle Net Configuration list.

15. What utility can be used to determine whether a client can locate a listener?

16. Which of the following is an example of an external naming method?

   a. host naming

   b. local naming

   c. Network Information Services (NIS)

   d. all of the above

17. The Tnsnames.ora file is used to specify the information regarding the settings on the client machine. True or False?

18. Which of the following is a prerequisite to using host naming for name resolution?

   a. The network must consist of at least fifty Oracle9i servers.

   b. The Oracle Connection Manager must be configured.

   c. There must be at least 12,000 database users configured for the network.

   d. The network must use the TCP/IP protocol.

19. Which parameter specifies the sequence of methods to be used during name resolution?

20. If a Microsoft Active Directory is being used to resolve net service names, which resolution method must be specified on the client?

14

HANDS-ON ASSIGNMENTS

 Before performing any of the assignments in this section, create a copy of the Tnsnames.ora and Sqlnet.ora files. These files should be restored after completing the assignments.

Assignment 14-1 Creating a Tnsnames.ora File

In this assignment, you delete the existing Tnsnames.ora file on your computer and then create a new file. Make certain you have created a copy of the file before attempting this assignment.

1. After creating a copy of the Tnsnames.ora file, delete the file from the \network\admin folder on your computer.

2. Attempt to connect to the database through SQL*Plus.

3. After receiving the error message indicating that the service name cannot be resolved, start the Oracle Net Configuration Assistant.

4. Provide the necessary information to use the local naming method for name resolution. (*Hint:* If you are uncertain regarding any of the database information, take a peek inside the copy of the Tnsnames.ora file previously created. Make certain you do *not* make any changes to this file.)

5. After completing the necessary steps in the Oracle Net Configuration Assistant, verify that a new Tnsnames.ora file has been created in the \network\admin folder.

6. After verifying that the file has been created, connect to the database through SQL*Plus. Exit SQL*Plus after you have confirmed your connection.

Assignment 14-2 Specifying Multiple Name Resolution Methods

In this assignment, you change the information regarding the name resolution methods stored in the Sqlnet.ora file.

1. Start the Oracle Net Configuration Assistant.

2. Select the appropriate option to add other resolution methods to the client's current Sqlnet.ora file.

3. Provide the appropriate responses to include the host–naming and directory naming methods to the client's name resolution configuration.

4. Change the resolution sequence so that host naming is always the first method used, followed by local naming, and then directory naming. Exit the Oracle Net Configuration Assistant.

5. View the contents of the modified Sqlnet.ora file.

Assignment 14-3 Using Net Manager to Change the Resolution Order

In this assignment, you use Net Manager to change the name resolution order specified in Assignment 14-2.

1. Start Net Manager.
2. Select the appropriate item from the Oracle Net Configuration list to view the current resolution order.
3. Use the Net Manager to change the resolution order so that the Tnsnames.ora file is referenced first, followed by the HOSTS file, and then the Oracle Internet Directory.
4. Save the changes you have made and exit Net Manager.

Assignment 14-4 Using the Oracle Net Configuration Assistant to Test a Listener Connection

In this assignment, you use the Oracle Net Configuration Assistant to make certain you can connect to the database.

1. Start the Net Configuration Assistant.
2. Select the appropriate option to test whether the client can locate the database's listener.
3. Complete the test. Make any login changes necessary to connect to the listener.
4. After successfully testing the client's configuration, exit the Net Configuration Assistant.

Assignment 14-5 Using the PING Utility

In this assignment, you verify that the machine housing the Oracle9*i* database can be reached.

1. Open an operating system command prompt window.
2. Issue the appropriate PING command to connect to the computer used by the Oracle9*i* server. If your database server is stored on the same machine as your client software, use the IP address 127.0.0.1 or the name of your computer in the PING command.

14

Assignment 14-6 Using the TNSPING Utility

In Assignment 14-5, you used the PING utility to verify that you could reach the host machine of the database. In this assignment, you verify that you can actually locate the listener for the database.

1. If necessary, open an operating system Command Prompt window.

2. Issue the appropriate TNSPING command to locate the listener. Use the name of the database in the TNSPING command to identify the database.

3. Use the IP address from Step 2 of Assignment 14-5 to re-issue the TNSPING command to reference the database by its IP address rather than by the database name.

4. Exit the Command Prompt window.

Assignment 14-7 Tracing Client-Side Connection Information

In this assignment, you will enable the tracing feature available within Net Manager.

1. Review the material regarding tracing contained in Chapter 17 of the *Oracle9i Net Services Administrator's Guide, Release 2 (9.2),* Part Number A96580-02, available on the Oracle Technology Network at *http://download-west.oracle.com/docs/cd/ B10501_01/network.920/a96580/troubles.htm#435078*. You will need to have a free OTN account to access the documentation area within the site.

2. Open Net Manager and enable tracing client-side information at **SUPPORT** level for the **TESTDB** database. Name the file **NetTracing** and designate the **UDUMP** directory of your Oracle9i installation as the tracing directory.

3. Save the changes you have made and exit Net Manager.

4. Open SQL*Plus and connect to the TESTDB database.

5. After connecting to the database, exit SQL*Plus.

6. Open the NetTracing text file with a text editor and review the entries that were made during the connection process.

7. Close the file.

Assignment 14-8 Tracing Client-Side Connection Information

In this assignment, you will enable the tracing feature available within Net Manager.

1. Review the material regarding logging contained in Chapter 17 of the *Oracle9i Net Services Administrator's Guide, Release 2 (9.2),* Part Number A96580-02, available on the Oracle Technology Network at *http://download-west.oracle.com/docs/cd/ B10501_01/network.920/a96580/troubles.htm#435078*. You will need to have a free OTN account to access the documentation area within the site.

2. Open Net Manager and enable the logging of client-side information for the **TESTDB** database. Name the file **NetLogging** and designate the **UDUMP** directory of your Oracle9i installation as the tracing directory.

3. Save the changes you have made and exit Net Manager.

4. Open SQL*Plus and attempt to connect to a database named **GENERIC**. When Net Manager is unable to locate this database, an error message will be recorded in the NetLogging file.

5. Exit SQL*Plus.

6. Open the NetLogging text file with a text editor and review the entries that were made during the connection attempt. Compare these entries with the contents of the NetTracing text file from Assignment 14-7.

7. After reviewing the NetTracing and NetLogging text files, use Net Manager to disable both tracing and logging.

8. Delete the NetTracing and NetLogging text files from the UDUMP directory.

CASE PROJECTS

Case 14-1 Identifying a Name Resolution Strategy

As presented in Chapter 13, the network at Janice Credit Union will be undergoing several changes to provide more services to its customers. This includes the need to support Internet access to the database and increased network volume. Several new workstations need to be added to the network for the new employees that are required. Draft a memo and outline your preference(s) for a name resolution strategy. You should include at least two naming methods, the resolution sequence, and your rationale for each. In addition, identify any assumptions you made when developing your strategy.

Case 14-2 Troubleshooting the Oracle9*i* Network

A user calls Carlos and states that he can no longer connect to the database. He had been connected to the database earlier before he left for lunch. But when he came back, he could not log into the database. Carlos also just came back and is not aware of any other user's having reported similar problems (at least there are no e-mails indicating problems).

If you were Carlos, what steps would you take to identify why the user is unable to connect to the database? Include any questions you would ask the user before actually trying to troubleshoot the problem.

14

15

ORACLE SHARED SERVER

**After completing this chapter,
you should be able to do the following:**
- Configure the Oracle Shared Server
- Change the value of the shared server parameters
- Monitor performance of the Oracle Shared Server through views
- Dynamically change parameter values

One of the scalability options provided by the Oracle9*i* server is the Oracle Shared Server (previously known as Oracle Multithreaded Server). This option allows the server to provide a larger number of concurrent connections without an increase in the hardware requirements. However, the scalability does not occur without a cost. Although the server can process more requests, this does not equate to an increase in the server's performance.

In an environment where the transactions are small and pauses occur between a user's requests, the increase in the response time of the server is barely noticeable. However, if user requests tend to generate a large result set, the increased response time can be significant and require the DBA to consider investing additional capital in the system. This is also true for long running queries (i.e., queries that require full table scans) because they do not have pauses during execution. In this chapter, you will examine the Oracle Shared Server, how it works, and its configuration requirements.

Becoming proficient in the Oracle Shared Server will allow you to maximize your server resources when additional hardware is not available. Your first step in becoming proficient is studying this chapter.

The Current Challenge in the Janice Credit Union Database

In Chapters 13 and 14, Carlos examined the server- and client-side configuration requirements necessary to expand the Oracle9i network environment. However, he still needs to determine how to get the existing server to handle the anticipated increase in volume. Recall that his budget is limited. Therefore, Carlos is considering using the Oracle Shared Server option to address the increased processing load that will be placed on the database server.

As an example of the usefulness of this option, consider that if reconfiguration of the server allows each shared server process twenty concurrent user connections, then this significantly increases the volume of user requests that can be addressed by the system. However, the downside is that there could be an impact on performance. This type of trade-off requires Carlos to look at the advantages and disadvantages of a shared server environment to determine whether the option should be implemented.

Set Up Your Computer for the Chapter

Before attempting the examples demonstrated in this chapter, make certain you create a backup of the database files and the database's Tnsnames.ora and Init.ora files. In the event an error occurs, these files can be used to restore the database.

Notes About Dual Coverage in this Chapter

In this chapter, both the GUI and manual approaches are demonstrated when changing the values assigned to the **shared server** parameters. The GUI approach consists of using the Database Configuration Assistant to change the database option to the shared server option, while the manual approach requires the user to manually edit the database's Init.ora file. Although the Enterprise Manager Console can also be used to change parameter values, this approach is not included because it was previously presented in Chapter 2.

The examples in this chapter are presented using the Windows 2000 Server operating system because a database server in a networked environment would most likely use that operating system. However, the examples can also be performed if Oracle9i is installed on a computer using Windows XP.

Server Processes

To understand the impact that a shared server environment can have on the database system, Carlos needs to examine how the different server environments function. Recall from Chapter 1 that the client makes a connection to the Oracle9i server to process a request.

Previously, the Oracle9*i* server was configured to provide a dedicated connection for each client. In a **dedicated server** environment, the user submits a request for a connection to the database. After the connection is established, the user uses this same connection throughout the session to request and receive data, as shown in Figure 15-1. Even if 100 users are connected to the database server, each user has a separate server process allocated for his or her exclusive use.

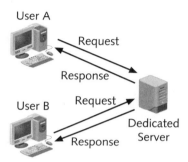

Figure 15-1   Dedicated server process

In a typical online transaction-processing system (OLTP), it is reasonable to assume that there are pauses that occur between the requests submitted by a particular client. Basically, the server processes a user request, then sits idle while waiting for the next request to be received from the user. This procedure can be made more efficient by allowing the server to process requests from other clients during this gap, or lag, between requests.

Oracle9*i* provides the option of configuring the database for shared server processes. With a shared server process, multiple users can use the same process to execute requests. However, rather than submit the request directly to the server, it is submitted to a dispatcher via a virtual circuit. A **virtual circuit** is a portion of shared memory allocated for each dispatcher.

Think of a **dispatcher** as a real estate agent. Each agent does not work exclusively with each client. Instead, the agent works with several clients, and clients may be in different stages of the purchasing process. With the shared server processes, the requests are received by a dispatcher and placed in a request queue. When the server is available, it selects the next request in the queue, processes the request, and then places the results in the response queue. The completed request is then retrieved by the dispatcher and returned to the client, as shown in Figure 15-2.

15

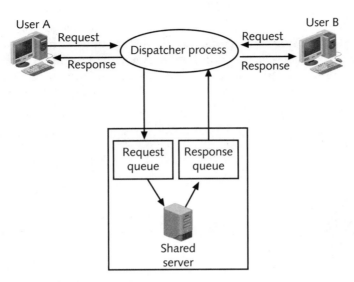

Figure 15-2   Shared server process

CONFIGURING THE SHARED SERVER

To configure the shared server option, the dispatcher for the database must be configured and an Oracle Net Server listener must be started. The procedure to start and control a listener was presented in Chapter 13. The dispatcher specifications are included in the database's Init.ora file using the DISPATCHER parameter. In addition, there are other parameters available that can be used to specify the number of dispatchers to start, the maximum number of dispatchers that can be started, and so on. The available parameters will be discussed in this and subsequent sections of the chapter.

Carlos has two approaches he can use to configure the shared server option. The simplest approach is to use the Database Configuration Assistant. The GUI interface allows Carlos to specify the shared server option and then perform the necessary parameter changes to use the option. The alternative is to directly edit the database's Init.ora file.

Because Carlos has never configured an Oracle Shared Server, in the following section, he will use the Database Configuration Assistant. The subsequent section will focus on manually editing the Init.ora file.

Configuring the Oracle Shared Server—GUI Approach

In this section, Carlos uses the Database Configuration Assistant to configure the shared server option. The responses provided by Carlos are used by the wizard to update the Init.ora file and then start the shared server process.

To configure the Oracle Shared Server using the GUI approach:

1. On the Start menu, select **All Programs**, **Oracle – OraHome92**, **Configuration and Migration Tools**, and then click **Database Configuration Assistant**.

2. When the Welcome window shown in Figure 15-3 appears, click **Next**.

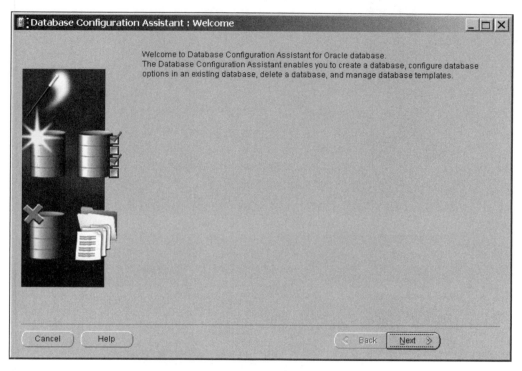

Figure 15-3   Welcome window

3. When the window in Figure 15-4 appears, click the second option button, and then click **Next**.

4. When the Databases window shown in Figure 15-5 appears, click the name of the database to be configured for the shared server option from the Available Database(s) list, and then click **Next**.

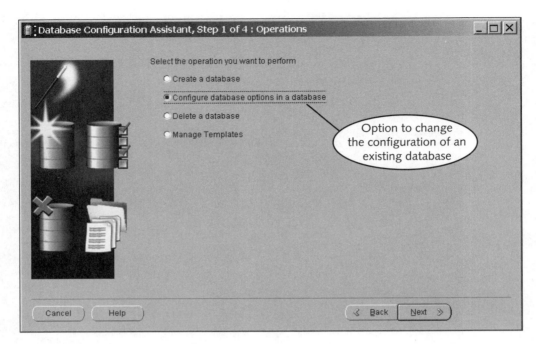

Figure 15-4   Operations window

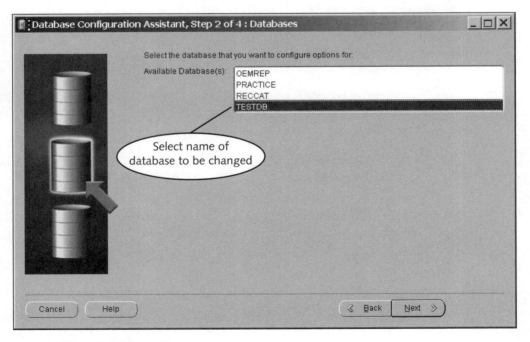

Figure 15-5   Databases window

5. When the Database Features window shown in Figure 15-6 appears, click the **Next** button because none of the existing features need to be changed.

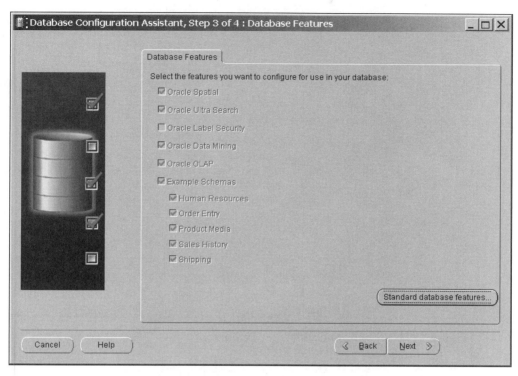

Figure 15-6   Database features window

6. When the Database Connection Options window appears, click the **Shared Server Mode** option button, as shown in Figure 15-7, and then click **Finish**.

7. A dialog box appears indicating that the database needs to be restarted before the changes can take affect. Click **Yes** to restart the database.

8. A dialog box appears indicating the type of operation to be performed. Click **OK** to accept the change.

9. The dialog box shown in Figure 15-8 is displayed to indicate the progress of the changes being made to the database. When the process is completed, a dialog box appears indicating that the configuration was completed successfully. Click **No** to indicate that no another operation is to be performed at this time.

15

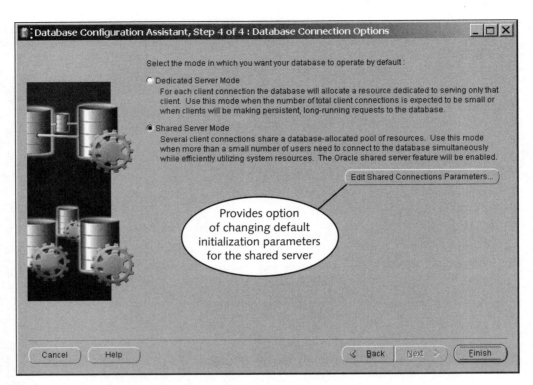

Figure 15-7   Database Connection Options window

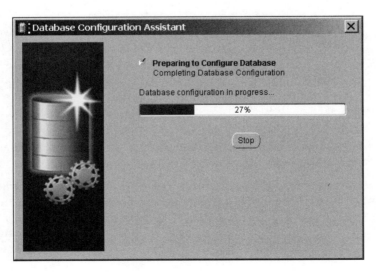

Figure 15-8   Status of configuration change

After the database has been restarted, it is configured with the shared server option. Take another look at the Edit Shared Connections Parameters button in Figure 15-7. In the step sequence, Carlos chose not to change any of the default values for the Oracle Shared Server. However, if Carlos had wanted to specify the number of dispatchers to be available at instance start-up, the maximum number of connections to be managed by each dispatcher, and so on, he could have clicked this button and entered the appropriate value(s), as shown in Figure 15-9. If a value is not specified, Oracle9*i* assumes the default value for each parameter.

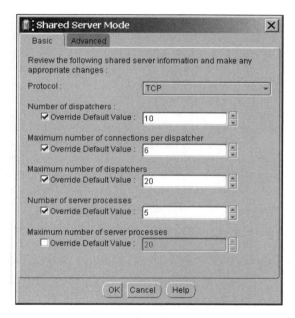

Figure 15-9   Basic Shared Server Mode options window

Configuring the Oracle Shared Server—Manual Approach

15

With this approach, Carlos is required to manually enter the necessary parameters into the database's Init.ora file. Recall from Chapter 2 that Oracle9*i* has two versions of the initialization file available: the binary spfile and the text pfile. Because the Oracle9*i* database references the spfile by default, Carlos first needs to shut down the database, create the text pfile based on the current contents of the spfile, edit the pfile, and then recreate the spfile before restarting the database.

To create and edit the Init.ora file to configure the TESTDB database to use the Oracle Shared Server option:

1. Log into SQL*Plus with a SYSDBA privileged account.

2. Type **shutdown** to shut down the database.

3. Type **create pfile from spfile;** to create the text version of the spfile.

4. On the Start menu of the operating system, select **All Programs**, **Accessories**, **Notepad** to start Notepad as the text editor to be used when editing the Init.ora file.

5. Open the Init.ora file for your database. Recall from previous chapters that the actual file name will be Init*sid*.ora where *sid* is the name of your database. By default, the file is located in the database folder of the home directory of your Oracle9*i* software installation.

6. At the end of the text file, type ***.DISPATCHERS = "(PROTOCOL = TCP)(DISPATCHERS = 3)"** to indicate that three dispatchers should be started on instance startup and that each dispatcher is for client requests via the TCP protocol. If you do not indicate the number of dispatchers, only one dispatcher is started by default.

If the initialization file already contains the DISPATCHERS parameter, ask your instructor for the correct parameter syntax for your specific system. In some cases, the SERVICE attribute of the options clause for the parameter may be required to identify the service name that the dispatcher registers with the listener. Consult the documentation available on the Oracle Technology Network at *http://otn.oracle.com* for the various attributes available for the DISPATCHERS parameter.

7. Save the modified Init.ora file and exit Notepad.

8. In SQL*Plus, type **create spfile from pfile;** to update the spfile with the changes.

9. Type **startup** to restart the database.

After the database has been restarted, the shared server option is enabled. However, in this example, only the DISPATCHERS parameter was specified; it is the only parameter that is required by the shared server. The parameter specifies that there are three dispatchers and they are configured to listen for requests via the TCP protocol.

Because each dispatcher can listen for requests only on one protocol, if the network uses multiple protocols, additional dispatchers must be configured. To configure multiple dispatchers, list the arguments for each dispatcher in the DISPATCHERS parameter. For example, to configure two dispatchers for the TCP protocol and one for the IPC protocol, Carlos would include the following parameter in the Init.ora file: ***.DISPATCHERS = "(PROTOCOL = TCP)(DISPATCHERS = 2)(PROTOCOL=IPC)(DISPATCHERS=1)"**. In the following section, optional parameters that Carlos can use to customize the shared server environment will be discussed.

Recall from Chapter 2 that initialization parameters can also be changed through the Enterprise Manager Console.

INITIALIZATION PARAMETERS

There are multiple parameters that Carlos can include in the Init.ora file to customize the shared server environment. Table 15-1 identifies the various parameters and their purpose.

Table 15-1   Shared server initialization parameters

Parameter Name	Purpose
CIRCUITS	Identifies the number of virtual circuits that are available for network sessions
SESSIONS	Identifies the maximum number of concurrent sessions that can be created
MAX_DISPATCHERS	Specifies the maximum number of dispatchers that can run at one time
SHARED_SERVERS	Specifies the number of shared server processes to be created when the instance is started
MAX_SHARED_SERVERS	Identifies the maximum number of shared server processes that can run at one time
SHARED_SERVER_SESSIONS	Identifies the maximum number of user sessions that can be processed concurrently by the shared server processes

In previous versions of Oracle, the shared server was referred to as the Multithreaded Server and its parameters had an MTS prefix. The MTS-prefixed parameters are still included in the Oracle9i software for backwards compatibility. If your database does not interact with non-Oracle9i clients, ignore the MTS-prefixed parameters.

Each of the optional shared server parameters can be entered into the initialization file, allowing Carlos to customize the server configuration. Carlos can use the CIRCUITS parameter to specify the total number of virtual circuits allowed for all incoming and outgoing network connections. Remember from earlier in the chapter that a virtual circuit is an allocation of shared memory. Therefore, Carlos is cautious about the number of circuits allocated, because the parameter affects the total size of the SGA. If Carlos does not include the CIRCUITS parameter in the Init.ora file, the value will default to the value assigned to the SESSIONS parameter if it has been specified.

The MAX_DISPATCHERS parameter is used to indicate the maximum number of dispatchers that can run concurrently. Recall that the DISPATCHERS parameter specifies the original number of dispatchers at instance start-up. As a rule of thumb, Oracle recommends that there be one dispatcher for each 1000 concurrent connections. The maximum number of dispatchers Carlos should specify is simply the maximum number of concurrent sessions divided by the number of connections supported by each dispatcher. However, there is one catch. The maximum number of processes a dispatcher can run

15

concurrently is actually operating-system dependent. Therefore, Carlos' choice in the operating system used by the machine housing the Oracle9i database server directly influences some of the values assigned to the shared server parameters. The default value assigned to the MAX_DISPATCHERS parameter is 5.

The SHARED_SERVERS and MAX_SHARED_SERVERS parameters, respectively, are used to indicate the number of shared servers to be started on instance startup and the maximum number of shared servers that can be running concurrently. If the SHARED_SERVERS parameter is added to the initialization file, then the parameter must be set to a value of at least 1 to enable shared server connections. A value of 0 would mean that no shared servers are started automatically on instance start-up. The default value for the MAX_SHARED_SERVERS parameter is twice the value assigned to the SHARED_SERVERS parameter or 20 if a value for SHARED_SERVERS is not specified.

The SHARED_SERVER_SESSIONS parameter indicates the total number of user sessions that can be concurrently connected to the shared server processes. For example, Carlos could assign a value of 500 to the SESSIONS parameter to indicate that only 500 user processes can connect to the shared server processes. However, he may want to reserve some of those sessions to be dedicated processes to use the Recovery Manager utility, or other utilities. If Carlos wants to reserve 30 sessions to be dedicated sessions, he can limit the number of user processes that are connected to the shared server processes to 470 by assigning that value for the SHARED_SERVER_SESSIONS parameter. The default value for this parameter is derived; it is based on the lesser of the value assigned to the SESSIONS parameter minus five or the value of the CIRCUITS parameter.

TUNING THE ORACLE SHARED SERVER

So, now that the various parameters have been identified, how does Carlos know what value to assign to each one? There is no hard and fast rule for the exact value of each parameter. The DBA needs to monitor the activities of the Oracle Shared Server over a period of time to determine whether enough dispatchers are available and whether there are enough shared servers. There are different data dictionary views available that can be referenced to determine the status of the dispatchers and other shared server options. A list of the most commonly used views is provided at the end of the chapter.

Using Views

After Carlos starts using the shared server option for the database at Janice Credit Union, he can use these views to determine whether adjustments need to be made to the existing configuration. For example, if the value displayed in the MAXIMUM_SESSIONS column of the V$SHARED_SERVER_MONITOR view is the same as the value assigned to the SHARED_SERVER_SESSIONS parameter, then the parameter's assigned value should be increased. In addition, if the value displayed in the

MAXIMUM_CONNECTIONS column of the same view is equal to the value assigned to the CIRCUITS parameter, then the value for this parameter should also be increased.

To determine whether the value for the SHARED_SERVER_SESSIONS and CIRCUITS parameters should be increased:

1. If necessary, log into SQL*Plus using a SYSDBA privileged account.

2. Type **select * from v$shared_server_monitor;**.

An example of the output Carlos received from his newly configured database is shown in Figure 15-10. Notice that most of the values are 0. Why? No connections have been made to the database after the shared server option was configured. However, notice that the number of server processes started at instance start-up was 4, as shown in the SERVERS_STARTED column. This was specified by the value assigned to the SHARED_SERVERS parameter. Of course the results displayed for your database will be different depending on what changes you have made.

Figure 15-10   V$SHARED_SERVER_MONITOR view

Of course, in a production system with several concurrent connections, the values displayed in these views would be much higher.

Note that the value in the SERVERS_HIGHWATER column indicates that at one time there were 5 shared servers started. How to dynamically change the number of shared servers running and the number of dispatchers is presented in the next section of this chapter.

Performing Calculations

Sometimes Carlos needs to perform calculations based on the values provided by a view to determine if maintenance is required. For example, Carlos may wonder whether more dispatchers should be started to handle the client requests. This requires calculations using data Carlos can find in the V$DISPATCHER view. This view provides the idle and busy time for each dispatcher.

Ideally, Carlos would want a dispatcher to be busy only 50 percent of the time. To determine this percentage, Carlos can simply divide the amount of time the dispatcher is busy (using the value in the BUSY column) by the total amount of time the dispatcher has been available, which can be calculated by adding the values in the BUSY and IDLE columns. The result is multiplied by 100 to determine the percentage of time that the dispatcher is actually busy.

To determine whether additional dispatchers should be started:

1. If necessary, log into SQL*Plus using a SYSDBA privileged account.

2. Type **select name, (busy/(busy+idle))*100 from v$dispatcher;**, as shown in Figure 15-11.

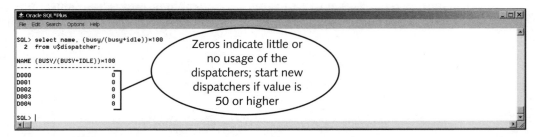

Figure 15-11   Monitoring dispatchers' busy rate

As mentioned, no activity has occurred after the shared server option was configured. Therefore, the busy rate shown in Figure 15-11 is 0 for each dispatcher. However, if the value returned for one or more of the dispatchers is 50 percent or greater, this would indicate that Carlos should start at least one additional dispatcher. After the additional dispatcher has been started, Carlos should monitor the busy time for all the dispatchers to determine whether additional dispatchers should subsequently be started.

DYNAMICALLY CHANGING SHARED SERVER PARAMETERS

The number of dispatchers or shared servers that are started can be changed using the ALTER SYSTEM command, as long as the values do not exceed those specified by the respective MAX parameters. For example, suppose Carlos queried the V$DISPATCHER view and decided to start an additional dispatcher. Notice in Figure 15-11 that there are currently five dispatchers started (0–4). To have a total of six dispatchers started, Carlos would enter the command **ALTER SYSTEM SET DISPATCHERS="(PROTOCOL=TCP) (DISPATCHERS=6)"**. Notice that the command specifies the total number of dispatchers, not the additional number that should be started.

Recall that when the V$SHARED_SERVER_MONITOR view was queried in Figure 15-10, there were a total of five shared servers that had been started while that instance was running. How can Carlos start additional servers after the instance is started? Again, Carlos would simply change the value assigned to the appropriate parameter, as demonstrated in the following section:

To dynamically start additional shared server processes:

1. If necessary, log into SQL*Plus using a SYSDBA privileged account.

2. Type **alter system set shared_servers = 5;**, as shown in Figure 15-12. You can close SQL*Plus if you are not continuing with the end of chapter assignments at this time.

Figure 15-12   Specifying the number of shared servers

After the command is executed, the specified number of shared servers is started. If Carlos decides to reduce the number of shared servers or dispatchers, he can simply reissue the ALTER SYSTEM command with the lower value assigned for the parameter.

Now that Carlos has learned to configure and monitor the Oracle Shared Server option, he is prepared to begin drafting the initial network and server configurations necessary to support management's customer support plan. Again, the shared server processes require modifications over time as more services are offered and the volume of transactions increases. Remember that as the volume of processing increases, performance will begin to suffer. At some point, the performance level will drop to an unacceptable level and Carlos will be forced to add more resources to the database system. However, until then, the shared server environment should provide the necessary scalability to support the initial increase in demand.

15

CHAPTER SUMMARY

- Use of the shared server option addresses scalability, not performance.

- In a shared server environment, a server process services requests from multiple user processes.

- A dispatcher is used to place client requests in the request queue and retrieve server responses from the response queue.

- The only mandatory parameter to configure the Oracle Shared Server is the DISPATCHER parameter.

- A dispatcher can be configured for one protocol.

❑ Initialization parameters specify the number of dispatchers and shared server processes to be started at instance start-up.

❑ Views can be queried to determine whether changes should be made to the Oracle Shared Server.

❑ Additional dispatchers and shared server processes can be dynamically started using the ALTER SYSTEM command.

SYNTAX GUIDE

Parameter	Description	Example
CIRCUITS	Identifies the number of virtual circuits that are available for network sessions	`circuits = 1`

DISPATCHERS	Specifies the number of dispatchers to start at instance start-up and the protocols they support	`dispatchers=`
		`"(procotol=tcp)`
		`(dispatchers=1)"`

MAX_DISPATCHERS	Specifies the maximum number of dispatchers that can run at one time	`max_dispatchers = 10`
MAX_SHARED_SERVERS	Identifies the maximum number of shared server processes that can run at one time	`max_shared_servers` `= 8`
SESSIONS	Identifies the maximum number of concurrent sessions that can be created	`sessions = 500`
SHARED_SERVERS	Specifies the number of shared server processes to be created when the instance is started	`shared_servers = 5`
SHARED_SERVER_SESSIONS	Identifies the maximum number of user sessions that can be processed concurrently by the shared server processes	`shared_server_` `sessions = 4`

Views	Description	Example
V$CIRCUIT	Provides information regarding each dispatcher, how much information has been transferred, and the current status of each circuit	`select * from v$circuit;`
V$DISPATCHER	Identifies the name and address of each dispatcher, current status, and the amount of busy and idle time for each dispatcher	`select * from v$dispatcher;`
V$DISPATCHER_RATE	Provides the current and historical statistics for each dispatcher, including the number of inbound and outbound connections, average number of bytes processed, and the transfer rates	`select * from v$dispatcher_rate;`
V$QUEUE	Identifies how long requests are waiting in the queue, number of items currently in the queue, and the total number of items that have been in the queue	`select * from v$queue;`
V$SESSION	Identifies the type of server being used by each connection	`select * from v$session;`
V$SHARED_SERVER	Identifies the number of requests processed by each shared server, current status, and the total idle and busy time of each shared server in hundredths of seconds	`select * from v$shared_server;`
V$SHARED_SERVER_MONITOR	Shows the number of additional shared servers started after instance start-up and the maximum number of circuits and sessions used during the instance	`select * from v$shared_server_monitor;`

15

REVIEW QUESTIONS

1. In a shared server configuration, the client interacts with the:

   a. SGA

   b. server process

   c. dispatcher

   d. initialization file

2. A circuit is the path used by a server process to interact with the user. True or False?

3. The _____ view can be queried to determine how long requests are waiting in the queue.

4. How do you start additional dispatchers in a shared server environment?

5. A dedicated server process uses a dispatcher to interact with one client at a time. True or False?

6. Which of the following parameters is mandatory when configuring the Oracle Shared Server?

   a. CIRCUITS

   b. SHARED_SERVERS

   c. SESSIONS

   d. DISPATCHERS

7. Which of the following views can be used to determine the total number of bytes processed by a particular dispatcher?

   a. V$DISPATCHER

   b. V$DISPATCHER_RATE

   c. V$CIRCUIT

   d. V$DISPATCHER_RATE_MONITOR

Refer to the following entries in the Init.ora file of the PRACTICE database when answering Questions 8–11.

```
DISPATCHERS="(PROTOCOL=TCP)(DISPATCHERS=4)"
MAX_DISPATCHERS=6
SESSIONS = 9
```

8. Which of the following is a valid statement that increases the number of dispatchers currently started in the instance?

   a. `ALTER SYSTEM SET DISPATCHERS="(PROTOCOL=TCP)(DISPATCHERS=5)"`

   b. `ALTER SYSTEM SET DISPATCHERS="(PROTOCOL=TCP)(DISPATCHERS=1)"`

c. `ALTER SYSTEM SET DISPATCHERS="(PROTOCOL=TCP)(DISPATCHERS=4+1)"`

d. `ALTER SYSTEM SET DISPATCHERS="(PROTOCOL=TCP)(DISPATCHERS=9)"`

9. Based on the parameters specified, what is the default number of allocated virtual circuits?

10. Based on the parameters specified, what is the maximum number of user sessions that can be concurrently processed by the shared server processes?

11. The _____ view can be used to determine the total number of shared servers that have been started during the life of the instance.

12. The Database Configuration Assistant can be used to specify the total number of dispatchers that should be created when the instance is started. True or False?

13. The shared server parameters are stored in the _____ file.

14. What task is performed after a dispatcher receives a request from a client?

    a. The request is placed in the request queue.

    b. The request is passed to the next available shared server process.

    c. The request is placed in the response queue.

    d. The request is sent to the default listener.

15. The response queue is located in which of the following memory areas?

    a. PGA

    b. UGA

    c. SGA

    d. DGA

16. Each dispatcher can only be configured to process requests from one protocol. True or False?

17. What parameter is used to specify the number of shared server processes to be started when the instance is started?

18. The _____ is a tool that can be used to specify the maximum number of dispatchers that can be started without the user needing to know the name of the appropriate parameter.

19. The default value assigned to the MAX_DISPATCHERS parameter is:

    a. 1000

    b. 100

    c. 20

    d. 5

20. Which shared server parameter directly affects the size of the SGA at instance start-up?

15

Hands-on Assignments

 If your database is currently configured to use the Oracle Shared Server option, restore the backup of the Init.ora and Tnsnames.ora files created at the beginning of the chapter before working on the following assignments.

Assignment 15-1 Configuring the Oracle Shared Server

In this assignment, you enable the shared server option and specify the maximum number of dispatchers that can be started.

1. On the Start menu, select **All Programs, Oracle – OraHome92, Configuration and Migration Tools**, and then click **Database Configuration Assistant**.

2. When prompted, select the necessary option to enable the shared server option.

3. Before restarting the database, use the Database Configuration Assistant to specify that the maximum number of dispatchers is 11.

4. Restart the database and then query the V$DISPATCHER view to verify that the correct number of dispatchers were started.

Assignment 15-2 Changing Parameter Values Through the Database Configuration Assistant

In this assignment, you are required to change the maximum number of processes that can be started by the Oracle Shared Server using the Database Configuration Assistant.

1. On the Start menu, select **All Programs, Oracle – OraHome92, Configuration and Migration Tools**, and then click **Database Configuration Assistant**.

2. Use the Database Configuration Assistant to change the maximum number of shared server processes that can be started to 15. Make certain the database is not changed back to dedicated server mode and that no other values are reset.

3. When prompted, restart the database and then verify the number of shared server processes that have been started by querying the V$SHARED_SERVER_MONITOR view.

Assignment 15-3 Changing a Parameter Value in the Init.ora File

In this assignment you modify the current value assignment to a parameter in the database's initialization file.

1. Create a pfile from the database's existing spfile.

2. Open the database's Init.ora file using a text editor.

3. Locate the MAX_SHARED_SERVERS parameter that was created in Assignment 15-2. Change the value of the parameter to 20.

4. Save the changed pfile.

5. Recreate the spfile and start up the database.

Assignment 15-4 Inserting a Shared Server Parameter into the Init.ora File

In this assignment, you modify the database's initialization file by specifying the number of server processes to be started at instance start-up.

1. Create a pfile from the database's existing spfile.
2. Open the database's Init.ora file using a text editor.
3. At the end of the initialization file, add the necessary parameter to make certain that four server shared processes are started at instance start-up.
4. Save the changed pfile.
5. Recreate the spfile and start up the database.
6. In SQL*Plus, type **show parameter shared_servers** to verify the value of 4 that has been assigned to the SHARED_SERVERS parameter.

Assignment 15-5 Adjusting the Number of Dispatchers

In this assignment, you are required to determine how many dispatchers are currently started in the instance and then reduce the number of dispatchers.

1. In SQL*Plus, query the V$DISPATCHER view to display the name, status, and busy rate of the current dispatchers.
2. If the busy rate of all the dispatchers is below 15%, then decrease the current number of dispatchers by one.
3. Redisplay the V$DISPATCHER view to verify that the correct number of dispatchers are running.
4. Use the SHOW PARAMETER command to determine the current value assigned to the DISPATCHERS parameter. Is the total number of dispatchers defined by the DISPATCHERS parameter the same as the number of dispatchers displayed in the output from Step 3 of this assignment?

15

Assignment 15-6 Manually Changing the Init.ora File to Configure the Database for Dedicated Server Processing

In this assignment, you are asked to identify, and then remove, the parameter or parameters in the database's initialization file that enables the shared server option.

1. Create a pfile from the database's existing spfile.
2. Open the database's Init.ora file using a text editor.
3. Delete the parameter or parameters in the initialization file that specify that the shared server option is to be used for the instance and save the modified file.
4. Recreate the spfile and then start the database.
5. Query the V$SHARED_SERVER view to verify that no shared servers have been started.

Assignment 15-7 Dispatcher Configuration

In this assignment, you will change the DISPATCHERS parameter.

1. Shut down the database and create a pfile from the database's existing spfile.

2. Open the database's Init.ora file using a text editor.

3. Change the DISPATCHERS parameter so there are three TCP/IP dispatchers and two IPC dispatchers. Add the necessary argument so that each dispatcher handles a maximum of five network connections. If necessary, visit the Oracle Technology Network at *http://otn.oracle.com* and search the Oracle9*i* database library for the definition of the DISPATCHERS initialization parameter.

4. Recreate the spfile and then start the database. Verify the number of dispatchers that were started for the instance.

Assignment 15-8 Using the Listener Control Utility (lsnrctl) for Dispatcher Information

In this assignment, you use the Listener control utility (lsnrctl) presented in Chapter 14 to obtain information about the dispatchers.

1. On the Start menu, click **Run**.

2. Type **cmd** in the Run dialog box to open a command-line window.

3. Type **lsnrctl services** at the operating system prompt, and press **Enter** to list the current listener services.

4. Identify which listeners are listening on behalf of dispatchers.

5. Identify the protocol of each listener.

6. Determine how many user processes have requested connection to the dispatchers.

7. Type **exit** at the operating system prompt to close the command-line window.

CASE PROJECTS

Case 15-1 Determining the Initial Settings for the Oracle Shared Server

Now that the basic configuration requirements for the Oracle Shared Server have been examined, Carlos needs to determine the initial values to be assigned to the Oracle Shared Server. At this point, Carlos expects that most client requests are received on Friday afternoons. Combining the connections made from ATMs, tellers, users via the Internet, and other sources, he expects the maximum number of connections to be requested during that timeframe will be 2120.

Based on this information, determine the most appropriate value to be assigned to the various shared server options that are available. In addition, specify any other information that should be considered while determining the values to be assigned to the parameters. Write your suggested values (and rationale for those values) in a memo addressed to Carlos. In the memo, include the additional information that is needed and why you believe it is important.

Case 15-2 Monitoring Routine

Carlos has asked you to develop a schedule for monitoring the activity of the Oracle Shared Server to determine whether any parameters need to be changed. Make certain you account for activity changes that can occur throughout a one-day period. If necessary, reference the parameter and view information provided on the Oracle Technology Network at *http://otn.oracle.com* to develop your schedule. (*Hint*: Consider using script files to retrieve statistics from views at regular intervals and have the results stored in text files for comparison.)

15

Glossary

archiver (ARC*n*) — An optional background process responsible for creating copies of filled online redo log files that can later be used to perform database recovery.

backup — A process of creating copies of database files that can be used to restore the database if recovery is necessary.

backup piece — A binary file that belongs to a specific backup set.

backup set — A backup file created by RMAN; usually consists of either data files, control files, or archived redo log files.

cancel-based recovery — An incomplete recovery process in which the stopping point for the recovery process is manually determined by the DBA.

change-based recovery — An incomplete recovery process in which the stopping point is based on the system change number (SCN) specified by the DBA.

checkpoint (CKPT) — A mandatory background process responsible for triggering the DBW*n* to write changed data to the data files. It also stamps the control file and data file headers with the system change number (SCN).

cold backup — Creating a copy of the database files while the database is closed.

complete recovery — All archived data or backed up files are used to update the contents of the database.

connect descriptor — Identifies the name of the desired service and network route information that the client uses to contact the database server.

control file — Small binary file containing the physical structure of the database. A database can have a maximum of eight copies of a control file; a minimum of one is required.

data block — Smallest logical storage structure that can be referenced by the Oracle9*i* database.

data dictionary cache — A portion of the shared pool containing information regarding database objects.

data file — Physical storage location for the actual data contained within the database.

data segment — A logical storage structure specifically used to store data for the database tables.

database — Used to store organized data. Consists of both a physical and a logical structure.

database buffer cache — A portion of the SGA containing the most recently used data blocks. When filled, the least recently used data blocks are overwritten.

database recovery — Resetting a database to a point immediately prior to a database failure.

database writer (DBW*n*) — A mandatory background process responsible for writing all changed data blocks from memory to the data files. A maximum of ten such processes can be used by any one database. By default, only DBW0 is started when an instance is started.

dedicated server — Each server process handles requests for a single user process.

directory naming method — Uses an LDAP-compliant centralized directory server to store the database server connection information.

disaster recovery plan — Identified procedures that ensure the database can be recovered in the event of a natural or man-made disaster.

dispatcher — A process that handles multiple user requests for connection to a shared server process.

dump file — Binary file generated by the Export utility to store exported data.

Export utility — Used to export data from an Oracle9i database into a dump file.

extent — A set of contiguous Oracle data blocks. Each extent belongs to only one segment.

failure — When the database cannot be accessed by users due to an error.

full backup — Includes all data blocks within a file.

host-naming method — The client obtains the database server connection information from a HOSTS file located on the network.

hot backup — Creating copies of the database files while the database is still open.

Import utility — Used to import data from a dump file previously created by the Export utility.

incarnation — A version of a database.

incarnation number — A value assigned to identify the different versions of a database.

incomplete recovery — All archived redo log files are not included in the recovery process.

incremental backup — Only changed data blocks are copied during the backup procedure, rather than the entire file.

index segment — A logical storage structure specifically used to store indexes.

instance — Consists of both background processes and memory structures.

instance failure — A type of non-media failure that occurs when an instance is shut down improperly. Such failure is normally resolved by SMON.

job command — A group of commands enclosed in curly braces submitted to RMAN at one time and executed in sequential order.

library cache — A portion of the shared pool containing the most recently used SQL and PL/SQL statements.

listener — A server process that accepts and processes client requests for connections to the database.

local naming method — Uses a local Tnsnames.ora file to map a net service name to the appropriate connect descriptor.

log switch — Occurs when the active redo log file is closed and any changes are recorded in the next available redo log file.

log writer (LGWR) — A mandatory background process responsible for writing all changed data blocks to the online redo log files.

LogMiner — A utility provided in every Oracle9i installation that can analyze the contents of the online and archived redo log files.

mean-time-between-failure (MTBF) — The amount of time that has passed between database failures. A goal of the DBA is to increase MTBF.

mean-time-to-recovery (MTTR) — The amount of time taken to recover a database after a failure has occurred. A goal of the DBA is to decrease MTTR.

media — The physical material used to store files.

media failure — A type of failure resulting from not being able to access one or more database files; requires recovery to be performed by the DBA.

media recovery — Restoring the database file or files that previously could not be accessed by the database.

net service name — The simple name, such as a host name without network location information, used to identify a database.

network — A group of computers or other devices connected together through a transmission medium.

non-media failure — A type of failure that does not result from a database file being inaccessible. Most non-media failures are corrected automatically by the background processes.

n-tier architecture — Other servers are included between the client and the database server to provide specific services for the network.

Oracle Net Services — A set of components available with the Oracle9i database software that is designed to address the issues of connectivity, manageability, and scalability.

pfile — A text version of the initialization file referenced by Oracle9i during instance start-up. Contains the values assigned to parameters used to configure the instance and database.

process failure — A type of non-media failure that occurs when an internal program error occurs or if the user does not terminate a session correctly. Such failure is usually resolved by the PMON.

process monitor (PMON) — A mandatory background process responsible for releasing resources and performing any necessary rollback operations after a failed user process has occurred.

program global area (PGA) — A memory structure. Actual contents vary based on instance configuration. Allocated when a server process is started, and deallocated after the process is finished.

protocol — A set of communication rules.

recovery — A term describing the process of returning a database to a desired state in the event of a failure.

Recovery Wizard — GUI interface provided through the Enterprise Manager Console to access Recovery Manager.

redo log buffer — A portion of the SGA containing changes made by INSERT, UPDATE, DELETE, CREATE, ALTER, or DROP operations.

redo log file — Contains the database changes generated by committed transactions. Each database must have at least two redo log files.

redundant array of independent disks (RAID) — A set of hard disks that enable mirroring and/or file redundancy to eliminate data loss due to disk failure.

repository — Stores metadata regarding the target database and backup and recovery operations.

resetting — Updating the Recovery Catalog after an incomplete recovery is performed of the target database.

resynchronizing — Updating the Recovery Catalog after the control file in the target database has been altered or after performing backup and recovery operations without a connection to the catalog.

scalability — The network's ability to expand.

segment — A logical storage structure consisting of one or more extents. Each segment can belong to only one tablespace. Segment types include data segments, index segments, undo segments, and temporary segments.

sequence-based — An incomplete recovery performed through RMAN that uses the log sequence number of a redo log file as the terminating parameter for the recovery process.

shared pool — A portion of the SGA consisting of the library cache and the data dictionary cache.

shared server — Allows several user processes to use the same server process.

single-tier architecture — A mainframe computer with dumb terminal connections; oldest form of a network.

spfile — A binary version of the initialization file referenced by Oracle9i during instance start-up. Contains the values assigned to parameters used to configure the instance and database. By default, the spfile is referenced before the pfile version of the initialization file.

SQL*Loader — Used to import data from an ASCII file; data can originally be from a non-Oracle database.

stand-alone command — A single command that is executed immediately in RMAN.

statement failure — A type of non-media failure that occurs because of a syntax error in a SQL or PL/SQL statement. No background process is necessary for recovery.

system global area (SGA) — A memory structure consisting of the shared pool, database buffer cache, and redo log buffer. Allocated when an instance is started.

system monitor (SMON) — A mandatory background process responsible for instance recovery at database start-up by "rolling forward," or updating the data files, with committed changes.

tablespace — A logical storage structure consisting of one or more segments. Can belong to only one database.

tablespace point-in-time recovery (TSPITR) — Procedure to recover a table; requires use of physical and logical backups and a clone database.

target database — Database that serves as the source for backup operations or the database to be recovered through RMAN.

temporary — A logical storage structure used to perform sort operations required by SQL statements.

time-based recovery — An incomplete recovery process in which the stopping point is based on the time provided by the DBA.

trace file — A text file generated by background processes or activated by the DBA to detect errors.

undo segment — A logical storage structure used to store the data necessary to perform rollback operations.

user error — A type of non-media failure that occurs when data or tables are incorrectly altered. Depending on the severity of the error, intervention by the DBA may be required.

virtual circuit — A portion of shared memory allocated to each dispatcher.

whole backup — A backup of the control file, data files, any valid archived redo log files, and any other files referenced by the database.

wizard — A step-by-step series of GUI prompts used to assist with the completion of a task.

Index